ALLE ZEIT WACH
SJ
1842

V.V. Roshchina   V.D. Roshchina

The Excretory Function of Higher Plants

With 95 Figures

Springer-Verlag
Berlin Heidelberg New York
London Paris Tokyo
Hong Kong Barcelona
Budapest

Dr. VICTORIA V. ROSHCHINA
Prof. VALENTINA D. ROSHCHINA
Russian Academy of Sciences
Institute of Cell Biophysics
Pushchino, Moscow Region 142292
Russia

ISBN-13:978-3-642-78132-2 e-ISBN-13:978-3-642-78130-8
DOI: 10.1007/978-3-642-78130-8

Library of Congress Cataloging-in-Publication Data. Roshchina, V.V. (Viktoria Vladimirovna)
The excretory function of higher plants / Victoria V. Roshchina, Valentina D. Roshchina, p. cm.
Includes bibliographical references and indexes.
ISBN-13:978-3-642-78132-2 1. Plants—Secretion. 2. Plant cells and tissues.
I. Roshchina, V.D. II. Title. QK725.R75 1993 581.1'4—dc20 93-4993

Softcover reprint of the hardcover 1st edition 1993

Typesetting: Macmillan India Ltd., Bangalore 25

31/3145/SPS-5 4 3 2 1 0 – Printed on acid-free paper

Contents

Introduction

Until the 1930s biological science had not paid sufficient attention to the excretory function of plants. The interest in the ability of plants to release various compounds arose in connection with two discoveries. In 1930, the embryologist Boris Tokin reported that in plant excreta there are substances which he called "phytoncides", because they possessed an antimicrobial activity. Later, Hans Molisch (1937), based on numerous observations, concluded that chemicals released by plants can participate in a competition between plants in phytocenosis. At the same time the first monograph *Die Stoffausscheidung der höheren Pflanzen* by Frey-Wyssling (1935) appeared, in which the excretory function of plants was considered, mainly from the anatomo-physiological viewpoint.

Almost half a century after these fundamental works, the excretion by plants was studied in various fields of biology: plant anatomy and cytology, biocenology, plant immunology, and pharmacology.

Anatomical and cytological data dealing with plant secretory structures are reviewed in detail in some monographs (Schnepf 1969; Vasilyev 1977; Fahn 1979; Buvat 1989; Denisova 1989). Many publications are devoted to the role of plant excreta in phytocenosis, mainly in connection with allelopathic interactions. They are summarized in the books of Chernobrivenko (1956), Grodzinskii (1965, 1991), Ivanov (1973), Rice (1974, 1984), Kolesnichenko (1976).

Great progress has also been made in the analysis of the chemical nature of plant excreta, especially using high-sensitive gas chromatography, combined with mass spectroscopy. First steps in this direction were made in individual works by G. Sanadze, V. Dadykin, and VD Roshchina from 1960–1975. Now, such studies have been intensively developed, as is seen from the numerous fundamental publications of the school of Professor Wollenweber (1972–1991). Advances in this direction have stimulated a new tendency in allelopathic research, now considered as a new direction, called allelochemistry. Knowledge of the natural products participating in the interactions between various plant species, plants and animals, and plants and microorganisms permits the use of these substances or their artificial analogs for plant protection, in agriculture and forestry. The chemicals are easily decomposed by microorganisms and do not pollute the surrounding media. Since 1959, phytopathologists, headed by Professor Verderevskii (1959), began to study plant excreta as a possible factor of plant immunity. The discovery of the phytoncidic character of released plant compounds led to intensive investigations of plants as the origin of the medicinal substances known as phytoncides. Success in the coordination of such studies is due to Professors Tokin, Drobot'ko, Aizenman, Vichkanova, Grodzinskii, who published monographs in the years 1980–1991.

For a long time, all the lines dealing with plant excretion developed parallel to each other. Now finally, the necessity for discussing and summarizing experimental data and theoretical approaches, especially in the field of physiology and biochemistry, has given rise to the present work.

Acknowledgments. Writing this book would have been impossible without the great and unselfish help of many specialists from different countries, who sent us their papers and permitted us to use their results. We are deeply thankful to all of them. We would also like to thank those who participated in the preparation of the manuscript: Ms. Margarita Ilyina, Olga Golovchenko, Swetlana Sidorova, and Tatyana Panarina, who participated in the translation of the text from the Russian, Ms. Lubov'Kozlova, who corrected and typed part of the text, and Vitalii Drozdov and Leonid Petrov for the photos.

Classification of Secretory Processes and Terminology

The secretory activity of plants is a manifestation of the fundamental property of living organisms: the ability to exchange substances and energy with the environment. Frey-Wyssling (1935) divided metabolites into recreta, secreta, and excreta, thus considering metabolic processes as the basis of the classification. According to his classification, the recreta are substances that, being absorbed by the plant from the environment, do not alter in the course of cell metabolism (water, salts, etc.) The secreta are products of assimilation and require the energy of cell metabolism for their formation and release. The excreta are the final metabolic products, which are not further involved in the metabolism. In accordance with these definitions, the secretion processes are illustrated by the scheme:

absorption	→	assimilation	→	dissimilation
|		|		|
recretion		secretion		excretion

This classification, however, cannot be considered ideally suited for plants, and as Tukey (1970) justly noted, many authors encounter difficulties when using it. In particular, compounds commonly classified with excreta (essential oils, oxalates, resins) can, as shown by radioactive labeling, be a key definition to this group. It is also difficult to classify the gaseous compounds that are accumulated in the internal air of plants and released with aqueous vapor under particular conditions (Novitskaya 1966; Roshchina 1971, 1973a) as well as many other compounds whose formation pathways and functions are either very poorly studied or absolutely unknown. Since it is difficult to classify new substances under the available groups, many botanists simply call all newly isolated substances secreta or excreta (Esau 1965).

When studying the role of secretion in plants, some scientists have tried to classify them according to their origin and mechanism of action. The classifications by Tokin, the author of the term "phytoncide" (Tokin 1980a), Grümmer (1955), and Grodzinskii (1965) are the best known. These classifications are outside the scope of the present work, as they have been repeatedly considered in the scientific literature.

After the Second All-Union Symposium on volatile biologically active compounds of biogenic origin, held in 1971 in the USSR. a new term "exometabolites" became widely applicable to denote organic substances released by higher and lower plants to the environment in the course of normal vital activity (Tambiev 1984). It has been proposed to call the products of lysis and cell damage, which also possess a well-pronounced physiological action, "extracellular products" (Tambiev 1984). The above terminology is not, however, adequate, since it is difficult to distinguish between exometabolites and extracellular products under natural conditions. As Tokin noted (1957, 1980a), there are no absolutely undamaged plant individuals in nature. Recently, the term "stress metabolites" has been used to indicate substances released under extreme influences.

Thus, as there is no stable classification for plant secretion in the literature, all the terms we have suggested may be used as possessing equal rights.

CHAPTER 1

Cellular Aspects of Secretory Activity in Plants

1.1 Significance of Secretory Processes for the Cell

The main physiological functions of the cell occur in the cytoplasm, including the organelles. This active part of the cell is separated from the cell wall by the plasmalemma and from the vacuole by an internal membrane, the tonoplast. An excess of metabolites, which can cause disturbances in normal reactions of the organism, is evacuated by diffusion or via the active mechanism of transport through the plasmatic membrane or the tonoplast. In the former case, the secreted substance is accumulated in the free cellular space occurring out of the plasmalemma-limited volume, in the latter case in the vacuole.

Free cellular space | Cytoplasm | Vacuole

Products of
⟵ metabolism ⟶

Thus, substances that could disturb cell homeostasis are removed from the sphere of active metabolism.

The intercellular secretion is an evolutionarily determined pathway for the elimination of excessive metabolites. This pathway appeared, probably, when terrestrial forms of life arose, in particular, giant forms of plants, which limited the ability of cells to excrete metabolites to the environment. The plants tended to reserve byproducts inside the cells: in vacuoles and in the free cellular space.

1.2 Compartmentation of Metabolites and Mechanisms of Their Secretion

Secreted products present a wide spectrum of chemical substances. Among them are, first of all, the products of basic (primary) metabolism: carbohydrates, proteins, hormones, which are necessary for plant growth and development. Additionally, there are products of secondary metabolism, which are believed to be synthesized in the cytoplasm and organelles as a result of unbalanced plant growth (Haslam 1985), leading to the appearance of excessive intermediates. The latter, as they are accumulated in the organism, are converted to secondary compounds.

Secondary metabolites are represented by numerous chemicals formed in the metabolic pathways from some few biogenetic precursors: acetyl ~ CoA, mevalonate, cinnamic acid, and amino acids (Fig. 1). The number of known terpenoids approaches 5000, of alkaloids 7000, of flavonoids 1000 etc., growing from year to year.

Since the secondary compounds are biologically active and can damage the vitally essential cell part, the cytoplasm, mechanisms must exist for rendering them harmless. There are two principal ways for plants to protect themselves from chemically reactive substances. One of them is their compartmentation in physically separated specific organelles. The other way is the chemical modification of these substances to relatively harmless ones, which does not exclude their localization in particular cell compartments.

Table 1 shows the sites of synthesis and compartmentation of secondary metabolites. As is seen, the products of intracellular secretion are for the most part compounds formed during photosynthesis, and their principal sites of synthesis are chloroplast membranes, although some secondary metabolites can also be formed in other organelles: endoplasmic reticulum, Golgi apparatus and its vesicles, microsomes, and leucoplasts. The basic compartment for water-soluble compounds is the vacuole, whereas for gaseous and lipophilic excretions the same function is played by the free space of the cell or the extracellular space.

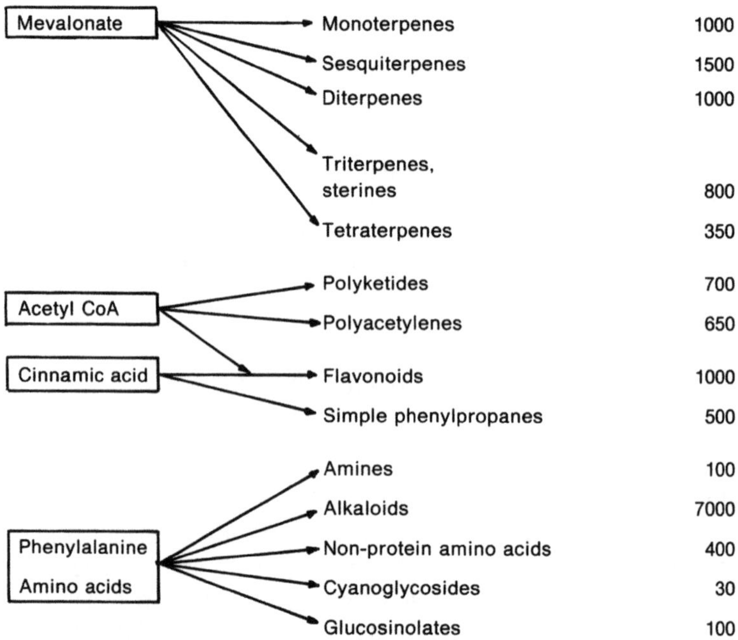

Fig. 1. Biogenetic origins of the main classes of secondary metabolites. (After von Proksch 1985)

Table 1. Cell compartmentation of some secondary metabolites

Compound	Site of formation	Store compartment
Alcohols	C (7)	FS (1)
Aldehydes	MS (7)	FS (1)
Alkaloids	Chl (13); C (12)	V (1, 8); ER (1); FS (1)
Cyanogenous Glycosides	ER (1, 10)	V (1)
Coumarins	ER (14); V (14)	V (1)
Ethane	Chl (9)	FS (1)
Ethylene	Chl (9); M (2)	FS (1)
Esters	–	FS (1)
Flavonoids	Chl (1, 4, 5)	V (1, 5); FS (1)
Gums	–	D (1); GA (1); FS (1)
Hydrogen	Chl (15)	FS (15)
Isoprene	Chl (1, 6)	FS (1)
Methane	–	FS (1)
Monoterpenes	L (11), IM (11)	FS (1)
Mucilages	GA (16)	V (16); FS (16)
Organic acids	M (8)	V (8)
Propane	Chl (9)	FS (1)
Resins	–	D (1); GA (1); FS (1)
Tannins	V (1)	V (1, 8); ER (1); FS (1)
Terpenoids	Chl (1); L (3)	V (1); C (1); FS (1)

(–) – site of synthesis is not known, Chl – chloroplasts, M – mitochondria, L – leucoplasts, ER – endoplasmic reticulum, FS – free space of cell, V – vacuole, D – dictyosomes, GA – Golgi apparatus and its vesicles, C – cytoplasm, IM – internal membranes, MS – microsomes. References in parentheses: 1 – Luckner et al. (1980); 2 – Vinkler and Apelbaum (1985); 3 – Gleizes et al. ((1983); 4 – Zaprometov and Kolonkova (1967); 5 – Charriere-Ladreix (1977); 6 – Sanadze et al. (1972); 7 – Grineva (1975); 8 – Matile (1978); 9 – Bochler-Kohler et al. (1982); 10 – Solomonson (1981); 11 – Bernard-Dagan et al. (1982); 12 – Deus-Neumann (1984); 13 – Wink (1985); 14 – Matile (1987); 15 – Mal'tsev et al. (1986); 16 – Esau. (1965).

Prior to being excreted from the cytoplasm, the metabolites overcome the cytoplasmic membranes: the plasma membrane when the substance is transported to the free space of the cell, or the tonoplast when it is secreted to the vacuole. The protein parts of the membranes have hydrophilic pores and are therefore permeable to water and water-soluble low-molecular weight components localized in the cytoplasm. Sugars, amino acids, and salts may be released in accordance with their concentration or the gradient of electrochemical potential. The same substances can also be transported against the concentration and potential gradients. In this case, ATP is involved in the process while triggering various active mechanisms: ionic pumps, systems with participation of carriers, and transport of nonpolar compounds. Since these processes require energy consumption, they are directly dependent on the metabolism. However, secretion is mainly provided by a special transporting mechanism called exocytosis, which is also an active form of secretion. Via exocytosis, particles and macromolecules that are unable to overcome the barrier formed by the plasmalemma or the tonoplast are excreted (Vasilyev 1977; Fahn 1979; Lüttge and Higinbotham 1979). Among these are polysaccharides, pectins, proteins, etc. These high-polymer substances are synthesized or assembled from particular precursors in the Golgi apparatus, which consists of individual dictiosomes. Each dictiosome presents a stack of five to seven cisternae. Secretory vesicles are separated from their borders (Fig. 2). Particular compounds are transported within the vesicles to the plasmalemma or the tonoplast, where they fuse with the membranes, and then the products are released into the free space (extracellular space) or the vacuole. Vesicles with a smooth surface transport large polymeric molecules. Dictyosomes also form vesicles with a rough surface, so-called coated vesicles, which seem to be widely distributed in plant cells (Coleman et al. 1988). In coated vesicles, specific large molecules (ligands), hormones, and transmitters are mostly transported. The coated vesicles differ from all the others by the presence of a coat consisting of a fibrillar protein clathrin (180 kdaltons) bound with a smaller polypeptide (35 kdaltons). It is believed that the clathrin coat protects such vesicles and their contents against lysis.

The vesicles being transported from the Golgi apparatus are well seen in an electron microscope (Fig. 2). For the vesicles to be formed, their movement and fusion with plasmalemma ATP are required. Exocytosis is induced by specific signals: hormones, Ca^{2+}, cAMP, as well as the factors depolarizing the membranes (Steer 1988).

It has been shown (Steer 1988) that an increase in the Ca^{2+} level promotes secretion of amylase by aleurone cells of *Hordeum* L. and of peroxidase by a suspension culture of *Spinacia* cells. It is proposed that ATP-dependent actomyosin-like proteins which need Ca^{2+} for contraction and ATP hydrolysis are involved in the transport of secretory vesicles. This is confirmed by experiments where the inhibitor of animal actin filaments, cytochalasin B, blocks the transport of secretory vesicles in plants. There is evidence for the participation of microtubules in translocation of particular elements of the Golgi apparatus and secretory vesicles (Kreis 1990). A particular role in the transport of secretory vesicles is played by guanosine triphosphate (GTP)-binding proteins, since the

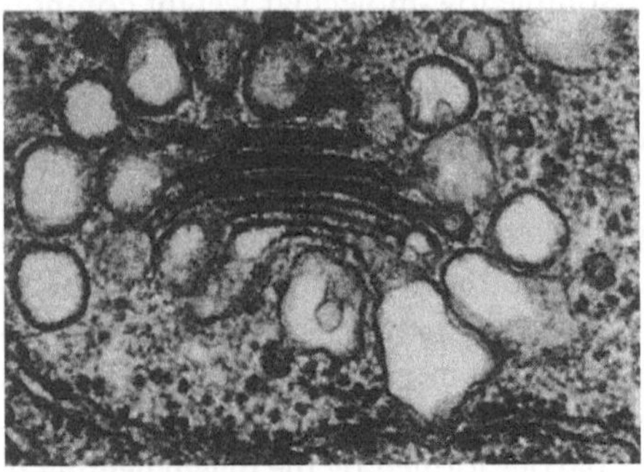

Fig. 2. Electron microscopy of dictyosomes of the pollen tube of *Nicotiana sylvestris* ×55000. The dictyosome consists of five cisterns and forms large secretory vesicles. (Kristen et al. 1988)

movement of vesicles after they have left the Golgi apparatus is inhibited by the unhydrolyzable analogs of guanosine triphosphate (Tooze et al. 1990).

Secretion of proteins by secretory vesicles can be well illustrated by three examples (Akazawa and Hara-Nishimura 1985):

1. extracellular secretion of amylase by seedlings of rice, *Oryza*, and of polyphenoloxidase by a cell culture of maple, *Acer*;
2. extracellular and (or) intracellular secretion of acidic phosphatase or peroxidase to the vacuole by *Acer* cells in culture;
3. intracellular storage of globulin as protein bodies by *Cucurbita* cotyledons.

In all three cases, proteins synthesized in the rough endoplasmic reticulum come to the Golgi apparatus, and within the vesicles detaching from these structures are transported to the plasma membrane or the tonoplast. Then the vesicles fuse with the membranes and the excreted proteins are eliminated to the free space or the vacuole. Probably, the proteins originating from the rough endoplasmic reticulum can be transported to the plasma membrane within transitory and secretory vesicles via the Golgi apparatus. Secretion of such a type takes place in the ovary glandular cells of *Aptenia cordifolia* (Kristen et al. 1988). Sometimes the protein within vesicles is transported directly to the plasmic membrane, by-passing the Golgi apparatus. This is the case in *Crocus* pistil stigma (Heslop-Harrison and Heslop-Harrison 1985), *Dionaea muscipula* digestive glands (Robins and Juniper 1980), and in marginal ligule cells of *Isoetes lacustris* (Kristen et al. 1988).

The secretion of noncellulose polysaccharides in higher plants proceeds as follows: synthesis in the Golgi apparatus, transport in the Golgi vesicles to the plasma membrane, and excretion via exocytosis. In such a way, secretion occurs in leaf trichomes.

Thus, there are three pathways for the vesicular transport of the hydrophilic micromolecules of plant cells (Kristen et al. 1988):

1. Golgi apparatus → plasma membrane (for noncellulose polysaccharides);
2. Endoplasmic reticulum → plasma membrane (for proteins);
3. Endoplasmic reticulum → Golgi apparatus → plasma membrane (for proteins that form only glycoproteides in the Golgi apparatus).

Depending on the kind of secretion cells and the secretory mechanism, three types of secretion are distinguished: merocrine, apocrine, and holocrine. During merocrine mechanism, the secretion is released from cells that remain alive after the secretory act. This type of secretion is classified into two subtypes: ecrine and granulocrine. In the case of ecrine secretion, the secreted substance consists of small molecules which can permeate through the membranes, as described above, by passive or active transport. The granulocrine secretion takes place when the secreted substance is water-insoluble or permeates the membrane with great difficulty. In this case, secretion is achieved via exocytosis.

Apocrine secretion occurs with damage to cells. Removal of the secretion is accomplished by excretion of a part of the cytoplasm, but in this case, only the nucleus-free portion of the cell is separated. An example of such secretion is the rupture of salt hair heads in some halophytes. Often it is unclear whether in such a secretion the cell is regenerated or dies off.

In the case of holocrine secretion, the cell degrades completely. The parts of the degraded cell are evolved together with the secretion. This type of secretion takes place when slime is secreted by the root cap cells.

1.3 Secretion into the Free Space of the Cell

The free space of a cell usually means the cell part that is outside the plasmalemma (Kursanov 1976). Since the cell wall and the plasmalemma are considered as a single complex which can be separated only with great difficulty (Salyaev 1985), secretion via the plasma membrane must be virtually secretion into the free space of the cell wall. Some space between the cell wall and the plasma membrane (periplasmic space) arises only as a result of secretory activity of the cell when the inclusion of the Golgi vesicle is ejected outside the plasma membrane, and in this case the plasma membrane may be separated from the cell wall by a considerable space filled with secretion (Salyaev 1985).

Secretion into the free space of the cell is particularly observed in specialized secretory cells, being the first stage of excretion from the cell. This process, however, is intrinsic in any plant cell, in particular, dividing meristemic cells form the cell wall from the components secreted to the free space. The major constituents of the cell wall are an amorphous matrix with a high water content, and the fibrillar system of the skeleton, made of cellulose. The matrix elements, hemicelluloses (essentially ureides) and pectins, are synthesized in the Golgi apparatus and then excreted through the plasmalemma by exocytosis (Frey-Wyssling 1973; Lüttge and Higinbotham 1979). The fibrillar system of the skeleton consists of cellulose (β-1,4-polyglucosan) and galactan (β-1,3-polyglucosan), which are synthesized outside the plasma membrane in the newly formed matrix of the cell wall. These elements are synthesized by a complex of enzymes localized in the plasmalemma (Frey-Wyssling 1973).

Recently, special attention has been paid to the synthesis of cell wall elements and their involvement in the biogenesis of the cell wall. It has been shown (Griffing et al. 1986) that polysaccharides are synthesized with the participation of two main enzymes: glucan synthase I (EC 24.1.12) and glucan synthase II (EC 24.1.34). Both enzymes have been found in membrane-coated vesicles. The former is found to be initially localized in the Golgi apparatus, the latter in the protoplasmic membrane and mitochondria. The membrane vesicles are assumed to separate from the cisternae of the Golgi apparatus, capturing the enzyme and transporting it to the cell wall, where the enzyme participates in the synthesis of β-1-4-bound glucan.

Many of the enzymes synthesized in the cytoplasm can be released into the free space of the cell by overcoming the membrane barrier. Some of them are then localized in the cell wall and participate in processes such as lignification, differentiation, and extension of the cell wall. Cellulose, β-1,3-glucanase, cellobiase, and some other hydrolases, malate dehydrogenase, acidic isoforms of peroxidase, and other enzymes have been found there. Thus, the cell wall may be considered as an "enzyme depot" (Salyaev 1985; Schloß et al. 1987).

Among the excreta released via the plasma membrane, there are products of decomposition or waste such as ammonia and hydrocarbons: ethane, ethylene, and propane. In particular, hydrocarbons are formed as a result of the peroxidation of chloroplast membranes (Böchler-Kohler et al. 1982). Generally, most volatile substances are released via the plasmalemma to the free space of the cell.

An example of secretion by specialized cells into the free space is the emission of monoterpenes. The synthesis of these substances occurs in large ameboid leucoplasts which have several inner membranes and their strome is devoid of ribosomes (Bernard-Dagan et al. 1982). Around the leucoplasts there is an envelope made of the endoplasmic reticulum, which is connected with their outer membranes, thus permitting terpenoids to migrate from the seat of their synthesis, the plastid outer membrane, to the seat of their accumulation, the free space of the cell. The synthesis and secretion of terpenes is presented in more detail in Chapter 3.6.

Phenols are known to be secreted into the free space of the cell as well. The compounds appear in sweed, *Brassica napus*, at the first stage of embryogenesis (36 h after the beginning of germination) either in the cytoplasm (in small vacuoles) or outside the plasmalemma. In the latter case, the secretion of phenolic substances serves as a barrier preserving from infection (Zobel 1989).

1.4 Secretion into the Vacuole

A completely developed plant cell has a large central vacuole which can occupy up to 90% of the cell volume (Nobel 1970). The vacuole is the seat of accumulation of water-soluble compounds which come there from the cytoplasm. The membrane surrounding the vacuole possesses selective permeability, so that it admits to the vacuole substances of a particular type. Physiologically, the substances accumulated in the vacuole belong to two different categories.

Substances of the first category are useful for the plant: sugars, amino acids, organic acids, etc., which are retained in the cell and can be further involved in the metabolism. The vacuole can contain up to 100 types of proteins. Among them are many hydrolases: acidic proteinase, acidic phosphotase, mannosidase, and galactosidase (Kenyon and Black 1986). Oxidoreductases are represented by the main isoforms of peroxidases (Schloß et al. 1987). Like all proteins, the enzymes are synthesized in the cytoplasm on ribosomes and are then transported across the tonoplast to the vacuole.

On the other hand, the vacuolar sap contains secondary metabolic products, flavonoids, alkaloids, etc., whose accumulation is caused by the limited excretion capacity of the cell. Thus, the central vacuole of the cell presents a kind of secretory system where substances of various chemical natures are isolated from the sphere of active transformations. The composition of the material secreted by the cytoplasm to the vacuole varies strongly, depending on plant species, phase of development, and organ (root, leaf, etc.) where the cell is located. The vacuole has a number of functions (osmoregulation, turgor maintenance, storage of assimilates, and secretion). The secretory function prevails in the mature cell when nonutilized metabolites are accumulated in the vacuole (Matile 1987).

The total concentration of substances, as well as the particular compounds, in the cell sap is higher than in the cytoplasm, so the movement of molecules is directed against the concentration gradient. As a rule, the transport of secreted substances across the tonoplast is carried out by active mechanisms: exocytosis and pinocytosis (Fineran 1971); in some cases, however, a passive process is possible (Buzuk and Lovkova 1986). The mechanism of transporting secondary substances across the tonoplast has been poorly studied except for phenols and alkaloids.

Phenolic compounds may also be synthesized in both the vacuole and chloroplasts. In the vacuole were found chalcone synthase, chalcone isomerase, and flavonoid glycoside transferase, and in the chloroplasts phenylanine ammonia-lyase, cinnamate 2-hydroxylase, and hydroxycinnamyl$\sim$CoA: quinnate hydroxycinnamyl transferase, and chalcone synthase (Hrazdina and Wagner 1985). Recently, new studies of phenol synthesis and the accumulation of these substances in chloroplasts and vacuoles have been undertaken. Under permanent illumination of the tea callus tissue, chloroplasts arose whose formation was accompanied by a significant increase in the amount of total soluble phenolic compounds and by the appearance among them of flavones along with catechins and proanthocyanidines (Zaprometov 1985). The flavones were represented by two aglycones (kaempferol and quercetin) and their five glycosides. It has been shown on isolated chloroplasts of *Populus nigra* that flavonoids are accumulated in the thylakoid membranes and in the external membrane of chloroplasts (Charriere-Ladreix 1977). Their accumulation in the vacuole is a secondary process, occurring due to transport from the chloroplasts. Some elements of the endoplasmic reticulum probably participate in the transport of phenols into the vacuole (Charriere-Ladreix 1977). In this case, phenolic substances being transported within secretory vesicles penetrate through the tonoplast into the vacuole by exocytosis. It is also possible that the movement of compounds across the vacuolar membrane proceeds via pinocytosis. This type of absorption in the central vacuoles of parenchymatous cells was shown by Fineran (1971). According to his scheme of the event, a particle or a molecule is absorbed on the tonoplast, which initiates invagination of the membrane, resulting in the formation of a vesicle which is found in the vacuole. After the membrane of the vesicle has disappeared, the adsorbed substance appears in the vacuolar sap.

As shown by Charriere-Ladreix (1977), phenolic compounds from chloroplasts are transferred in the form of aglycones. Once having penetrated to the vacuole, they undergo glycosidation by binding with a sugar residue, which diminishes the toxicity of phenolic compounds. The transformation of aglycone to glycoside increases their solubility in water and simultaneously reduces their solubility in biomembrane lipids, thus hampering the back outflow of phenolic compounds from the vacuole through the tonoplast.

Besides phenols, alkaloids, and terpenes, some hydrocarbons synthesized in chloroplasts and leucoplasts can also then be accumulated in the vacuole. Some recent studies have been devoted to the accumulation of alkaloids in the vacuole (Deus-Neumann and Zenk 1984, 1986; Müntz 1984; Buzuk and Lovkova 1986). The synthesis of alkaloids takes place in the cytoplasm (Deus-Neumann and Zenk 1984) or can sometimes occur in plastids or in some individual vesicles (Hartmann 1985; Wink 1985). Initially, the conceptions of the accumulation of alkaloids in vacuoles were based on the fact that the vacuolar sap has an acidic reaction (low pH). According to this model, alkaloids can quickly penetrate through the tonoplast by diffusion in the lypophilic layers. In the acidic medium of the vacuole, alkaloids are protonated and hence captured as cations, for which the tonoplast is either poorly permeable or absolutely impermeable (Matile 1984). Moreover, phenols and other components of the cell sap can immobilize alkaloids by forming complicated complexes with them. This mechanism, known as "ion trap", has been evidenced by classical observations with neutral red, whose absorption mechanism is believed to be analogous to that of alkaloids (Matile 1984). However, this phenomenon is not considered to be a widespread one, as follows from the experiments with isolated vacuoles (Deus-Neumann and Zenk 1986). It has been shown that isolated vacuoles accumulate only alkaloids specific for a given plant. The vacuoles of plant species that do not synthesize these alkaloids are unable to accumulate them. Mende and Wink (1987) studied the uptake of the alkaloids lupinine (1-oxymethylquinolisidine) and atropine by protoplasts and isolated vacuoles from tissue culture cells of *Lupinus polyphyllus*, *Spinacia*, *Galium*, and *Atropa belladonna*. Lupinine was taken up only by vacuoles of *Lupinus*, and atropine only by analogous organelles of *Atropa*. The vacuoles of other plants studied absorbed neither lupinine nor atropine. Some specificity is alkaloid uptake by isolated latex vacuoles from *Chelidonium majus* and *Papaver somniferum* was also observed in later works (Hauser and Wink 1990; Roberts et al. 1991). Based on similar experiments, a concept of high-taxon-specific carriers of alkaloids, which function at the expense of metabolic energy and are localized in the tonoplast, has been created (Deus-Neumann and Zenk 1986). However, there is evidence that the specificity in the alkaloid uptake by vacuoles depends not only on the taxonomy of the plants which synthesize the substances. In particular, Sato et al. (1990) have demonstrated that the absorption of alkaloid berberine by protoplasts and vacuoles isolated from cells of tissue culture of *Coptis japonica* and *Thalictrum flavum*, which hardly or only poorly synthesize this compound; but when being absorbed by protoplasts, this alkaloid was not accumulated in the vacuole. In another case, cells in tissue culture of *Thalictrum minus* excreted berberine and were unable to take up the exogenous alkaloid (Sato et al. 1990). Probably the absorption by vacuoles was due not only to taxonomic specificity, but also to the ability to excrete this alkaloid.

Some authors (Buzuk and Lovkova 1986; Roberts et al. 1991), using alkaloids of different structures, have shown that the rate of penetration of these substances into vacuoles varies strongly. Matern in his review (1987) believes that the sterical parameters of a molecule to be transported are essential for the transport of secondary metabolites into the vacuole and their fixation there. The advantages of the stereoisomeric model is that it explains the penetration into the vacuole of basic, acidic, or neutral substances and their selective accumulation. The evidence for the validity of this model could be the isolation of conformation- and configuration-specific carriers from the tonoplast.

The rate of alkaloid uptake by vacuoles is actually high (0.03 mg/mg of vacuolar protein) (Deus-Neumann and Zenk 1986), which may be due to both active and passive mechanisms of transport through the tonoplast. The participation of both mechanisms has been shown by using specific blockers (Buzuk and Lovkova 1986). Some factors point to the participation of the proton mechanism of H^+ -ATPase in the active transport of alkaloids. In particular, vesicles isolated from the latex of *Papaver somniferum* were unable to absorb morphine from the external solution (Pham et al. 1989) if ATP and Mg^{2+} were absent in external media, or absorbed only poorly (Roberts et al. 1991). The absorption of lupinine by *Lupinus* vacuoles was activated 30- fold by Mg^{2+}, ATP, and KCl, and inhibited by DCCD (dicyclohexylcarbodiimide), which usually blocks the activity of H^+-ATPase (Mende and Wink 1987). Hauser and Wink (1990) have shown that the protoberberine and benzophenanthridine alkaloids naturally accumulated by latex vacuoles of *Chelidonium majus* penetrate these isolated organelles via diffusion rather than with participation of carriers. The kinetics of the uptake is rather characteristic for a simple diffusion; but among seven studied heterologous alkaloids of various structure (nicotine, vinblastine, 9,10-dihydroergocryptine, strychnine, colchicine, lupanine, 13-hydroxylupanine), dihydroergocryptine and vinblastine, whose movement is stimulated by ATP, had the highest rate of accumulation by isolated vacuoles. Thus, except for the case of the two latter alkaloids, ATP had no influence on this process or decreased it slightly, although the alkaloid transport occurred against the concentration gradient. It is believed to be due to chelidonic acid, whose concentration in vacuoles reaches 661 mM, whereas in latex vesicles only 58 mM. Chelidonic acid forms complexes with these alkaloids and in this manner can prevent their back-diffusion from the vacuolar compartment. This is a possibility for the alkaloids to be accumulated within the vacuoles via a passive mechanism.

The isolated latex vacuoles from *Papaver somniferum* took up the alkaloids morphine, codeine, thebaine, nicotine, noscapine, papaverine, and caffeine. The rate of the process is dependent on the maintenance of tonoplast pH and on ATPase, which generates this pH (Roberts et al. 1991). Exogenous ATP stimulated morphine accumulation. Sequestration of the alkaloids taken up appears to involve protonation and anion-cation stabilization with participation of meconate and sulfate, important acids of the vacuoles. The specificity of alkaloid uptake had no clear correlation with pK or lipophilicity. A possible channel mechanism of alkaloid influx more related to alkaloid shape is suggested. According to the view of Roberts et al. (1991), it is more likely than the mechanism of protein carriers.

Among the alkaloids to be secreted, there are physiologically active and sometimes toxic substances. It remains unclear why the cells or organelles are resistant to them. Recently, it has been proposed (Roos and Luckner 1986) that it is due to an asymmetrical architecture of the membranes, which provides that their external and internal sides possess different sensitivity to specific metabolites. This was demonstrated in experiments with vacuoles from the latex of *Chelidonium*, where binding of the isoquinoline alkaloids, sanguinarine and chelerythrine, was observed. Both compounds induced lysis of isolated vacuoles if their concentration was higher than the organelles were able to accumulate. The vacuolar membrane as more stable if the alkaloids were accumulated only in the vacuole, but was damaged when the same substances were accumulated outside the vacuole.

Penetration of other secondary metabolites through the tonoplast has been poorly studied. In order to gain some insight into the physiological role of the secondary metabolites taken up by the vacuole, it is essential to know that they are accumulated irregularly and only in particular plant species, and that they can be released from the cell only after damage to the latter.

Although most of the secondary metabolites are stored in nonplasmatic compartments of living cells, the lipophilic substances can form lipid drops inside the cytoplasm. Among the compounds are carotenoids, resins, and essential oils. In these drops, other lipophilic substances such as alkaloids can be dissolved as well (Roos and Luckner 1986). Similar drops are also found in vacuoles.

1.5 Idioblasts

The secondary function mentioned above is inherent in all plant cells. However, in plants there are specialized cells (idioblasts) in which the secretory function prevails. The idioblasts are scattered among other tissues of vegetative and reproductive organs of the plant and may differ substantially from other cells of the same tissue in their form, structure, and content (Esau 1965). The specialized secretory cells accumulate mineral salts, resins, tannin, oils, etc. Some of them contain mucilage and are called mucilage-secreting cells.

The isolated cells become secretory and should be called idioblasts where non-utilized products are kept inside. Usually there are no intercellular spaces around them, and the secretion is within the idioblast as long as this cell lives.

The secretory products are stored either within the vacuole or in the cytoplasm (Buvat 1989). The latter case is known for oil cells like those found in leaves of *Citrus limon* and *Laurus nobilis*.

The depositions of mineral salts in idioblasts are mainly calcium oxalate, potassium carbonate, and silicon oxides. Most frequent are depositions of calcium oxalate, which is found in many plant families, and the number of plants where such crystals are detected is becoming ever greater. Calcium crystals may be single (rhombohedrons or octahedrons) or have a complex structure in the form of concretions (druses, spherites, raphides, etc.). Sometimes the calcium oxalate inclusions occur in the form of small pyramidal crystals, the so-called "crystalline sand" (Cody and Horner 1985). The form of the crystals, as some researchers believe (Cody and Horner 1985), reflects the properties of the environment in which the crystal grows, i.e. those of the cell vacuole. Therefore, the crystals are indicators of the intracellular microconditions. It is also believed that the crystal form is a taxonomic feature (Esau 1965).

The cells accumulating calcium oxalate differ even at earlier stages of development from neighboring parenchymatous cells in their ultrastructure: they are rich in cytoplasm and have a high content of cell organelles. The crystals are formed in special chambers surrounded by membranes and found in vacuoles. The oxalate transport through the tonoplast and the chamber membrane is probably realized via vesicles originating from the elements of the endoplasmic reticulum and dictyosomes. It has been shown experimentally (Schötz et al. 1970) for *Oenothera*. The active Ca^{2+} transport with participation of the ionic pumps localized in the membrane is also possible (Vasilyev 1977). After completion of crystal formation, a polysaccharide envelope is formed around the crystal-bearing vacuole. This vacuole can be lignified or suberized. Finally, it is connected with the cell wall by a short stalk (Fig. 3). In the case of suberization, the idioblast dies off and presents a dead structure. In other cases, the protoplast remains alive, though the cytoplasm and its organelle content reduce drastically.

It is proposed that in photosynthesizing organisms for oxalate synthesis, glycolic acid is utilized, which is formed in the course of photosynthesis (Frank and Jensen 1970). Franceschi (1987) studied the formation of the crystals in *Lemna minor* using [^{14}C]-labeled easily reversible precursors of oxalic acid. Labels from [^{14}C]-glycolic, glyoxylic, oxalic, and ascorbic acids are incorporated into crystals. The participation of glycolate and glyoxylate in the metabolic pathway of oxalate formation has been shown. However, they do not originate from the process of photorespiration, as was believed earlier (Frank and Jensen 1970). Indeed, calcium oxalate crystals are formed only in very young tissues which possess a low photosynthetic activity (Franceschi 1987). Under favorable conditions, crystals appeared as fast as within an hour, and the rate of the glycolate transport from mature tissues, where photorespiration takes place, cannot be high enough to provide growth and formation of crystals. Besides, in plants growing in the dark, there were more crystal idioblasts with calcium oxalate (297 ± 32 per mm^3 of frond surface) than in plants growing in the light

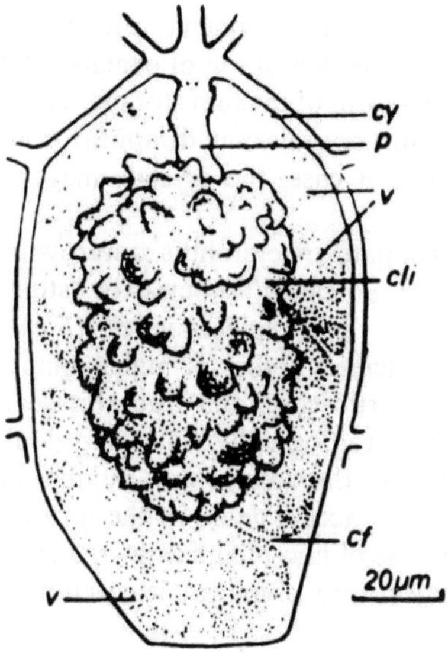

Fig. 3. Lithocyst in the upper epidermis of *Ficus elastica* leaves. The lithocyst has the cystolyth (*cli*) around the pedicel (*p*); *v* large vacuole with cytoplasmic filaments (*cf*) across; *cy* cytoplasm. (Hiltz 1951, after Buvat 1989)

(78 ± 19 per mm^3 of frond surface). Hence, it should be assumed that photorespiration cannot be the main source of glycolate for crystal formation.

In different plants, calcium oxalate crystals vary depending on localization and time of formation. Analysis of cells containing calcium oxalate crystals in kidney bean, *Phaseolus vulgaris*, and *Canavalia ensiformis* grown in Hogland's solution or in the soil shows that in kidney bean leaves, calcium is deposited in the adaxial part of the vascular bundle sheath, whereas in *Canavalia*, it is stored in the epidermis (Zindler-Frank et al. 1988). In kidney beans, crystal-bearing cells do not differ morphologically from the other cells of the sheath, whereas in *Canavalia* they are transformed to highly specialized idioblasts. The average content of calcium per dry mass is close in both species and equals to 1 – 0.8% in leaves.

In experiments with leaves of plum tree, *Prunus salicina*, and cherry tree, *P. avium*, the rate of calcium oxalate deposition has been shown to depend on the species and the phase of plant development (Sanchez-Alonso and Lachica 1988). As a rule, the calcium oxalate content in the plum tree is higher than in the cherry tree. Formation and accumulation of calcium oxalate proceeds parallel to the formation and maturation of fruits. The crystal-bearing idioblasts appear near the first leaf veins of *Gleditschia triacanthos* when the young leaf is at the beginning of its development, and then as the leaf becomes older, crystals of Ca oxalate accumulate also in other tissues. The total amount of calcium oxalate (in mg per 1 g of dry mass) in leaves of *Telfairia hooker* (Cucurbitaceae) is equal to 19.2 in plate, 18.8 in petiole, and 7.0–7.4 in stem and tendrils (Okoli and McEuen 1986). The formation of calcium oxalate is believed to be an effective adoptive mechanism for maintaining a low calcium level in the cytoplasm (Franceschi

1987). Probably this is very important because many plants are unable to absorb Ca^{2+} efficiently from the surrounding medium. The deposition of oxalate salts in crystals is a stable, nontoxic, and nondiffusing form of calcium storage. There is evidence that, under certain conditions, calcium crystals may disappear from vacuoles if the cells still remain alive. In such a case, the crystals may be considered as reserve material.

The best known concretions of calcium carbonate are cystoliths, outgrowths of the cell wall impregnated with this substance. So-called lithocysts or "stony cells" containing cystoliths are usually generated by basic epidermic cells (*Ficus*) or by epidermal hairs (*Humulus, Cannabis*). Calcined bodies of higher plants have been studied in leaves of the Moraceae, Urticaceae, and Cucurbitaceae families by using energy-dispersive X-ray microradiography (Okazaki et al. 1986). The amount of $CaCO_3$, as calculated from the volume and number of cystoliths, was 0.4 mg/cm^2. Calcium carbonate is accumulated in the form of cystolith-like crystals in cells localized at the base of microhairs (Fig. 4) and as heavy incrustations within the midrib and major veins of mature leaves (Okoli and McEuen 1986).

The ontogeny of lithocysts is well described for *Ficus elastica* by Buvat (1989) (Fig. 3). A young epidermal cell, which is going to differentiate into a lithocyst, is first individualized by the thickening of the cell wall, which grows toward the inside of the cell to form a pectocellulose appendage called a pedicel. The pedicel is surrounded by the cytoplasm which forms it. After pedicel formation, the cytoplasm begins to elaborate around its end a pectocellulosic weft, which is soon loaded with silica or calcium. Often, the amorphous calcium carbonate with silicon incorporations is formed. There is evidence that microtubules play a certain role in the formation of cystoliths and lithocysts. Thus, Galatis et al. (1989) showed that destruction of microtubules of *Pilea cadierei* by colchicine gives rise to defects or abnormalities in the formation of cystoliths. Moreover, the development of an idioblast (lithocyst) as a whole is severely deformed by the inhibitor of microtubules colchicine.

The silicon oxide can also be deposited in the cell wall and interior of the idioblast cell. In *Phalaris canariensis*, grown on a nutrient solution containing 50 mg/l SiO_2, silicon was accumulated in the epidermis and external cortical layer of root cells, in the epidermis of the stem and hairs of the thorn of the leaf lamina (Hodson 1986). In cereal plants, the idioblasts containing silicon in the epidermis of the lower palea (glume) in C_3 plants (*Avena, Triticum*, etc.) were found to differ from those in C_4-plants (*Sorghum*, etc.). In particular, the number of idioblasts in plants of C_4-species was greater than in those of C_3-species, and the shape of cells was different. Silicon-containing cells of C_3-species were spherical, elliptical, or christ-shaped and sickle-shaped, whereas those of C_4-species were shaped like dumbbells or crosses (Takeoka et al. 1979). The physiological significance of this phenomenon is not understood.

A peculiar type of silicon-containing cells are so-called pairing cells in cereals. These are two adjacent cells, one of which has a suberized, and the other a silicified cell wall. A cell that contains silicon in the cell wall dies off early, its

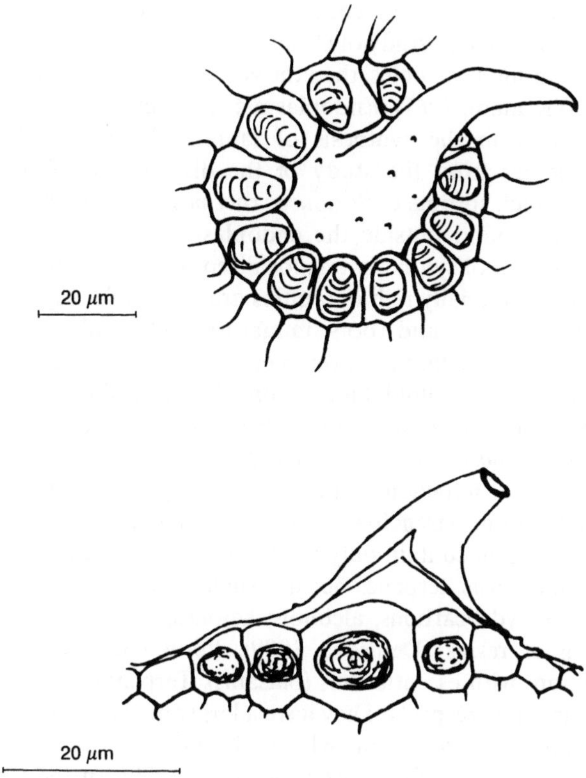

(a)

(b)

Fig. 4a, b. Cystolith-like structures and adjacent epidermal cells in *Telfairia occidentalis* (Cucurbitaceae). **a** Surface view with warty-walled macrohair. **b** Towards one edge of the hair base. (Okoli and McEuen 1986)

cavity is soon filled with silicon. The cell whose walls undergo suberization is rich in cytoplasm and has a well-developed tubular agranular endoplasmic reticulum charateristic for all secretory cells. Miroslavov (1974) believes that SiO_2 is accumulated in the secretory cell, which then evolves silicon into the adjacent cell by means of an active mechanism. The ecological significance of this process is probably protection against herbivorous animals.

Besides mineral concretions, organic substances are accumulated in idioblasts, among them phenols. Usually tannins can be stored in both specialized and normal plant cells. Tanniniferous cells are so frequent that it is difficult to refer to them as secretory cells (Buvat 1989). Usually they occur both in cells with a decreased proliferating activity and in tissues with a high physiological activity like those of leaf mesophyll. Tanniniferous cells are seen as cherry-red stains after coloration by neutral red.

Tannins are complicated mixtures containing esters of glucose and m-digallic acid. Electron microscopy study of *Oenothera* cells demonstrated that tannins are first localized in small vacuoles, then the vacuoles become larger and occupy the whole cell (Wiermann 1981). Endoplasmic reticulum, Golgi bodies, and plastids can take part in the synthesis of tannins (Wiermann 1981). Confirmation for this comes from the study of the distribution of phenol substances in apexes of branch sprouts of *Sambucus racemosa* (Zobel 1986). The precursors of phenol were found outside the endoplasmic reticulum, and a portion of the enzymes of phenol synthesis was localized on the membranes of the organelle (Zobel 1986). Especially rich in tannins are leaves of many plants, xylem, phloem, periderm of stems and roots, immature fruits, and seed coat. Probably none of the tissues lacks tannins, even meristematic cells (Esau 1965). Tannins are stored in vacuoles in colloidal form. Sometimes, tanniniferous cells contain two types of vacuoles, with and without tannins, as is shown for the motor swelling organs of *Mimosa pudica* (Buvat 1989).

Like tannins, substances of terpenoid nature may be deposited not only in specialized secretory but also in parenchymal and epidermal cells. The osmophores of the flower petals often contain only essential oils, which are mixtures of liquid terpenoids. In idioblasts, terpenes occur, usually in combination with other compounds such as hydrocarbons, alcohols, ketones, and a complex of substances in the form of resins (Denisova 1989). Resin consists of solid terpenoids often in mixture with essential oils (balsams). Terpenoid cells differ from nonsecretory cells in many respects. They have a larger nucleus, and a well-developed agranular endoplasmic reticulum, which is believed to play a key role in the biosynthesis of terpenes (Vasilyev 1977; Denisova 1989). In these cells there are many mitochondria which provide energy for the processes of both synthesis and secretion of the substances into the vacuole. Terpenoid cells also possess a developed plastid apparatus consisting of numerous leucoplasts (Vasilyev 1977). In specialized secretory cells, terpenoids are first accumulated in small amounts inside the protoplasm and then enter the vacuoles. Cell walls of such idioblasts become rapidly suberized, and the secretion contained in the cell is separated by a cellulose and suberic envelope. In most plants, idioblasts of adult organs lack living matter, and the cellulose wall impregnated with suberin extends as an excrescence towards the inside of the secretory cell to form a stalk or cupule, by which the oil-containing vacuole (oil-sack) is attached to the cell wall (Postek and Tucker 1983; Bakker and Gerritsen 1990). Sometimes, in oil-containing cells (for instance, in *Magnolia grandiflora*), this stalk is lacking, and the suberine layer develops later and does not extend inward in the oil cell (Postek and Tucker 1983). Oil cells look homogenously “shiny” under the light microscope, unlike mucilage cells, which occur with or replace them (Baas and Gregory 1985). The latter lack a cellulosic suberized layer and cupule.

Oil idioblasts occur widely in the plant world. They are found in representatives of more than 20 families: Magnoliaceae, Annonaceae, Cannellaceae, Hernandiaceae, Illinacee, Lauraceae, Myristicaceae, Piperaceae, Araceae, Aristolochiaceae, Saururaceae, Labiatae, Onagraceae, Rutaceae, Simarubaceae,

Valerianaceae, etc. (West 1969; Bakker and Gerritsen 1990). In some plants, oil idioblasts have been studied in detail. In *Asarum europaeum* L. (Aristolochiaceae family), essential oils are in large idioblasts of isodiametric shape spread in parenchymal tissue. Oil cells are localized in the epidermis. A drop of oil is enclosed in a thin-wall pocket attached to the stalk (a small core impregnated with cutin). On the average, they amount per 1 mm^2 to 97 for petals, 19 for leaf lamina, and 44 for petioles and rhizomes (Suslova and Sharygina 1968). Suslova and Sharygina (1968) also found that the greatest amount of essential oil is contained in rhizomes and roots, reaching 2.13–3.69% of the absolute dry mass, whereas in leaf laminae it is only about 0.1%. According to Ozarowski (1956), besides terpenoids, essential oil of the same species involves asarone (30–50.2%), asaryl aldehyde (2–3%), methyleugenol (15–20%), resin (10–12%), diasarone, and sesquiterpene alcohol α-pinene (1–2%). Unlike *Asarum* cells, single idioblasts of Magnoliaceae contain mainly sesquiterpene lactones (Caniato and Cappelletti 1984). Some examples of the substances are shown in Fig. 5. Thus, terpenoid-containing cells may differ in the composition of their secretion, depending on the species and the organ of the same species.

Besides low-molecular mass terpenoids, idioblasts may also include their polymers. For instance, idioblasts containing polyisoprene in the form of gutta occur in the bark of the stem of *Eucommia ulmoides* (Tian Lan-xin et al. 1990). Gutta is a high-molecular weight hydrocarbon $(C_5H_8)_n$, with elementary composition similar to that of rubber. Both substances are polymers of isoprene C_5H_8, and differ as stereoisomers. Gutta-containing structures are filamentous cells. They are formed at the stage when the procambium differentiates into the primary sieve-like elements of the protophloem. Thus, gutta-containing idioblasts are derived from meristematic bark cells. Gutta particles in these cells are intensively synthesized and accumulated in the cytoplasm where the organelles are degraded. In mature gutta-containing cells where the cell cavity is filled with gutta, the nucleus and other organelles disintegrate and leave the intact fibrous cell wall (Tian Lan-xin et al. 1990).

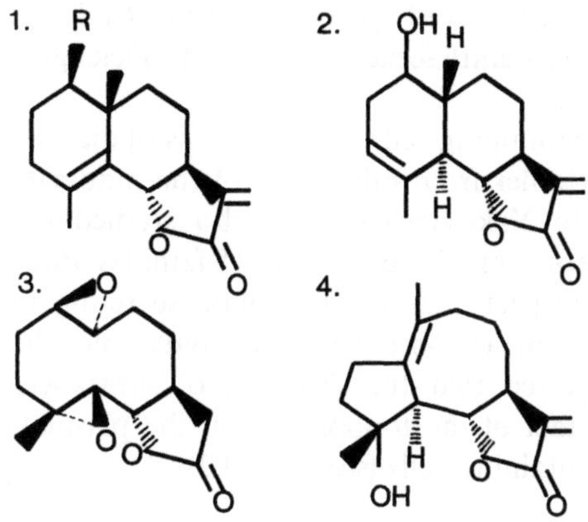

Fig. 5. Sesquiterpene lactones in oil idioblasts of a plant belonging to the Magnoliaceae. *1* Magnolialide (costunolide-1,10-epoxide); *2* santamarine; *3* michenolide; *4* micheliolide

Typical for some groups of plants is the accumulation of alkaloids in idioblasts. Eilert et al. (1986) found single idioblasts containing acridone alkaloids in roots and suspension culture of *Ruta graveolens*. A distinguishing feature of idioblasts is the aggregation of small vacuoles in the center of the cell, which show fluorescence characteristic of acridone alkaloids, and are stained with permanganate. Neumann and Müller (1972) reported that in *Sanguinaria canadensis*, excess of the alkaloid sanguinarine and its related benzophenanthridine alkaloids is stored in idioblasts as well.

It should also be noted that idioblasts are the site of accumulation for steroids, glycosides, and saponines. Water-soluble furastanol glycosides are present in leaves of *Dioscorea deltoidea*. They are localized in idioblasts situated on both sides of the leaf. In sections treated with Ehrlich reagent (1% n-dimethyl aminobenzaldehyde in methyl alcohol-hydrochloric acid mixture) they look as crimson-colored cells (Fig. 6) (Gurielidze et al. 1988).

In plants belonging to families Cactaceae, Crassulaceae, and Orchidaceae, polysaccharides are accumulated in idioblasts, mainly in the form of slime which most frequently is secreted into the space between the cell wall and plasmalemma, gradually forcing the protoplast towards the center of the cell. In this process, the volume of both the cytoplasm and vacuole is reduced. Finally, the interior of the cell can be entirely filled with slime and the cell dies. Probably, the secretion of slime occurs also into the vacuole (Vasilyev 1977). Mucilage-containing cells are met in leaves of genera *Alcea* (Bykova and Yakovleva 1991), *Salix* (Mariani et al. 1988), *Tilia* (Yakovleva 1988). Species of genus *Alcea* (family Malvaceae) vary in the amount of polysaccharides from 0.2 for *A. nikitinii* to 3.10% of dry mass for *A. flavovirence* (Bykova and Yakovleva 1991). The mucilage of their cells differs in two zones: the zone of dense slime (just near the cytoplasm of the mucilage cells), and the zone of loose slime. These zones leave contact with each other, unlike species of the family Tiliaceae such as *Tilia platyphyllos*, *T. cordata*, and others, where the mucilage zones are separated by the cell wall. Mucilage of slime cells contains fibrils collected in sheaths of various forms unlike slime-releasing cells. More often, mucilage cells are observed in epiderma and mesophyll as single cells, and, rarely, are combined in groups. Fibrillar mucilage is also observed in seed epidermis cells in *Collomia grandiflora* (Polemoniaceae) and *Ruellia* (Acanthaceae) (Schnepf and Deichgraber 1983a, b).

Besides these specialized mucilage-containing cells, the epidermis of dicotyledon leaves contains cells that do not differ from other cells of the induviate tissue, but secrete slime. According to Yakovleva (1988), who studied the ultrastructure of slime cells in the epidermis of 35 species from 15 families, three types of cells can be distinguished with respect to localization of the secretion, in the cell cavity or intercellular spaces or in both the cell cavity and the intracellular spaces. The author suggested that the secretion of substances outside the cell proceeds through contact of the tonoplast with the plasmalemma, which is also typical for salt-containing cells (see Sect. 3.2).

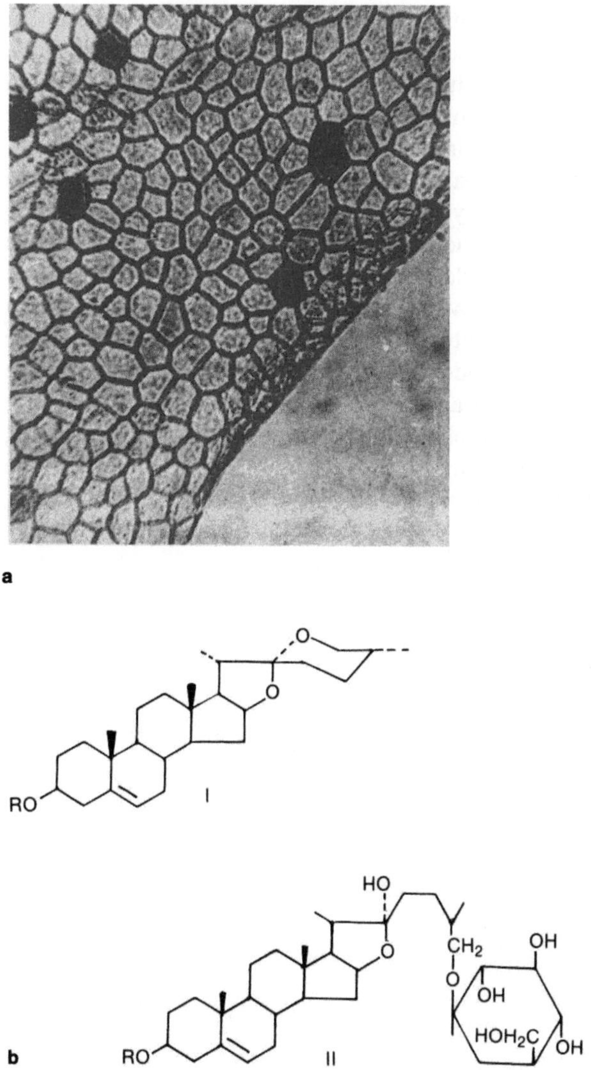

Fig. 6. a The longitudinal section of leaf epidermis of *Dioscorea deltoidea* (×28a). Dark cells are idioblasts with oligofurastanosides. **b** Formulas of steroid glycosides. *1* Oligospirostanoside; *2* furanostanolic glycoside. (Gurielidze et al. 1988)

Slime-secreting cells of different plants have a similar structure, inherent in all secreting cells as long as they live. They are rich in cytoplasm, have a well-defined Golgi apparatus, and numerous mitochondria with well-developed crista (Vasilyev 1977). The chemical composition of slime is variable in different plants (see Sect. 3.4).

Slime cells are located in different organs and tissues, and the amount of secretion in them is variable. The accumulation of mucilage in leaves of *Thuja occidentalis* occurs in idioblasts, whereas in *Taxus faccata* it proceeds in vacuoles of mesophilic cells (Distelbarth and Kull 1985). The amount of slime undergoes seasonal oscillations. In winter it is maximal, in spring it is reduced, and in summer it is minimal (Distelbarth and Kull 1985). Idioblasts with slime are formed primarily in the aerenchyma of different organs as well as in the epidermis of stems and leaves. The mucilage cells are found mainly in generative organs such as petals, floral receptacle, staminal tube, and style.

Idioblasts from staminate and ovulate cone sporophylls of the Mexican cycad *Zamia furfuracea* contain neurotoxin 2-amino-3(methylamino)-propanoic acid (Duncan et al. 1991, 1992; Vovides 1991). This toxin is sequestered in the larval phase of the weevil pollinator and taken into the pupa case. Cycads are also toxic for mammalia. Their idioblasts are mainly located in parenchyma, except *Cycas rumphii* and *Stangeria eriopus*, where they are found in the epidermis and/or hypodermis (Vovides 1991).

CHAPTER 2

Intratissular Secretion

Idioblasts are single secretory cells. They are secretory structures of a simple type. The evolution of secretory structures is proposed (Fahn 1979) to go from scattered cells, idioblasts, to organized intratissular canals, ducts, and cavities covered by epithelial cells and to end on the glands located on the plant surface. Two main types of secretory tissues can be distinguished depending on origin and localization: secretory tissues that are located within the plant and secrete substances into the intra- or intercellular spaces of plant, and secretory structures that are located on the plant surface and secrete compounds directly out onto the plant surface. The first type of secretion will be dealt with in the present chapter and the second will be described in Chapter 3.

The internal tissues of the plant are pierced by a network of intercellular spaces filled either with gases to form an air-bearing system, or with various secretions to form a system of intratissular secretory channels (Kucera 1985).

2.1 Air-Bearing System of Plants

Almost all plant tissues have a system of intercellular spaces filled with air. Well-developed are the intercellular channels in the leaf blade and photosynthesizing stem parenchyma, the mature parenchyma is less rich in them, and meristematic tissues usually contain no intercellular spaces. Studies of permeability of plant organs for gases entering under pressure indicated that in plants there are two systems of intercellular spaces, a continuous and an intermittent one. Continuous intercellular spaces in plants are large in size; therefore, air and gases are free to pass from one part of the plant to another. The leaves of angiosperms are characterized by a widely branched system of intercellular spaces. Air spaces account for 10–60% of the plant volume and are the basic pathways of diffusion of gases, mainly oxygen, carbon dioxide, and water vapor (Jeffree et al. 1986). A developed system of intercellular spaces is found in plants growing in a habitat with poor gas exchange. Thus, herbaceous marsh and aquatic plants have a special air-bearing tissue aerenchyma, the formation of which is due to oxygen deficiency. This is confirmed by studies of the aerenchyma from the root bark of orchard grass, *Dactylis glomerata*, and moisture-loving species of fescue, *Festuca* (Focke 1985). Some mesophytes such as *Phaseolus vulgaris*, *Vicia faba*, *Zea mays*, etc., also have a developed system of intercellular spaces. In these plants, gases move not only along the central air-bearing cavity but also along the intercellular spaces of parenchyma and phloem of stem.

Intercellular spaces are the internal medium of the organism and they play an important role in its vital activity. Although the intercellular spaces are in contact with the external air via numerous stomata and lenticels, the gas composition in them differs strongly from the atmospheric air. It is more variable, since it depends on the metabolism and gas excretion of cells surrounding the intercellular spaces.

Cellular walls which line intercellular lacunas feature a high biochemical activity (Salyaev 1969). They contain enzymes (see Chap. 1.3) and, consequently, numerous biochemical processes may arise that lead to the formation of different organic substances which can be found in the intercellular spaces of plants. In some periods of life, the components of the cell wall participating in the formation of the plant skeleton may be partially hydrolyzed and involved in secondary metabolism, which, in turn, influences the chemical composition of the intratissular spaces of plants. No less important is the property of the cell wall contacting with the inner gaseous medium of the plant saturated with water vapors, to sorb vapors and gases and to desorb them into the internal medium of the plant. In other words, cell walls have the properties of polyvalent adsorbents with different types of bonding to the absorbed compounds (Salyaev 1969). Thus, in the internal gaseous medium of the plant there may occur rather complicated transformations which are due to specific features of the intercellular ducts and the walls of cells surrounding them.

The study of the internal gaseous medium of plants is primarily of great importance because the contents of intercellular spaces are an indication of the general metabolic state of the plant. Since the intercellular spaces communicate with each other, and gaseous products from different parts of plants are present in them, these compounds (among them the hormone ethylene) may direct and regulate the general metabolism and be involved in it at definite stages.

2.2 Internal Gases

2.2.1 Carbon Dioxide

The studies of gas composition in the internal medium of plants began as far back as two centuries ago. Much attention was focused on the gas composition of leaves. Since leaves are adapted to intensive aeration, it was believed that the gaseous medium of leaves is close in composition to atmospheric air. Later, this proved true only in relation to nitrogen. The amount of oxygen and carbon dioxide in internal media of plants varies strongly, and not seldom differs significantly from their concentrations in the atmosphere. The level of CO_2 in leaves of plants is often higher than in the surrounding air. For instance, in sugar beet, *Beta vulgaris* var. *altissima*, the CO_2 amount can vary from 1 to 7% (Rubin and Panasenko 1956). In leaves of the plants belonging to the Crassulaceae family, the concentration of carbon dioxide achieves 0.15–2.5%, depending on the species (Cockburn et al. 1979). Extremely high concentrations of CO_2 were found in leaves of woody plants during a hot dry summer (Roshchina 1975).

The amount of CO_2 in the internal air of leaves reflects the influence of many internal and external factors. The most important internal factors are the rates of photosynthesis and respiration. Also of some importance may be dark fixation of CO_2, decomposition of organic acids in the light (Soldatenkov 1971), and the entrance of CO_2 through roots (Kursanov et al. 1952).

Billings and Godfrey (1967) found a high level of carbon dioxide in intercellular spaces of stems of herbaceous plants of mountain meadows and swamps. A great amount of CO_2 was found in intercellular spaces of root systems. Burström (1959) reported that roots of wheat, *Triticum*, grown on nutrient solution, have a branched network of intercellular spaces filled with gas. In meristemic tissues located at the top of the root, the internal gaseous phase consists almost entirely of pure carbonic acid. MacDougal and Working (1933) were the first to summarize the data on gas content in stems of woody plants. They found that in stems, air occupies a considerable part of the volume. Heart wood of trees contains more gases than sapwood does. In stems of leafy woody species, more gases are accumulated than in coniferous plants. The air contained in stems of trees and bushes differs in composition from atmospheric air. It has a greater amount of carbon dioxide, many times exceeding that in atmosphere. For instance, the concentration of CO_2 in wood of pine, *Pinus*, is 200 times and in walnut, *Juglans regia*, 600 times that in the atmosphere (Yatsenko-Khmelevskii 1954). In eastern poplar, *Populus deltoides*, the amount of carbon dioxide in the stem may reach 26% (Shaedle 1975) and in 5–7-year-old sprouts of leaf-bearing species 23% (Roshchina 1975).

CO_2 is unevenly distributed throughout the plant. This gas formed by living cells may be accumulated not only in intercellular spaces but also in vessels and tracheids. As a result, a considerable part of the xylem may be filled with air. On the contrary, the external sapwood contains a great amount of water and thus accumulates the gas to some extent. The greatest amount of gas is contained in the heartwood, which serves as a reservoir for the evolved gases. The intercellular spaces of the phloem communicate with the external air via pores and therefore accumulate less CO_2 (MacDougal and Working 1933).

Seasonal changes in the amount of gases in stems of woody plants have been studied by some investigators (MacDougal and Working 1933; Chase 1934). It was found that CO_2 concentration is high during vegetation, and in winter is very low. Seasonal changes in carbon dioxide content in sprouts of woody plants have been investigated by Roshchina (1973b,1974c, 1975). Using gas chromatography, the author studied a great number of trees. Along with the CO_2 content, the intensity of respiration and humidity of sprouts was analyzed (Fig. 7). The amount of CO_2 in the internal air was found to depend on season and taxonomy of the plant and was always many times higher than in the surrounding air.

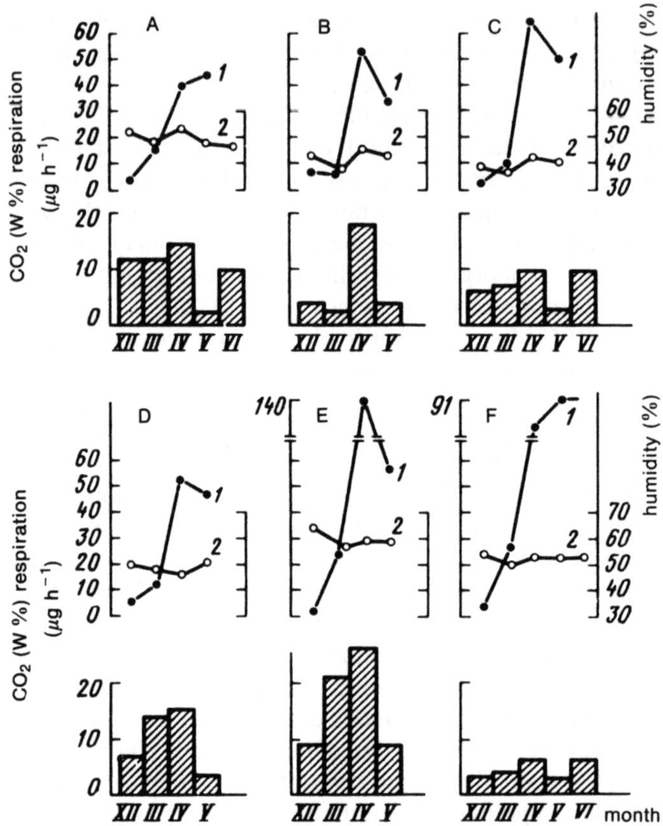

Fig. 7A–F. Seasonal changes of CO_2-concentration, respiration, and humidity in 4-6-year-old shoots of trees. **A** *Populus balsamifera*, **B** *Betula verrucosa*, **C** *Cotinus coggygria*, **D** *Larix decidua*, **E** *Pinus strobus*, **F** *Tilia cordata*. (Roshchina 1974c)

During the winter period, the concentration of gas did not change significantly, since the communication between plant and environment was minimal. In spring, the CO_2 level was highest and coincided with the onset of sap flow, increase in respiration rate, and awakening of physiological activity. In some plants, the increase in carbon dioxide concentration in the internal atmosphere was found to correlate with humidity. During leaf formation, the CO_2 level decreased. It seemed that carbon dioxide accumulated during the rest period, moved up the stem, and was utilized by young leaves during photoassimilation. The possibility of carbonic acid to be transported with the transpiration flow of woody plants and CO_2 reutilization during photosynthesis was shown by Zelawski and Riech (1970) and confirmed by Shaedle (1975). In spring, before opening of the leaves, there arises a CO_2 concentration gradient between bark and wood, leaves, and stems (Roshchina 1975), which provides for diffusion of gases from the bark to the conducting pathways and from stems to leaves.

The accumulation of carbon dioxide is concerned with the physiological state of plant and changes under unfavorable external factors. According to Novikov and Solov'ev (1973), the gas content of fungi-attacked wood of spruce fir *Picea excelsa*, features a high CO_2 concentration (to 21%) and a low O_2 concentration (to 0.4%), whereas in sapwood of healthy spruce fir, the CO_2 concentration at a depth of 1.5–2 cm is from 1.1 to 2.9% and that of O_2 from 15.5 to 19.4%. In tissues of plants frozen in ice, the concentration of carbon dioxide is considerably increased. Under these conditions, the CO_2 amount in winter wheat was very high, up to 44% (Rakitina 1970). It should be noted that quantitative results of gas analyses of the internal atmosphere of plants require critical interpretation because they depend on how the probes were taken. Special experiments showed that the composition of the extracted gases is due to the rarefaction at which the probe is taken. At great rarefactions, the gases are released from the liquid medium where the ratio of atmospheric gases, due to different solubility, is different from that in the gaseous phase. The method of vacuum extraction used by many investigators makes it possible to determine only the total amount of CO_2 and O_2 and does not permit one to separate free gas from the gas dissolved in cell sap. Langer et al. (1984) developed a method for analyzing internal gases without the use of vacuum. The collector of gases is a perforated capillary covered with a gas-permeable membrane from which gas enters the mass-spectrometer. It can be expected that this would contribute to the progress in studies of the inner gaseous medium of plant.

Little is known about the significance of internal gases for the physiological activity of the plant. However, their possible effects can be judged from the results of the experiments with plants placed in artificial gaseous medium. There are many books devoted to the effect of carbon dioxide on photosynthetic reactions (Punnet 1965; Govindjee and Rensen 1978; Larigauderie et al. 1986), on the synthesis of pigments (Verzilin et al. 1985; Khavari-Nejad 1986), processes of aerobic and anaerobic respiration (Grineva 1975; Siriphanich and Kader 1986), state of stomata, intensity of transpiration (Stalfelt 1967; Pallaghy 1971), turgor pressure, buffer equilibrium (Siriphanich and Kader 1986), movement of the cytoplasm and chloroplasts (Gärtner 1970), cell permeability and absorption of mineral salts (Kramer and Jackson 1954), metabolism of organic acids (Soldatenkov 1971), protein synthesis (Chong and Chang 1975), and many other processes. Based on the results of these works, it can be assumed that carbon dioxide present in inner tissues of plants is an efficient regulator of metabolic processes and, finally, of plant growth and development.

Of great importance is the interaction of carbon dioxide with ethylene. Sisler and Wood (1988) described in their review how the effects of CO_2 and ethylene depend on the concentration of the interacting components. Low concentrations of CO_2 enhance the effect of ethylene, for instance, when ethylene is used to stimulate the elongation of aquatic plants or germination of seeds. High concentrations of both CO_2 and ethylene counteract each other. This becomes evident when these agents are used to regulate the fall of leaves and fruits, suppression of growth, senescence of flowers, ripening of fruits, and induction of enzymatic activity. For instance, carbon dioxide stimulates elongation of roots of *Raphanus sativus* inhibited by ethylene (Radin and Loomis 1969), eliminates the suppression by ethylene of cuttings of flowering gold crests, *Dianthus* (Smith and Parker 1966), retards the fall of petiolcs of leaves of *Phaseolus vulgaris*, *Gossypium*, and other plants stimulated by ethylene (Abeles 1972) (see also Sect. 2.2.2).

To summarize, the responses of plants of CO_2 depend on the concentration of ethylene in the gaseous and aqueous media. Carbon dioxide is probably the antagonist not only of ethylene but also of nitrogen.

In 1987, the first work appeared (Stuhlfaulth et al. 1987) in which the effect of elevated concentrations of CO_2 on the secondary metabolism of plants was discussed. The author presented data indicating increased formation of cardiac glycosides in *Digitalis lanata*. A threefold increase in CO_2 concentration, as compared to the control, resulted in an enhanced amount of digoxin in 60% of the plants. Since secondary compounds prevail among excreta, this work appears to be of special interest for investigators.

As the question may have be of practical significance, the researchers were interested in the carbon dioxide content of the inner tissues of plants in relation to plant resistance to unfavorable conditions of the environment. Rakitina (1970) showed the existence of a strong correlation between the extent of damage to plant shoots frozen into ice and the level of gas exchange. Willaman and Brown (1930) also reported on a relationship between CO_2 concentration and the winter hardiness of plants. The most frost-resistant cultivars of the appletree, *Malus domestica*, were found to accumulate less CO_2 than the less resistant ones. Recently, some authors (Chang et al. 1983) have been able to show the dependence of the floodtolerance of plants on gas content. A non-tolerant batata, *Ipomoea batatas*, was characterized by a high concentration of CO_2 and low amount of O_2 in the roots. Conversely, in flood-tolerant appletrees, the gas atmosphere in the roots changed little under anoxia. Physiological disturbances were observed in flood-tolerant plants in anoxia. In particular, the decline in ATP level and in the activity of oxidative phosphorylation was great, concurrently with an increase in the rate of ethanol synthesis. Similar results were obtained by Good and Patrick (1987) in experiments on flooding of the 1-year-old flood-tolerant water oak, *Quercus nigra*, and flood nontolerant red ash, *Fraxinus pennsylvanica*. Inside the ash roots, the concentration of O_2 was greater and that of CO_2 was less than in the roots of the oak. Alcohol dehydrogenase, responsible for the conversion of alcohol into acetaldehyde, was more active in the ash. Thus the studied species differed by their ability to maintain oxidative processes for long-term periods of flooding and to prevent accumulation of toxic products in the tissues.

The increase in CO_2 concentration in plants is accompanied by approximately the same drop in O_2 concentration, due to which tissues may suffer anoxia. This would result in a decrease in the respiration level. Respiration was found to diminish directly with the fall in O_2 concentration from 20 to 2% (Grineva 1975). Further decrease of oxygen in the medium (less than 2%) leads to a disturbance of normal respiration and activation of the alcohol fermentation process. Under conditions of limited oxygen access, the rate of oxidative phosphorylation is retarded and ATP production is blocked. Thus biosynthetic processes would lack their energy source. These data indicate that the gas composition of the plant is very significant for the functioning of cells per se and of their particular structures.

2.2.2 Ethylene

Ethylene is one of the simplest organic substances, whose properties have been better studied than for other gases in plants. It is constantly generated in plant tissues during normal metabolism. It is found in excretions of fruits with pulpy pericarp (Rakitin 1967), of withering inflorescences (Nichols 1966), of germinating seeds (Abeles 1972), of isolated roots, (Radin and Loomis 1969), leaves (Debata and Murthy 1983), and other organs and parts of plants. Beyer and Morgan (1970) demonstrated, by means of vacuum extraction, the occurrence of ethylene in the inner air of plants. Considerable quantities (1 ml/l) of ethylene are accumulated in the air space of the water-logged parts of rice, *Oryza* (Raskin and Kende 1984). The gas medium also included 3% O_2 , 6% CO_2, and 91% N_2. The amount of ethylene in the atmosphere of intercellular spaces of lemons was calculated by Rakitin and Rakitin (1986). The gas atmosphere also contained 11% CO_2 and 20% O_2. Gases from intercellular spaces were obtained at 650 Hg mm, intratissular gases at 5 Hg mm (the authors evidently give residual pressure). The ethylene concentration in mature lemon fruits was higher (0.061) $\mu l\, l^{-1}$) than in immature ones (0.021 μl/l). In the intercellular spaces of the latter the amount of ethylene (0.05 μl) was two fold greater than in the inner tissues. Conversely, intratissular concentration of CO_2 in both mature and immature fruits was much higher than in the intercellular space.

The ethylene concentration does not remain unchanged, it is dependent on the environmental and internal conditions. The highest ethylene concentration was found during leaf abscission. At a small distance from the zone of leaf abscission, the amount of this gas increases 6–14-fold (Jackson and Osborn 1970).

The site of ethylene synthesis in higher plants has not yet been ascertained. This is because ethylene-releasing plant tissues once homogenized, cease to produce the gas. However, gentle treatment of cells made it possible to obtain subcellular fractions capable of ethylene synthesis. In the 1980s, some papers appeared (Mattoo et al. 1977) where the cellular wall-cellular membrane complex was suggested to be the site of ethylene synthesis. According to other authors (Guy and Kende 1984), about 80% of total ethylene is produced in the vacuole, as indicated by the experiments with vacuoles isolated from protoplasts of *Pisum sativum* and *Vicia faba*. Since one of the immediate ethylene precursors, 1-aminocyclopropane-1-carboxylic acid (ACC), is formed in the cytoplasm, the authors (Guy and Kende 1984) believe that ACC is translocated from the cytoplasm to the vacuole, where it is transformed into ethylene. Probably the vacuole is not the only site in the cell where ethylene is synthesized. It is likely that ethylene is also formed in chloroplasts, since isolated chloroplasts produce ethylene from methionine in the presence of pyridoxal phosphate, ferredoxin, and the electron donor (Koenze and Elstner 1976).

At present, the main pathway of ethylene synthesis in plants has been established. Many data on the enzymes and regulation of gas biosynthesis have been accumulated. The current state of the problem is summarized in Fig. 8. According to the ideas of some authors (Adams and Yang 1981; Yang and Hoffman 1984), ethylene is formed from methionine (MET) through S-adenosyl methionine (SAM) and 1-aminocyclopropane-1-carboxylic acid (ACC). As a result of methionine cycle functioning, the resynthesis of sulfur-containing amino acid takes place: SAM → 5-methylthioadenosine → 5-methylthioribose → 5 methylthioribose-l-phosphate → 2 keto-4-methylbutyril → methionine. The key reactions in ethylene synthesis are the conversion of SAM into ACC catalyzed by ACC synthase, and the production of ethylene from ACC

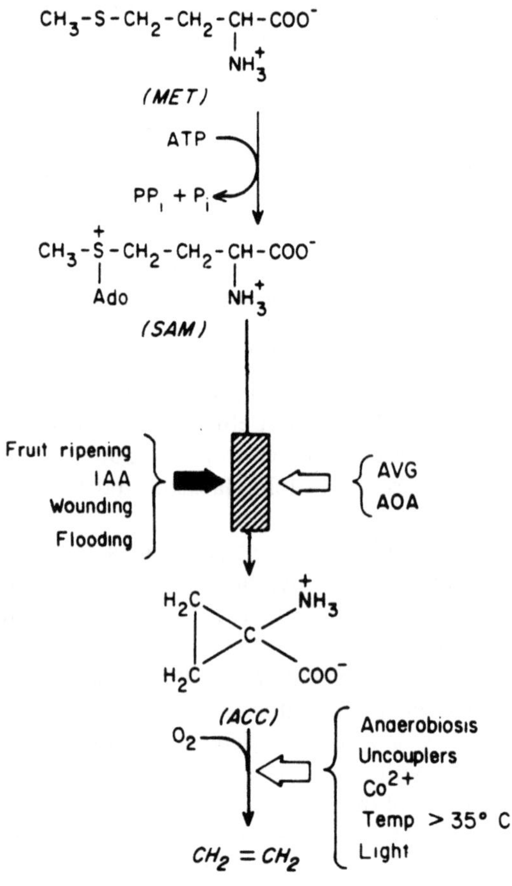

Fig. 8. Biosynthesis of ethylene and its regulation. *ACC, Ade, Ado, AOA, AVG, Met,* and *SAM* stand for 1-aminocyclopropane-1-carboxylic acid, adenine, adenosine, aminooxyacetic acid, aminoethoxyvinylglycine, methionine, and S-adenosyl methionine, respectively. (hatched box) stage of regulation; (black arrow) induction of the enzyme synthesis; (open arrow) inhibition of the reaction. (After Adams and Yang 1981; Yang and Hoffman 1984)

with the participation of the ethylene-forming enzyme (EFE). Kende in his overview (1989) presents data on the isolation and purification of the enzymes involved in the process of ethylene biosynthesis and gives their characteristics. The basic enzyme which limits the ACC → C_2H_4 stage was isolated and purified from the pericarp of tomato, *Lycopersicum esculeutum*, pumpkin, *Cucurbita*, etiolated seedlings of mung bean, *Phaseolus aureus*, and characterized. The ethylene-forming enzyme (EFE) is firmly bound to the membrane and can be destroyed by homogenization. The properties of this enzyme were studied in vivo on the preparations of isolated vacuoles capable of converting amino cyclopropan-l-carboxylic acid to ethylene.

The biosynthesis of ethylene is regulated by various growth and environmental factors. For example, during the process of ripening, ACC synthase production and enhancement of ethylene synthesis are induced by increased EFE activity (Adams and Yang 1981; Yang and Hoffman 1984). On the whole, mechanical injury, chilling, drought, flooding as well as pathogens and auxins, stimulate ethylene production. Since CO_2 concentration in plants is constantly varying, of particular interest are the works where the effect of this factor on the ethylene-forming enzyme (EFE) is studied. Philosoph-Hadas et al. (1986) revealed the dependence of EFE activity on the CO_2 content in disks of *Nicotiana tabacum* leaves. The enzyme activity shows the increase at CO_2 concentration from 0 to 2%, and does not change at CO_2 concentration from 2 to 3%, increasing again with the rise of CO_2 up to 5% in the dark and 10% in the light. Thus CO_2 stimulates the conversion of ACC to ethylene, which is a further indication of the regulatory role of the gas. Among recent publications, of interest are those dealing with the effect of cytokinins on ethylene biosynthesis in the leaves of *Gossypium* (Suttle 1986), where stimulation of ethylene production by natural and synthetic cytokinins was attributed to an increased biosynthesis of ACC and its conversion to ethylene. Using the disks of spinach leaves as test objects, Philosoph-Hadas et al. (1989) studied the effect of gibberellic acid, abscisic acid, and benzyladenine on ethylene biosynthesis. Each of the acids was found to enhance ethylene production. Phytohormones affected the production of endogenous ethylene as well as that of l-amino cyclopropane-1-carboxylic acid (ACC). The biosynthesis as blocked by the substances inhibiting the production of the ethylene precursor, ACC, or ethylene per se (see Fig. 8). In the latter case, EFE was also blocked (Yang and Hoffman 1984).

The amount of ethylene produced by the plant tissue is not great and varies over a wide range. The rate of ethylene excretion by leaves of *Ranunculus* in the light is 0.4 nmol g^{-1} fresh mass h^{-1} (Horton 1985). Fruits also release ethylene in an amount from 2.2×10^{-3} (*Citrus sinensis*) to 16 nmol g^{-1} fresh mass h^{-1} (*Passiflora*) (Jensen 1965).

Ethylene is a very reactive compound, whose chemical properties are mainly connected with the intercarbon double bond. According to the modern conception, ethylene is a hormone that controls various processes of plant growth and development, in particular, germination of seeds, lengthening and thickening of the stem (Abeles 1986), growth by elongation (Satler and Kende 1985), sexual differentiation of flowers (Byers et al. 1972), ripening of fruits (Adams and Yang 1981), and tissue aging (Abeles 1972). Moreover, it stimulates aerenchyma formation in flooded roots (Jackson 1989; Justin and Armstrong 1991). These phenomena reflect the effect of ethylene on some physiological and biochemical reactions: inhibition of photosynthesis (Pallas and Kays 1982; Sguler et al. 1985), reduction of membranous chloroplast systems (Shimokawa et al. 1978), decrease in the rate of respiration (Chin and Frenkel 1977), change in protein synthesis (Tucker and Latis 1984), membrane permeability (Farquhar and Mayak 1984), auxin transport (Lyon 1970), translocation of sugars (Jackson and Osborn 1970), the activity of adenosinetriphosphatase (Olson and Spencer 1968), cytochrome C-oxidase (Apelbaum et al. 1984), peroxidase (Imaseki 1970), and other enzymes and processes.

The mechanism of action of ethylene at the molecular level is far from being understood. However, the site-targets for it have been already established. Thus, during fruit ripening or fall, ethylene in the cell-targets binds to specific protein receptors and stimulates de novo synthesis and excretion of enzymes that break cellular walls as cellulase (EC 3.2.1.4). The ethylene-binding component of membranes has recently been isolated from plant tissue. It is a metal-containing enzyme located mainly in the endoplasmic reticulum and, to a lesser extent, in the plasmalemma (Sisler and Yang 1984). It remains unclear how the signal from ethylene bound to the receptor protein of membranes is transferred to the nucleus, where specific genes are triggered.

By its action, ethylene is an antagonist of CO_2 (see also Sect. 2.2.1). Analysis of the molecular structure of CO_2 and ethylene led to the conclusion (Abeles 1972) that, due to structural similarity, CO_2 can be a competitive inhibitor of ethylene. Many physiological effects may be explained by these peculiarities of ethylene and carbon dioxide. The universally accepted practices of apple storage at elevated CO_2 concentration is based on the ability of the gas to inhibit ethylene action. Thus the effect of ethylene on plant cells and an opposite effect of carbon dioxide probably belong to normal physiological regulation in the plant.

2.2.3 Other Volatile Compounds

In addition to ethylene, the inner air of the plant may have other organic incompletely oxidized compounds such as alcohols, aldehydes, and hydrocarbons. Besides the high concentrations of CO_2, the inner air of some woody plants is shown (Roshchina 1975) to contain also ethanol vapors, as revealed by gas chromatography (Fig. 9, Table 2). Calculations show that the content of CO_2 in plant shoots ranges from 11.3 to 19.9 of relative units. The presence of ethanol vapors in the inner air of the plant is an indication of insufficient supply of oxygen to the tissues and activation of the anaerobic pathway of respiration (Chap. 4.3.4).

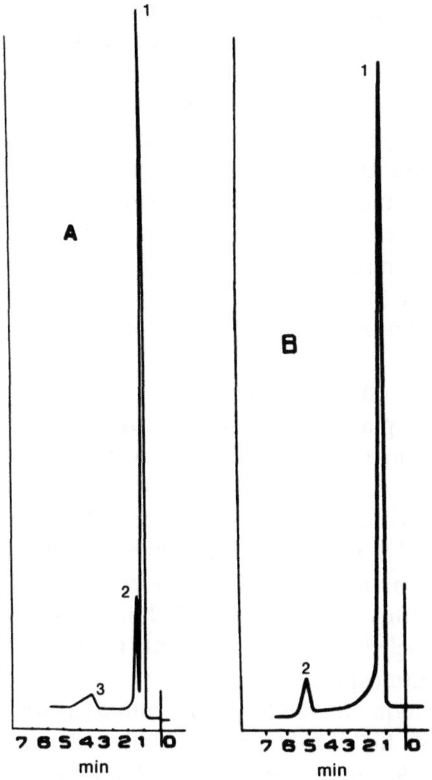

Fig. 9. Chromatogram of internal air from shoots of *Populus balsamifera* on polysorb (**A**). (Roshchina 1975) and *Tilia cordata* on tricresyl phosphate (**B**). (Roshchina 1974a, c) **A** *1* O_2 and N_2; *2* CO_2; *3* ethanol; **B** *1* air; *2* butilic aldehyde

Table 2. Amount of carbon dioxide and ethanol (relative %[a]) in internal air collected from 5–7-year-old shoots of woody plants. (Roshchina 1975)

Plant	Carbon dioxide		Ethanol	
Cotinus coggygria	17.3	0.05	2.18	0.49
Populus balsamifera	15.25	2.05	1.06	0.05
Fraxinus pubescens	12.5	0.75	1.91	0.47
Quercus robur	11.3	0.60	1.57	0,70
Betula verrucosa	12.95	0.79	1.22	0.36
Larix decidua	19.9	2.11	2.08	0.65

[a] Determined by the division of the measured area of each gas chromatographic peak by the sum of areas of all picks multiplied by 100.

Aldehydes are intermediators in the plant metabolism and can also be observed in the inner air. For instance, butyric aldehyde (butanal) was found in the inner air of 5–7-year-old plant shoots of 20-year-old poplar, *Populus balsamifera*, and *Cotinus coggygria* (Roshchina 1974 a,c) in amounts from a tenth to a thousandth of relative units. To our knowledge, there are no works reporting the presence of butanal in the inner air of plants; however, this aldehyde was found in volatile excretions of cowberry, *Vaccinium vitis idae*, batata, *Ipomoea batatas*, and red beet, *Beta vulgaris* f. *rubra* (Dadykin et al. 1967; Skvortsov and Smirnova 1972). Moreover, butanal is involved in gas excretions of grape *Vitis*, and black currant, *Ribes nigrum*, as well as benzaldehyde (Schauenstein et al. 1977). Some other aldehydes with a great number of carbon atoms have been found in the inner air of plants. Hamilton-Kemp and Andersen (1984) have vacuum-extracted volatile substances from leaves and stem-calms of wheat, *Triticum*, and have identified in their composition the saturated aldehyde nonal and unsaturated C_9-aldehydes as the main components of internal media of plants. Relatively high concentrations (up to 13–18% $\mu g\ kg^{-1}$) of such aldehydes as acetal, propanal, and hexanal were observed in fruits of the genera *Malus*, *Pyrus*, and *Fragaria* (Jensen 1965).

All the above compounds are normal metabolites. However, the phenomena of excretion of CO and methane by some plants are somewhat surprising. In 1917, carbon monoxide in an amount from 1.2 to 12% was found in the gas composition of air cavities or lacunas of brown algae, *Nereocystis leutkeana* (Langdon 1917). Many years later, it was shown that CO is excreted by the leaves of some higher plants (Delwiche 1970). It was suggested that in the light CO is formed as a result of degradation of polyphenols. Under the same conditions, CO can also be assimilated. In this case, the label ^{14}CO incorporates into serine or glycine. It was suggested (Delwiche 1970) that plants might have an important role in purification of the atmosphere from CO.

The air of intercellular spaces may contain other volatile compounds as well. Abell and Hursh (1931) were able to extract highly inflammable gas, presumably methane, from stems of two oak species. Recently, the release of CH_4 by lake plants has been reported (Dacey and Klug 1979). Gas lacunas of yellow pond-lily, *Nuphar luteum*, occupied from 20 to 40% of the root and rhizoid volume and about 15% of petiole volume, and their gas composition involved 37% CH_4, 54% N_2, 1% O_2 and 6% CO_2. In spring when the first upper surface leaves appear, methane in air bubbles is released into the atmosphere. As the gas composition of the inner air space is approximately similar to that of near-bottom sediments, some authors (Dacey and Klug 1979) believe that it is the gases dissolved in demersal deposits that may easily diffuse into the roots and rhizoids in order to reach the petioles and leaves of yellow pond-lily. Unlike CO, methane probably has no toxic effect on plants. In experiments with hydroponic cultures grown in the atmosphere with a high concentration of methane and sufficient oxygen supply, it was found (Arthur et al. 1985) that the increase in methane concentration up to 20% had no harmful effect on *Solanum lycopersicum*.

2.2.4 Transport of Internal Gases

There have been numerous attempts to establish the paths of gas movement in plants. The investigators were most interested in the oxygen supply of the inner tissues of plants. Brown and Escombe (1900) suggested that the major route of inner gas migration should be a diffusion regulated by stomata. According to Leuning (1983), diffusion of small quantities of gases from the leaf can be controlled by water evaporation inside the leaves. The gas flow inside the air spaces, as well as gas transport into and from the leaf, occurs by diffusion determined by the gradient of gas concentration. Individual gases diffuse independently into a stationary gas medium. However, diffusion is not the only path of gas movement. Plants in humid zones and marshes have a system of gas lacunas and porous air-bearing tissue called aerenchyma, which makes the movement of gases impossible as a solid stream. For example, air containing ethane as a marker enters from the outside into the lacuna of young leaves of *Nuphar luteum* against a low (<300 Pa) gas pressure gradient and moves downwards along the petioles of leaves to roots at a rate of 50 cm min^{-1} (Dacey 1980). Probably it is not only the difference in the gas concentration, but also the total balance of gas pressure involving thermal and hydrometric components, that is important here.

Owing to a well-developed surface and the system of intercellular spaces, pores, and stomata, the inner tissues of plants growing in unfavorable conditions are readily accessible to the outer atmosphere and have an aerobic type of respiration. Using a radioactive oxygen, $^{15}O_2$, in the experiments with *Vicia faba*, Evans and Ebert (1960) were able to show that O_2 is rapidly transported from the aboveground parts into the roots. The possibility of aeration of the cells of the roots by oxygen transported from the leaves was also shown by other researchers (Soldatenkov and Chirkova 1963). It is generally accepted that oxygen diffuses by the gradient of partial pressure from the surface of the plant to different inner tissues. Transport occurs via central air intercellular space, intercellulars of parenchyma and phloem of the stem (Soldatenkov and Chirkova 1963). However, the question of oxygen transport into roots cannot be unambiguously solved for all plant species. Of much importance here are the conditions of growing. Poor aeration is observed in the case of flooding or water freezing, etc. Since the roots often undergo the influence of such factors, the question of their supply with oxygen from the aerial part is especially acute. One should distinguish between the natural adaptation to root anaerobiosis and resistance to temporary water-logging.

The normal existence of marsh and water-flooded plants in the habitat is provided by the protective role of the leaf in combination with anatomo-morphological peculiarities of the structure of the aerial parts. The marsh plants possess a system of air-bearing cavities and lacunas representing a continuous route along which the air passes from leaves to roots. The gradient of oxygen concentration in tissues of aquatic plants was calculated in experiments with *Cladium mariscus* (Vallance and Coult 1951). The oxygen concentration was distributed as follows: illuminated leaves > dark-developing leaves > upper stem > lower stem > roots. Aquatic vascular plants have various types of ventilation system (Fig. 10 A, B), where gas flow is controlled by the gas pressure difference between the leaves of various age and atmospheric pressure. The difference in the pressure between leaves of various ages is determined by the diameter of their pores. According to the kinetic gas theory (Kennard 1938), the diameter of the smallest pores must be no less than or equal to 0.1 μm, so that one could observe the entrance of air into the pores in young leaves and its translocation to old ones. According to Dacey (1981), the size of pores in the leaves of *Nuphar luteum* does not correspond to the stomata size (> 1 μm). The gas pressure in the intercellular space of lotus, *Nelumbo nucifera*, is higher than in the surrounding air (Fig. 10C, D). The adoxal side of the leaf has two regions with different characteristics (Dacey 1987). There are regions with a relatively great diameter of the pores in the center of the leaf just above the site of the leaf attachment to the petiole. Gas exchange between the rest of the leaf (99% of the surface) and the atmosphere is limited by pores with a diameter of less than 0.03 μm. Air enters the leaf through the peripheral part and is released into the atmosphere through a highly porous region in the center of the leaf lamina (Dacey 1987). In the case of such ventilation of the leaf, the oxygen shifts downwards to the roots where anaerobic conditions are created, while CO_2 arising from respiration of roots moves upwards to photosynthesizing leaves (Dacey 1981; Dacey and Klug 1982a, b). Of much interest are experiments performed by Grosse and Mevi-Schütz (1987), who investigated migration of gas in aquatic vascular plant *Nymphoides peltata*.

In the model system, ethane was sucked into the lamina plate of the youngest leaf and then transported along the petiole. Under sunlight, gas diffusion was accelerated by 1200% due to the temperature difference between the leaf and

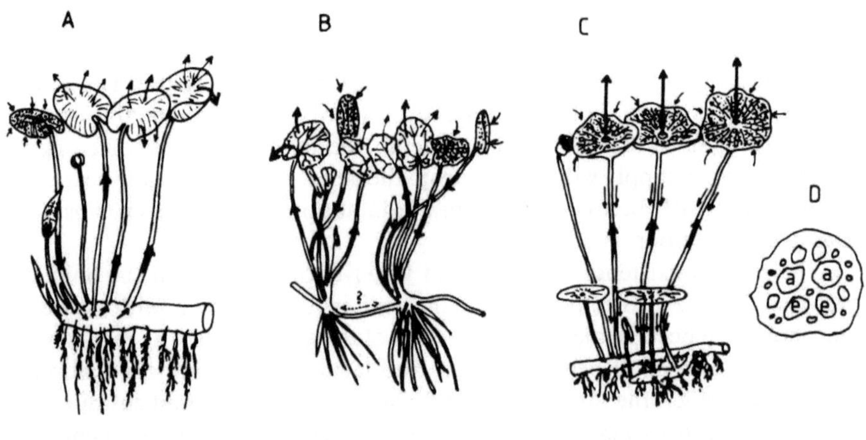

Fig. 10A–D. Transport of internal gases in aquatic vascular plants. The gas movement is dependent on the different size of the pores through which the gas enters the leaves and is released. The ventilation system is connected with the formation of gas pressure ΔP. $\Delta P = P_1 - P_2 > 0$; $P_1 > P_2 = P_0$, where P_1 is the gas pressure in young leaves, P_2 in older leaves, P_0 atmospheric pressure. **A** *Nuphar lutea*. **B** *Nymphoides peltata*. **C** *Nelumbo nucifera*. **D** Section of leaf aerenchyma of *Nelumbo nucifera* with small (entry) and large (release) gas channels. (Grosse and Schröder 1986)

surrounding air. In these experiments ΔT was 1.7 K. The temperature difference resulted in a pressure rise inside the aerenchyma of the young leaves. Acceleration of gas transport could be demonstrated by heating the lamina of the youngest leaf. Thus, the light energy was needed here for the temperature rise in the leaf and the establishment of the temperature gradient between the leaf and surrounding air. The process was independent of photosynthesis. This experiment showed the possibility of improving the oxygen supply of the roots in aquatic plants by means of the temperature gradient.

In water-logged herbaceous plants which do not constantly grow in water, for example in *Oryza sativa*, the inner oxygen route was enough for their normal life, as opposed to the seedlings of *Cucurbita maxima* (Vartapetyan et al. 1974). When rice was water-flooded for a long time, the composition of gases in aerenchyma of the internodes was investigated at various time periods (Stünzi and Kende 1989). The gradient of O_2 concentration from the apical part of the plant (near the water surface) to the basal part of the shoot was observed. CO_2 concentration changed opposite to that of O_2. With plant flooding, the level of ethylene in the air lacunas increased and reached 1 μl within 72 h. It was also shown (Setter et al. 1987) that O_2 concentration inside the air lacunas of rice during the day depended on the depth to which the shoots were submerged. Near the surface, O_2 concentration was 16–20% and at a depth of 0.8 and 1.8 m it was 10 and 5%, respectively. On the contrary, CO_2 concentration in rice lacunas increased with depth from 1–3% near the water surface to 5–10% at a depth of 1.8 m. Vartapetyan et al. (1974) came to the conclusion that the possibility of roots existing under anaerobic conditions could be explained not only by the improved transport of oxygen into them, but mainly by the peculiarities of their metabolism at anaerobiosis. The capacity of plants to coordinate the defensive reactions with the help of a regulatory system plays a key role in their resistance to anaerobiosis.

Among above ground parts of the plant, leaves are the most important in oxygen supply of the root system. Chirkova (1988) showed that in kidney bean, *Phaseolus vulgaris*, broad bean, *Vicia faba*, tomato, *Lycopersicon esculentum*, and potato, *Solanum tuberosum*, the unfallen leaves supported the activity of the root system for 6–7 days, whereas in the marsh plants such as bead-ruby *Maianthemum*, and bulrush, *Scirpus* L., the roots did not even die within 14 days when the leaves were functioning. Leaves without oxygen lose their ability to preserve the roots from dying. In plants resistant to a deficit of O_2, the products of anaerobic metabolism are not accumulated. Upon root anaerobiosis, these compounds rise, together with the transpirational flow, into the aerial part of the plant, and are removed through the lenticels and leaves. The most effective way of detoxication in their excretion via the root system into the surrounding medium. Such a mode of evacuation of the fermentation products depends on membrane permeability. Decrease of ethanol concentration in the tissues is also connected with its utilization in respiration as a substrate. An integral characteristic of tolerance to the external factors (including anaerobiosis) is the resistance of the plant membrane.

By using $^{15}O_2$ it was revealed that in rice, *Oryza sativa*, with well-developed aerenchyma, radioactive oxygen is easily diffused into the root, while in barley with a less developed system of intracellular spaces, this process proceeds with more difficulty (Barber et al. 1962). Application of heavy oxygen in experiments with corn made it possible to demonstrate its route within the plant into the roots and its outflow into the near-the-root medium (Jensen et al. 1964). It should be noted that in maize roots oxygen deficiency leads to aerenchyma formation and ethylene synthesis. (Atwell et al. 1988). Rice and some marsh plants are the most capable of medium aeration around their roots. The ventilation system of lobelia, for example, includes large lacunas for gas storage or reservoirs, from which the air is transported along the plant (Sand-Jensen and Prahl 1982). Gas transport in woody plants is not so well investigated as in herbaceous ones. However, oxygen movement from the aerated parts into the roots of woody plants such as: *Betula pubescens*, *Salix atrocenerea*, *Salix alba*, *Populus petrowskiana*, and *Fraxinus pensylvanica* was demonstrated (Hook et al. 1972). Due to gas transport, some woody species like *Alnus glutinosa* can grow on wet or water-flooded soils (Grosse and Schröder 1984; Grosse et al. 1990). Vertical gas flow occurs via intercellular spaces of wood and core parenchyma. The rate of gas diffusion was estimated by the ethane movement from the stems into the roots in the gas chamber where the plant was placed (Grosse and Schröder 1984). This value reached 1.41 $\mu l\,min^{-1}$.

Well-developed spongy tissue of roots as in alder, *Alnus* L., and the formation of aerenchyma in aquatic vascular macrophytes promote adaptation to the oxygen supply of the roots in the plants growing on wet soils or submerged into the poorly aerated sediments of lakes (Grosse and Schröder 1986). This accelerates the air route within the plant and makes O_2-transport to the more distant, water-logged parts of the plant more easy. Gas movement occurs due to the difference in pressures. In the intercellular system of dissimilating tissues, the higher water solubility of CO_2 leads to pressure decrease. The pressure gradient induces the air route within the plant. In plants growing on marsh and wet soils, additional supply of the roots with oxygen takes place under the pressure occurring due to the purely physical process of gas thermoosmosis. This thermoosmosis is the gas flow caused by a temperature difference across the bulk phase of a porous partition in particular, leaf tissue with intercellular pores as in the aquatic plant *Nuphar lutea* etc., as is shown in Fig. 10, or a porous cambial region of the tree lenticels with a pore size smaller than the mean free path length of the gas molecules (Grosse and Schröder 1986; Grosse and Mevi-Schütz 1987; Grosse et al. 1990). Thermoosmotic gas transport depends on the temperature difference between the porous aerated parts of the plant and the surrounding atmosphere. Oxygen supply of roots via this mechanism is described for some woody species of genera *Alnus* and *Betula* grown on wetlands (Grosse et al. 1990).

The elevated oxygen concentration in the tree roots should promote nitrogen assimilation and modulation by the aerobic nitrogen-fixing actinomycete *Frankia* (Mekurina et al. 1990).

Not only the leaves of the trees, but also the stem stomata and lenticels play a certain role in the oxygen supply of the roots (Chirkova and Gutman 1972; Chirkova 1988). For the lenticels, gas exchange through them occurs due to their anatomy, characterized by porous tissue which is formed by the small masses of loosely compacted (filling) cells (Esau 1965), often rising above the surface of the bark on the stems (sometimes, to a lesser extent, on the roots) of woody plants, and has abundant intercellular spaces. The deep layers of woody trunks are aerated via air pores of medullar rays propagating inside from the lenticels. Plugging of lenticels leads to root death (Chirkova and Gutman 1972). The possibility of radial transport of gases in the tree trunk, even through the layer of cambial cells, was revealed in the 1930s (MacDougal and Working 1933). The tissues of conducting and nonconducting phloem are aerated due to numerous intercellular spaces distributed radially through the medullar rays. Air spaces of the pith rays are connected with the environment through numerous lenticels in the bark and periderm. Thus, the aired system of phloema is structurally well developed so that it provides free gas exchange with the surrounding aerobic medium (Hook and Brown 1972).

Aeration of xylem depends, to a large extent, on cambium permeability, which varies significantly among different woody plants (Hook and Brown 1972). For example, *Fraxinus excelsior* has well-pronounced intercellular spaces passing through the cambium, whereas in other plants, such as the genera *Populus* and *Platanus*, the intercellulars of the initial rays of cambium are either lacking or so small that it is impossible to visualize them.

The cambium of those species which are able to be hydrophytes is more permeable for air than that of mesophytes (Hook and Brown 1972). Transport of carbon dioxide from the roots to the aerial parts of the plants was studied with the help of radioactive carbon (Kursanov et al. 1952; Vartapetyan et al. 1974). It was shown that transpirational flow plays a certain role in the movement of CO_2. The common scheme of aeration for woody plants should be considered as follows:

↗ Non-hydrophytes where the cambium prevents free gas exchange within tissues

O_2 → stomata and lenticels → tissues of bark and phloem are aerated via a number of intercellular spaces with radial and longitudal orientation

↘ Hydrophytes where intercellular spaces permit gas exchange from cambium to xylem

The transport of the phytohormone ethylene is also a question of great interest. Many investigators considered that ethylene transport may occur between different parts of the plant. Jackson and Campbell (1975) showed that in tomato, *Lycopersicon esculentum*, radioactively labeled gas penetrated from the roots into the stem and lenticels, so that the inner concentration of ethylene in the roots and aerated parts of the plant increased when the root system was water-logged. With the help of ^{14}C-ethylene, it was demonstrated that the gas moves rather quickly and remains unmetabolized. There is reason to believe that the transpiration flow does not participate in this translocation because the rate of ethylene transport decreased when the tree stems were ring-barked near the ground (Jackson and Campbell 1975).

These data are in good agreement with the conception that gases can diffuse from one part of the plant into another via the systems of intracellular spaces as was shown for oxygen, carbon dioxide, and nitrogen. However, ethylene is comparatively soluble in water (better than O_2 and N_2 and worse than CO_2) and its molecule is non-polar enough to penetrate through membranes. Thus, the possibility of ethylene transport in water solutions is quite probable. At the same time, in the experiments with *Vicia faba*, ethylene was not delivered to various parts of the plants in amounts sufficient for the gas to be physiologically active (Zeroni et al. 1977). This can be explained by the fact that the site of ethylene action in the plant is near or very close to the place where it is synthesized.

2.3 Intratissular Secretory Structures

Intratissular channels and reservoirs filled with various nonvolatile secretions are considered to be the inner secretory structures subdivided according to their origin into schizogenous and lysigenous. Schizogenous receptacles represent intercellular spaces filled with secreting substance and surrounded by the living epithelial cells. Resin ducts belong to them. Lysigenous structures developed in the place where the cells disappeared after accumulation of substances such as essential oils. The methods of development of the secretory channles differ in various plants and even in various parts of one and the same plant (Outer and van Veenendaal 1986). In some tissues, secretory channels are first formed schizogenously and then at later stages lysigenously, while in the other tissues only lysigenously or only schizogenously. It is suggested (Outer and van Veenendaal 1986) that the schizogenous mode of the secretory channel development is more ancient than the lysigenous one.

Laticifers representing living cells with latex in vacuoles are also believed to be intratissular secretory structures. It should be noted that laticifers and resin ducts do not exist simultaneously in one and the same plant species, except for those belonging to the family Anacardiaceae (Outer and van Veenendaal 1986) and therefore could be taxonomic features. Some secretory tissues could be formed in response to the outer stimuli: mechanical damage, introduction of pathogens and exogenous growth substances, probably ethylene, as well. For example, resin channels (ducts) can arise in such a way (Fahn 1987).

2.3.1 Secretion of Resins

Resins are formed either in the process of normal metabolism or as a result of wounding. They usually take part in further metabolism and are situated in the resin ducts and laticifers of various parts of the plant: roots, stems, mainly in wood, leaves, and seeds.

Resin ducts are wide-spread in the families Umbelliferae, Araliaceae, Compositae, Anacardiaceae, and Coniferae (Dell and McComb 1978 a; Outer and van Veenendaal 1986; Maksymovych and Ledbetter 1987). They appear as normal elements of wood in the genera *Pinus, Picea, Larix*, and *Pseudtsuga*, or develop as a result of damage in the genera *Abies, Cedrus, Tsuga*, and *Pseudolarix* (Yatsenko-Khmelevskii 1954; Chavchavadze 1973, 1979; Fahn 1988b). In conifers (Coniferae), resin ducts are located in the wood and bark of the trunk, branches, needles, and cone scales. Woody vertical and horizontal ducts are connected with each other, forming interlaced net. Resin ducts are also present in the root systems of *Culcasia, Cercestis, Furtadoa, Homalomena, Phillodendron*, and many other plants. However, there are cases when resin ducts are lacking in the roots, as in the genus *Schismatoglottis*, but are observed in the leaves of the same plant.

Classification of resin ducts and reservoirs (Chavchavadze 1973) is based on the principle of their creation: traumatic, pathological resin cavities and resin ducts formed in response to wounding, and normal resin ducts known as usual wood elements (Figs. 11, 12). Such traumatic structures appear schizogenously and are subdivided into schizogenous cavities – small intercellular spaces on the joint between parenchyma cells; small resin pockets – small intercellular spaces within epithelial cells; resin pockets – large intercellular pocket-like cavities; a resin cysts – protruded intercellular cavities with poorly marked epithelium which, as a rule, do not form a continuous sheet.

According to Chavchavadze (1973, 1979), ordinary normal resin ducts usually differ from the pathological ones by their lesser number per square unit and more perfect structure. Epithelial continuous layer always covers a surface inside the duct. However, the dead layer consisting of one or several cell rows and closely connected with the outer walls of the epithelium is sometimes discreted. The accompanying parenchyma usually represents one or two incomplete or complete cell rows, sometimes three or four.

Resin ducts arise as schizogenous intercellular spaces due to the disjointing of parenchyma cells. [It should be noted that according to Denisova (1989), the anatomical term "schizogenous duct" is more correct than "resin duct" because similar structures are often filled with other products]. After several divisions, these cells from the inner coat form the epithelium of developing resin ducts and are able to excrete resin. At present, the enzymes participating in this process are

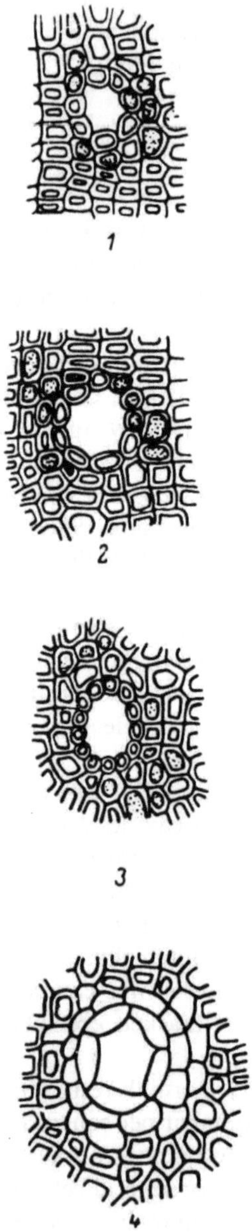

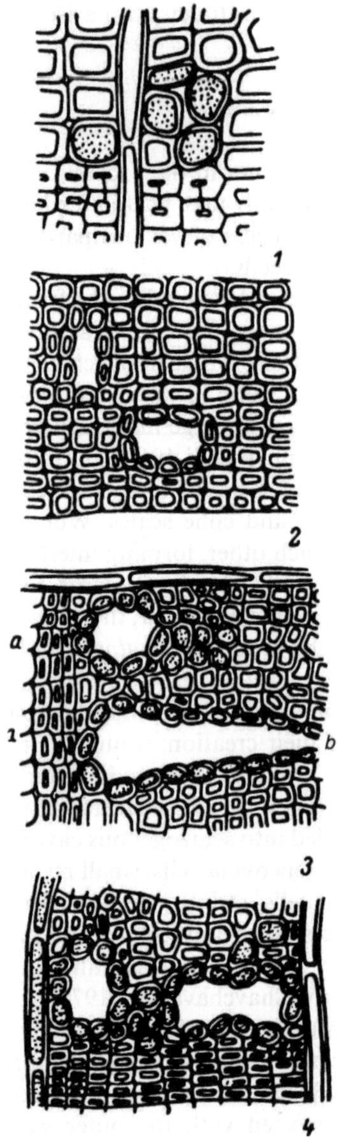

Fig. 11, 1–4. Normal vertical resin ducts. **1** *Keteleeria fortunei*, **2** *Pseudotsuga macrocarpa*, **3** *Larix laricina*, **4** *Pinus pumila*. (After Chavchavadze 1973, 1979)

Fig. 12, 1–4. Traumatic resin reservoirs and ducts. **1** Schizogenous cavity in *Abies firma*, **2** Resin duct pocket in *Metasequoia glyptostroboides*, **3** Vertical (*a*) and horizontal (*b*) resin cysts in *Cedrus libani*, **4** Vertical resin duct in *Abies venusta*. (After Chavchavadze 1973, 1979) being intensively investigated. Such enzymes, destroying cell walls, were extracted from the cell wall of mesocarp tissue of immature fruits of almond, *Prunus dulcis*. Their activities correlated with the development of resin ducts in these tissues (Morrison et al. 1987a). In particular, activities of polygalacturonase and 1,3-β-D-glucanase increased at the schizogenous stage of duct formation or even earlier, whereas α-galactosidase, β-galactosidase, α-arabinosidase, and α-mannosidase became more active at the later stage. Analysis of cell wall methylation showed that many specific glycosidic bonds, destroyed by the above-mentioned enzymes, were found in the cell walls of the mesocarp before resin duct formation (Morrison et al. 1987).

Secretions filling resin ducts are weak-volatile plant products insoluble in water but soluble in organic solvents. Diterpene derivatives, mainly cyclic acids, known as resinous acids, are those most often found in resins (Fig. 13). These acids, with empiric formula $C_{20}H_{30}O_2$, represent about 4/5 of the resinous excreta of the conifers. Additionally, resinous alcohols, phenols, tannins, and hydrocarbons are included in resin composition. Resin is often diluted in essential oil and these solutions are called balsams. The liquid balsam filling resin ducts of wood and secondary bark of the tree trunk in the genus *Pinus* is often known as resin or oleoresin. Oleoresins of pine, *Pinus*, spruce, *Picea*, cedar, *Cedrus*, larch, *Larix*, fir, *Abies* include 40–65% diterpene or resinous acids (levopimaric, pimaric, palustric, abietic, dehydroabietic, etc). and 20–35% monoterpene hydrocarbon with the general formula $C_{10}H_{16}$. Volatile components of oleoresins are α-and β-pinenes, carene, camphene, β-phellandrene, and limonene. About 5–20% oleoresin is usually represented by sesqui- and diterpene hydrocarbons and their derivatives.

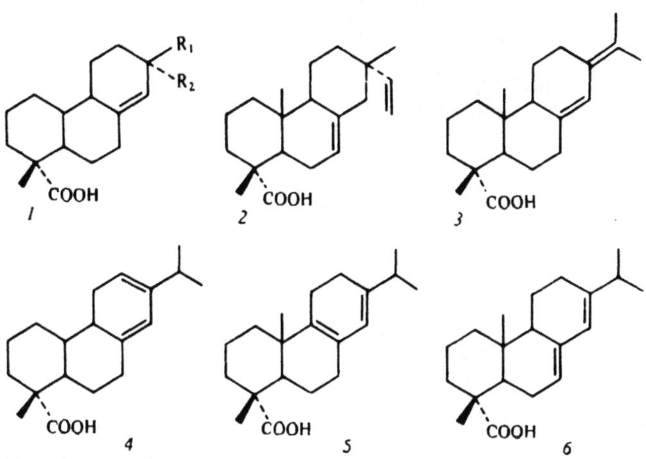

Fig. 13, 1–6. Resinous acids. **1** Pimaric acid: R_1 vinyl; R_2 methyl; dehydropimaric acid: R_1 ethyl; R_2 methyl; sandoracopimaric acid: R_1 methyl; R_2 vinyl, **2** Isopimaric acid, **3** Neoabietic acid, **4** Levopimaric acid. **5** Palustric, **6** Abietic acid

Flavonoids which can be converted into lignans, the compounds consisting of phenylpropane units connected by C–C bond between the middle carbon atoms, are also present in resin composition. According to Dell and McComb (1978a), fatty acids, some of which resemble waxes, are also often found in resins.

Compounds specific to certain families are also present in plant resins (Figs. 14–15). In particular, in resin ducts of the plants belonging to the family Asteraceae, benzopyrans and benzofurans are found (Fig. 16). The benzopyran/benzofuran ratio varies from 1:1 up to 1:10. In the composition of resin from European spruce, *Picea abies*, growing in Japan, phenylpropane derivatives are identified (Omori et al. 1983). Some of the components are known as biologically active and can act as repellents due to their toxicity for herbivorous insects. Pyran, furan, and aromatic rings are the chemical basis for this action. Other resin components, such as trachyloban-19-oic and 16-kauren-19-oic acids, are able to kill larvae of various insects (Elliger et al. 1976) or, like pyran and furan derivatives, to protect plants against herbivorous animals. Insecticidic (antitermitic) action is characteristic for the resin components of the tropical

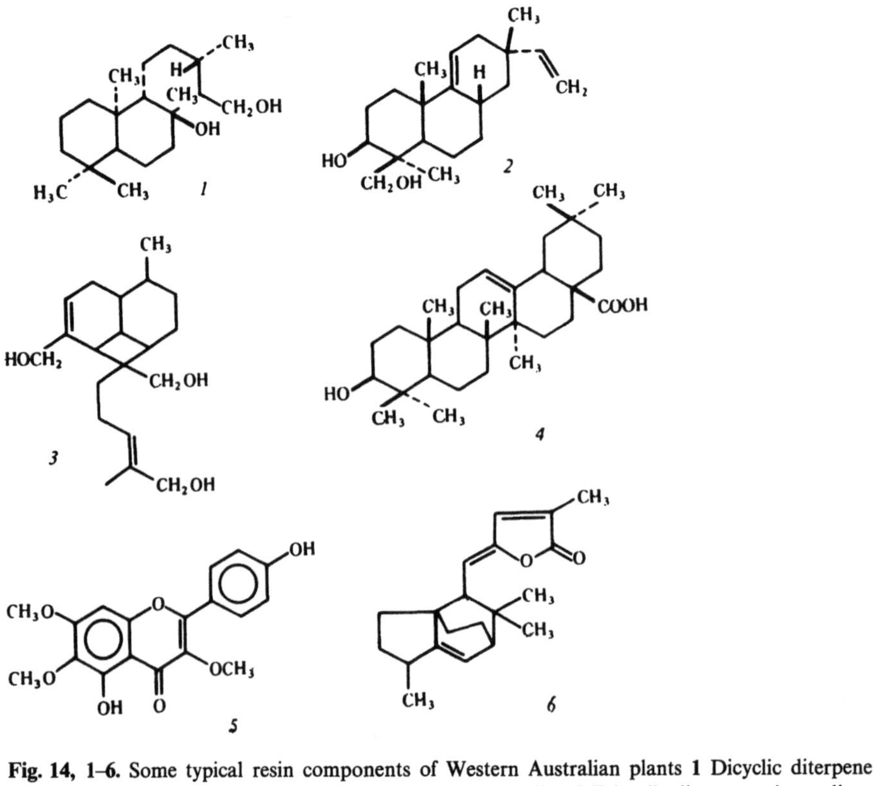

Fig. 14, 1–6. Some typical resin components of Western Australian plants **1** Dicyclic diterpene pimaradiene from *Ricinocarpos muricatus* eperuane-8β, 15-diol. **2** Tricyclic diterpene pimaradiene from *Newcastelia viscida* **3** Tricyclic diterpene from *Eremophila decipiens* **4** Oleanolic acid, **5** Penduletin from *Newcastelia viscida*, **6** Eremolactone from *Eremophila·fraseri*

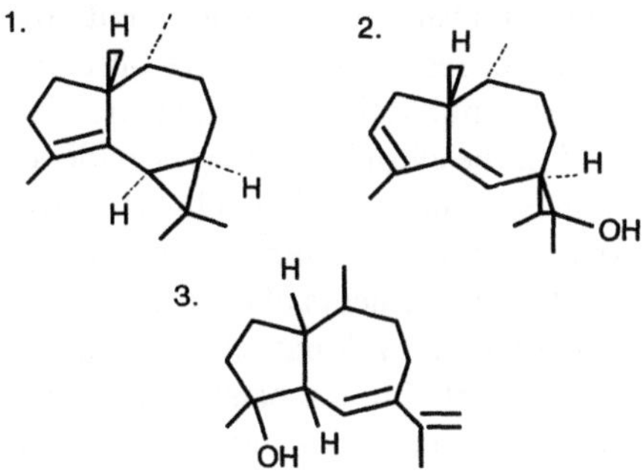

Fig. 15. Sesquiterpenes in resin of the tree *Dipterocarpus kerrii*. **1** Main component α-gurjunene, **2** and **3** Their derivatives

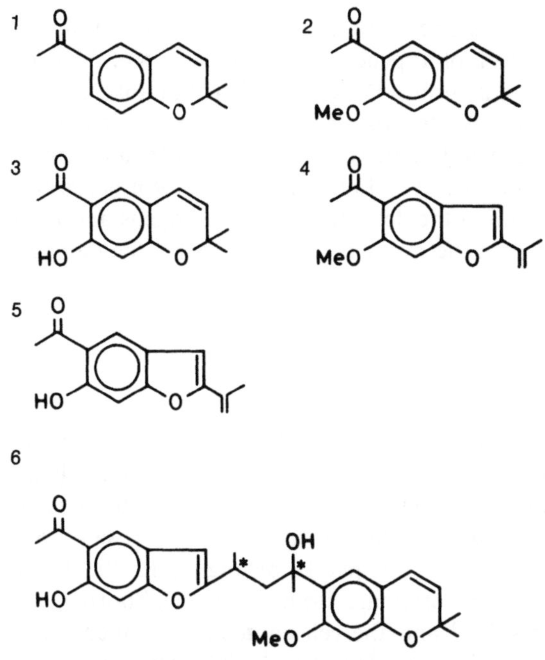

Fig. 16, 1–6. Benzopiranes (**1–3**) and benzofuranes from *Encelia* (**4–6**). 6-double stereoisomers

plant *Dipterocarpus kerrii*: sesquiterpenes, α- and γ-gurjunenols, and α- and γ-gurjunenes (Fig. 15) (Richardson et al. 1991).

Resin is synthesized in epithelial cells surrounding the resin duct. Secretory cavities or resin ducts should be isolated from the paths of gas diffusion since in some species such as *Eremophila* (Dell and McComb 1978a), the secreting terpenes can be oxidized easily.

The question of where the site of resin synthesis is inside the secretory cells is not yet completely solved. It has been proposed (Dell and McComb 1978b) that resins are synthesized in leucoplasts surrounded by the endoplasmic reticulum from which resin drops move out to be secreted into the resin duct. However, according to Fahn (1987), at earlier stages of resin duct development, the osmophilic drops, whose presence is indication of resin synthesis, are observed not only in these plastids but also in the endoplasmic reticulum, Golgi vesicles, mitochondria, on the nuclear membrane, and in the cytoplasm. This is evidence that resin synthesis may occur in various parts of the epithelial cells. It is also possible that different resin fractions are synthesized in different cell parts. The biosynthetic pathway of terpenoids, the essential constituents of resins, seems to be common to all similar compounds and starts with mevalonic acid (see Section 3.6; Fig. 50). The formed secretion, resin, is transported to the plasmalemma via the vesicles of the Golgi apparatus and endoplasmic reticulum and then excreted by exocytosis (granulocrine secretion). Other pathways of secretion are also possible.

The secretion in the resin duct is under great pressure (10 atm and higher), which causes an explosive liberation of terpenes after the break-up of the canal by mechanical damage to the plant surface (Vasilyev 1977) Probably, exuding resins cannot be considered as a waste arising in the process of metabolic exchange. Terpenes can be permanently renewed in plants and then destroyed and involved again in metabolism (Vasilyev 1977). Moreover, resin performs a protectory function, while preventing the development of pathogens and parasites, both in wounded plants when the resinous secretion covers the damaged tissue, and in intact ones.

2.3.2 Secretion of Latex

As distinct from the resin ducts, the reservoirs with latex, called laticifers, are living cells. Two types of laticifers are known: articulated and non-articulated. A nonarticulated laticifer is a giant cell which, when arising in the embryo, is not divided more but grows continuously, elongates, and branches (Esau 1965). Such laticifers permeate all organs of the plants belonging to the family Euphorbiaceae. Articulated laticifers originate from many neighboring cells whose cell walls have been dissolved and the cell contents combined into a continuous branched system. Laticifers are found in 12500 plant species belonging to 900 genera and 20 families of, mainly, dicotyledons such as Compositae, Papaveraceae, Campanulaceae, etc. (Fahn 1979; Nemethy and Calvin 1983). The presence of laticifers in natural conditions can be easily recognized by the latex efflux from damaged plant tissue. The latex in laticifers is under pressure which, in particular in *Hevea brasiliensis*, reaches 10 atm, thus providing for its intensive elimination. In wounded banana plants, the rate of latex exudation is from 400 to 50 $\mu l\ min^{-1}$ within the first 4 min after mechanical injury, then the amount of excreted latex decreases in time (Melburne et al. 1990).

A laticifer cell becomes secretory immediately after it appears. The duration of functioning of laticifers varies from species to species, being, as a rule, shorter than the life of the plant itself (Denisova 1989). The protoplasm and other organelles in a laticifer cell remain alive; the cytoplasm is localized near the walls, whereas the rest of the space is occupied by the vacuolar sap. Further development of this cell is characterized by degradation of cellular organelles. Small vacuoles in some plants do not combine, as a rule, to one large vacuole, so that the laticifers at later stages consist essentially of a great number of small separated vacuoles (Wiermann 1981). The structure of such laticifers and the vacuoles isolated from them is shown in Figs 17 and 18. For the latex of *Musa* sp., along with common cellular organelles, Kallarackal et al. (1986) described lipid globules, 5 to 35 μm in size, surrounded with a membrane and possessing a lysosomal activity, so-called lutoids, and cytoplasm fragments of granular structure. Similar inclusions are proposed for the latex of other plants.

The latex that flows after cutting a laticifer presents an emulsion. The dispersion medium of this emulsion is water (50–82%), whereas the dispersed phase consists of globules of water-insoluble substances: polyisoprene hydrocarbons, triterpenols and sterols, fatty and aromatic acids, carotenoids, phospholipids, proteins, and other substances (Wiermann 1981; Nemethy and Calvin 1983) The globules of these substances are suspended in the cell sap, so the emulsion is similar to milk. The latexes of different plants can be very different in appearance and chemical composition (Prokofyev 1948). In most cases, the latex is milky-white or yellow-green, but there are species (family Apocynaceae) in which the latex is like water in appearance, either more or less transparent, or in other plants yellow, orange, or even red. In Papaveraceae, the latex in genus *Papaver* is white, in the genus *Agremone* lemon-yellow, in *Chelidonium majus* orange (Prokofyev 1948).

Among the polyisoprene hydrocarbons of latex are resins, rubber, and gutta. The rubber has an empirical formula $(C_5H_8)_n$ and is a product of isoprene polymerization with a wide spectrum of molecular masses varying between 10^5 and 4×10^6, which is equivalent to polymers consisting for about 1500 to 60 000 isoprene residues (Goodwin and Mercer 1983). Gutta, like rubber, is a polymer consisting of such isoprene residues. Rubber and gutta are *cis*- and *trans*-1,4-polyisomers, respectively (Archer 1980).

Rubber Gutta

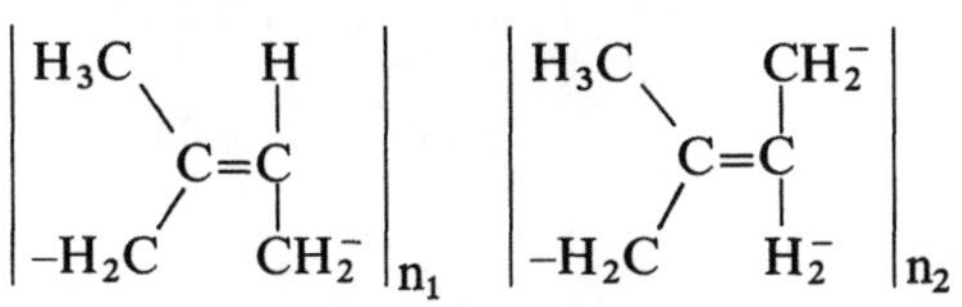

The latex of *Achras sapeta* contains a so-called chicle, a mixture of *trans*-(66%) and *cis*-(34%)-polyisoprenes, which is used for preparing chewing gum.

The production of rubber is peculiar to some families such as the Moraceae, Euphorbiceae, Apocynaceae, Asclepiadaceae, and Compositae. Rubber inconsiderable quantities is accumulated only in tropical plants of the genera *Hevea* (family Euphorbiceae), *Ficus* (family Moraceae), and *Parthenium argentatum*

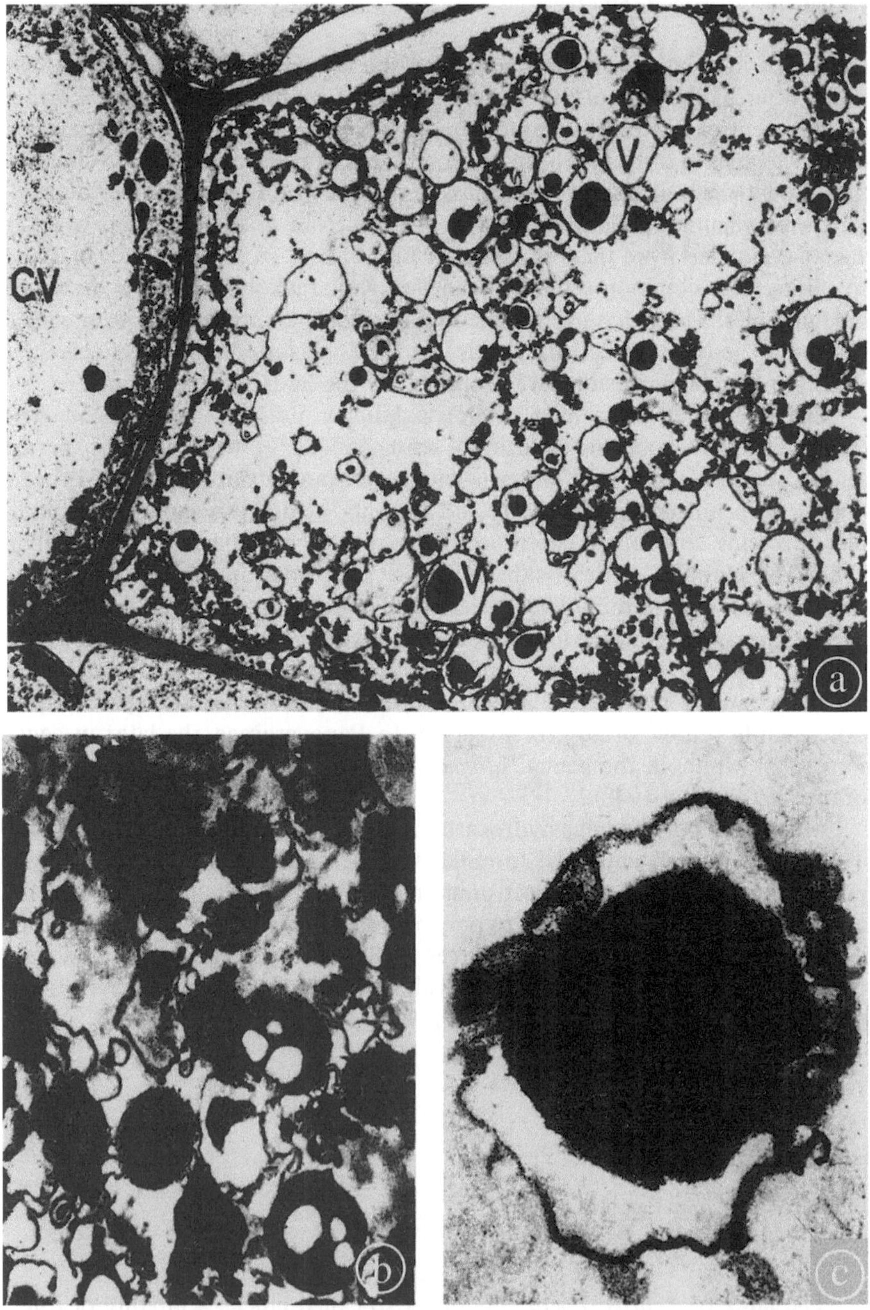

Fig. 17, a–c. Morphological appearance of latex vacuoles in situ and isolated from *Chelidonium majus* cells. **a** Cross-section of an old laticifer (first internodium of a plant at the age of 4 weeks) showing numerous small vacuoles (*V*) frequently with electron-dense inclusion bodies. The adjacent cells are characterized by large central vacuoles (*CV*). (6750 : 1). **b** Section through a pellet of isolated latex vacuoles (12000 : 1). **c** Well-preserved isolated vacuole showing the single surrounding membrane and the inclusion body (50000 : 1). (Matile et al. 1970)

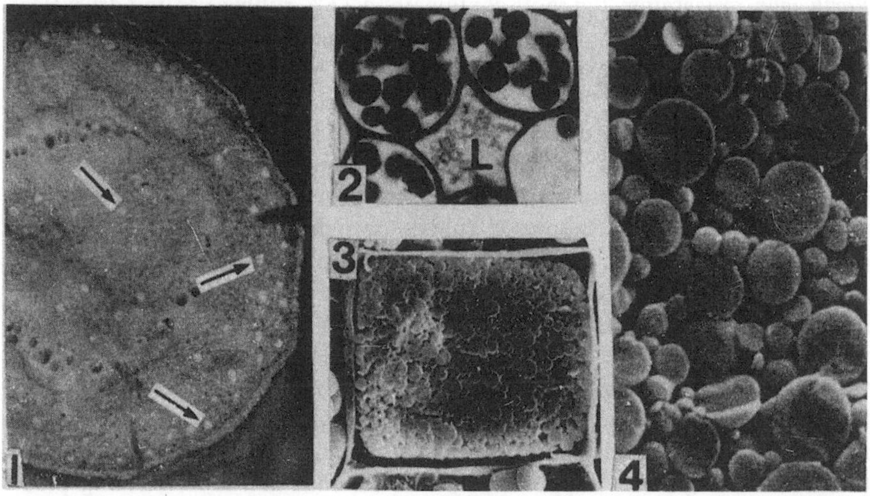

Fig. 18, 1–4. Light and scanning electron microscopy (SEM) micrographs of laticifers in *Calystegia silvatica* (Kit) Griesb. **1** Laticifers in stem. An intact fresh stem frozen directly in nitrogen "slush" and cross-fractured. Under the stereoscopic microscope, the laticifers show as white "spots" (*arrows*) among the tissues. × 18. **2** Enlarged view of a laticifer (*L*) and surrounding cells in a transverse section of the cortex (× 400). **3** The cortical laticifer cell in a fresh rhizome by SEM (× 770). **4** In situ latex of laticifer **3** by SEM (× 3900). (After Condon and Fineran 1989a)

(family Compositae). The latter is a shrub growing in the arid zones of North America. In Europe, the main rubber-producing species belong to the family Compositae and are found in the Ukraine and the Asian republics of the former USSR. Among them are Kazakhstan tau-saghyz (*Scorzonera tau-saghyz*) and kok-saghyz (*Taraxacum kok-saghyz*), whose roots contain up to 40 and 22% rubber per unit dry weight, respectively. Unlike the Asian species, the Crimean dandelion or crim-saghyz *Taraxacum hybernum* accumulates only 5–6% rubber by the second year of vegetation. Nearly the same amount of rubber is produced by leaves of *Asclepias cornuti* (family Asclepiadaceae) introduced to Europe from America. However, the Mexican silver guayule *Parthenium argentatum* acclimated in Europe synthesizes less rubber (10 to 12%) than in its native land (Prokofyev 1948). Other species may also contain rubber, but in quantities insufficient for commercial use. Among them are golden rod, *Solidago virga* (family Compositae) and dogbane, *Apocynum sibiricum* (family Compositae).

Of all the rubber-containing plants, mainly *Hevea brasiliensis* is commonly used for industrial purposes. High-yielding cultivars of *Hevea*, from which about 700 kg rubber per 1 acre in a year are collected, have been obtained in Malaysia. The rubber of *Hevea* is produced and accumulated by the ring-shaped laticifers of the bark. Since there are anastomoses between neighboring laticifers, the latex can flow out from the vast bark zone after tapping.

Rubber production by plants depends on light intensity; nevertheless, roots are the main source of this polymer in some species. For instance, in guayule,

Parthenium argentatum, the roots contain up to 7.6% rubber per unit dry weight. Along with the stem bark, the roots also can be a main source of rubber whereas the young shoots and leaves of this plant form as little as 0.5% rubber (Macrae et al. 1986).

Backhaus (1985) has reviewed studies devoted to rubber biosynthesis which reports that production of polyisoprenoid occurs in parenchymatous cells. Initially, rubber cells arise in the cytoplasm, then they can be detected in the vacuoles of these cells. Dispersed rubber particles of 5 to 6 μm in size are protected by a thin film consisting of adsorbed protein and phospholipid. As rubber is a product of industrial significance, the pathways of its biosynthesis have received greater attention than those of other latex components. The main precursors of rubber are sugars, acetate, pyruvate, and mevalonate. The pathway of their transformation is shown in Fig. 19. The direct precursor of rubber is isopentyl pyrophosphate, which serves as the origin for synthesis of all known natural terpenes and terpenoids. The possible pathway of rubber production, presented in Fig. 21, is supported by the fact that in rubber-yielding plants

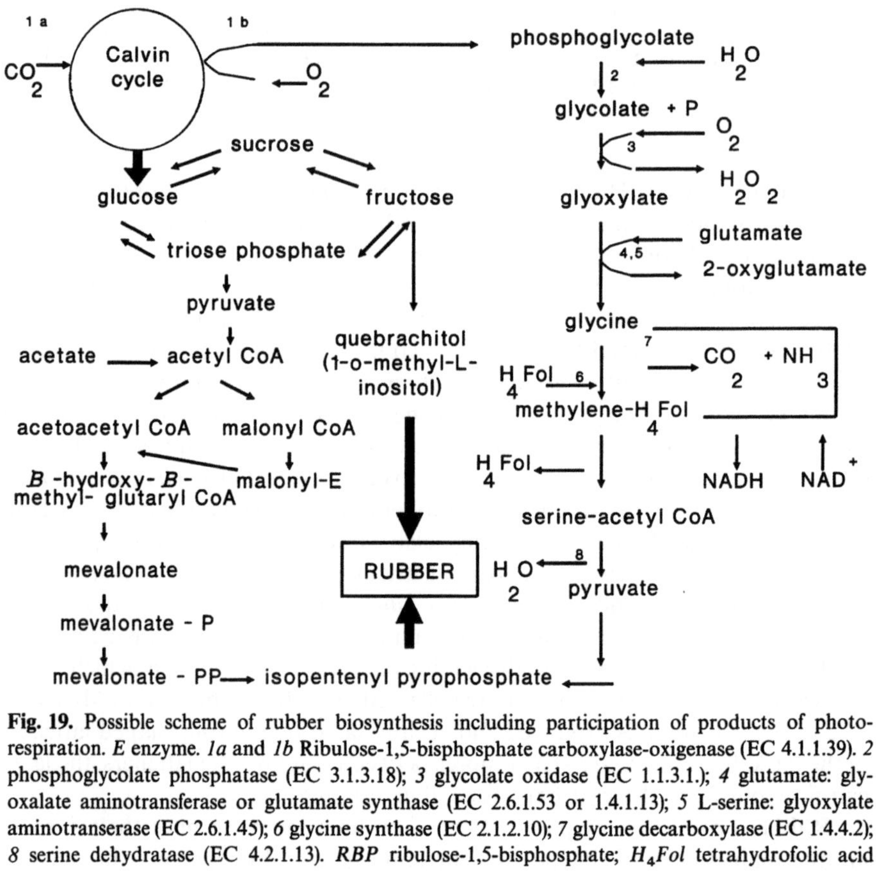

Fig. 19. Possible scheme of rubber biosynthesis including participation of products of photorespiration. *E* enzyme. *1a* and *1b* Ribulose-1,5-bisphosphate carboxylase-oxigenase (EC 4.1.1.39). *2* phosphoglycolate phosphatase (EC 3.1.3.18); *3* glycolate oxidase (EC 1.1.3.1.); *4* glutamate: glyoxalate aminotransferase or glutamate synthase (EC 2.6.1.53 or 1.4.1.13); *5* L-serine: glyoxylate aminotranserase (EC 2.6.1.45); *6* glycine synthase (EC 2.1.2.10); *7* glycine decarboxylase (EC 1.4.4.2); *8* serine dehydratase (EC 4.2.1.13). *RBP* ribulose-1,5-bisphosphate; H_4Fol tetrahydrofolic acid

appropriate enzymes are found. Indeed, the latex of *Hevea* contains a full enzyme system for rubber synthesis: enzymes essential to the transformation of acetyl ~CoA according to the following steps: mevalonic acid → phosphomevalonate → pyrophosphomevalonate → isopentyl pyrophosphate, a precursor of rubber. The biosynthesis of natural rubber is described in detail by Paterson-Jones et al. (1990). Besides the substances involved in the above pathways, some other products can participate in rubber synthesis, in particular those formed in photorespiration, whose role in rubber production has been investigated by Indian scientists (Reddy et al. 1987). A possible pathway based on this study is presented in Fig. 19.

The rate of rubber synthesis depends on many parameters. For the latex of *Hevea brasiliensis*, it amounts to 5 mg min^{-1} ml^{-1} latex (Gray 1987). The rate of rubber synthesis increases with water deficiency. In *Parthenium argentatum*, the activity of the *cis*-1,4-polyisoprenyl transferase involved in the final stage of rubber biosynthesis increases by 20–50% under water stress when the leaf water potential falls from 0.5 MPa to 2.5 MPa. Therewith rubber production in water-stressed plants also rises considerably in percent of dry mass from 5.08 to 6.35% in roots, from 4.18 to 6.01% in young stem, and from 3.65 to 4.89% in old stem, whereas in leaves there are no changes (Reddy and Das 1988). Moreover, rubber synthesis, in particular in guayule, *Parthenium*, increases at relatively low temperatures: 15 to 18 °C in the day and 10 to 13 °C at night (Downes and Tonnet 1985). This phenomenon is believed to be connected with the water content of tissues. However, it is not observed in all plants. Rubber synthesis is stimulated by ethylene. Upon ethylene treatment (200 mg of ethrel per tree) of *Hevea* stem incisions, the activity of the H^+ pump of tonoplast ATPase increased, which induced alkalization of the cytosol and enhancement of rubber synthesis (Gidrol et al. 1988).

The latex of many plants belonging to the families Euphorbiaceae and Compositae and growing in the temperate zone contains triterpenol esters instead of rubber. Indeed, euphol and sitosterol have been found in the acetone extract of dry latex from branches of *Euphorbia ruspoli* (Passannanti et al. 1985). Sometimes terpenoids and their alcohols are met in such high quantities that they can be used as a disinfectant. Formulas of some terpenoids from latex of plants belonging to the family Euphorbiaceae are given in Fig. 20.

The nitrogenous components in latex are represented by amino acids, proteins, alkaloids, and amines. The most common amino acid is 3,4-dihydroxyphenylalanine (DOPA) (Fig 21). According to Haupt (1976), the amount of DOPA in the latex of *Euphorbia lathyris* is estimated to be about 17% of fresh mass or 68% of the soluble nitrogen in the milky sap. In isolated latex, 17 mg of DOPA g^{-1} fresh mass is produced from ^{14}C-tyrosine within 2 min, whereas leaves, stems, and roots form less than 1 mg g^{-1} fresh mass within 6 h (Haupt 1976).

In latex-containing vacuoles, the enzymes ribonuclease, acidic phosphatases, α-amilase, and polyphenoloxidases exhibit high activity (Matile 1976). It is important to note that vacuoles isolated from the latex of *Chelidonium majus*

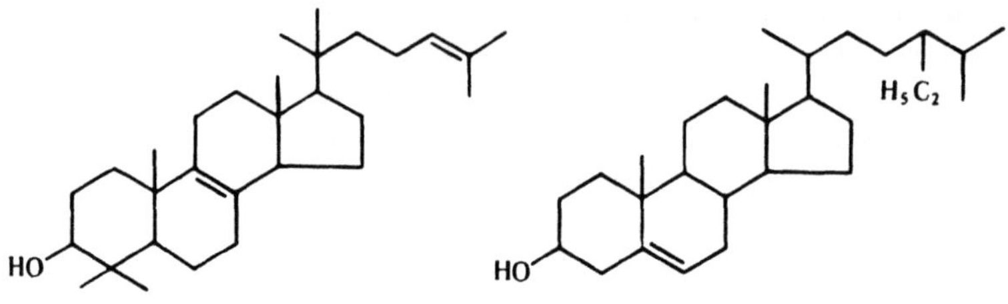

1 2

Fig. 20, 1,2. Triterpenes in latex. **1** Cyclic triterpene euphol. **2** Sitosterol

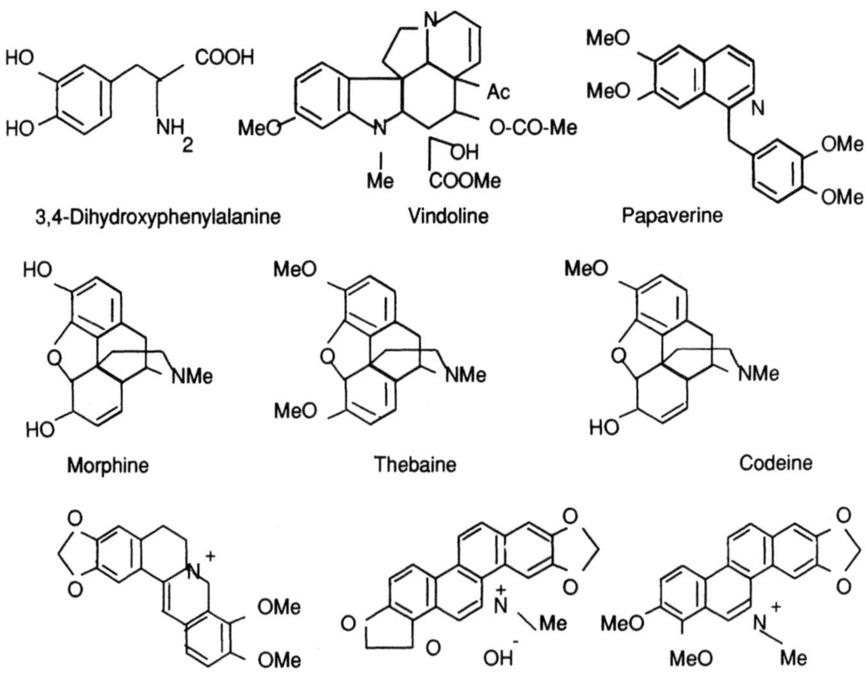

Fig. 21, 1–9. Nitrogenous components of latex

contained enzymes inherent in lysosomes: hydrolases, acidic proteases, RNAases, and phosphatases (Matile et al. 1970). Proteases have also been isolated from the latex of eight species of Euphorbiaceae (Lynn and Clevette-

Radford 1988). The functional groups of the active centers of these enzymes are serine and histidine. The molecular mass of the proteins varies between 33 000 and 117 000 daltons. In vacuoles, there are also enzymes of glyoxalate metabolism, the cycle of tricarboxylic acids and enzymes of alkaloid biosynthesis (Roberts 1986). The latex of *Synadenium grantii* from the family Euphorbiaceae contains acetylcholinesterase known as an enzyme of acetylcholine hydrolysis in animal cells (Govindappa et al. 1987).

The protective function in latex is fulfilled by lysozymes, proteins that hydrolyze the bonds of bacterial polysaccharides in microorganism envelopes. Four lysozymes have been isolated from the latex of *Asclepias syriaca* (Lynn 1989). They have a molecular mass of about 28 kdaltons and differ in amino acid composition. Lysozymes are also found in the latexes of other plants: *Papaya carica* (Howard and Glazer 1969), *Hevea brasiliensis* (Tata et al. 1983), and the rubber plant *Ficus elastica* (Glazer et al. 1969). The lysozymic activity of the proteins decreases significantly on treatment with histamine (20–50 mM), an inhibitor of plant lysozymes.

The representatives of the family Euphorbiaceae are characterized by the presence of lectins in the latex. These are carbon-binding proteins not yet well studied. Lectins are able to bond and precipitate specific polysaccharides and glycoproteins located on the cell surface. Due to the high affinity for antigene-glycoproteins, lectins induce their agglutination or even agglutination of whole cells. Similar proteins with a total molecular mass from 60 to 67 kdaltons and molecular masses of monomers from 27 to 38 kdaltons are found in 17 species belonging to the Euphorbiaceae (Lynn and Clevette Radford 1986a). Every type of latex may contain 5 to 13 isolectins. Lectins of plants perhaps function as defensive proteins, protecting them against the invasion of parasitic microorganisms (Sharon and Lis 1989). As defensive antifungal protein, hevein from the latex of the rubber tree *Hevea brasiliensis*, is known, and has been isolated and purified (Parijus et al. 1991).

Recently, the protein calmodulin binding Ca^{2+} has been identified in the latex of *Euphorbia lathyris* (Piazza et al. 1987 a, b). The amount of Ca^{2+} in the latex of this plant is very high: about 7.5% of dry matter, so that it is necessary to bond the excess calcium by calmodulin. The latter is thought (Piazza et al. 1987 a) to be a regulator of the triterpene biosynthesis in latex because calmodulin antagonists inhibit the incorporation of mevalonate into triterpenes and esters of fatty acids. A similar function is characteristic for animal calmodulin participating in cholesterol biosynthesis.

The latex of cell vacuoles includes a number of alkaloids (Fig. 21). In the review of Matile (1987), some data on the content of various isoquinoline alkaloids in the celandine, *Chelidonium*, are summarized. This latex contains the highest proportion of dihydrocoptisine: 1100 mg ml^{-1} (254.5 mM), and lesser amounts of sanguinarine: 332 mg ml^{-1} (126.5 mM), berberine: 218 mg ml^{-1} (35.2 mM), and chelerythrine: 174 mg ml^{-1} (60.6 mM). Chelidonic acid

(497 mM) (Fig. 22) and phenolic compounds (624 mM) are also found in vacuoles of laticifers. The chemotaxonomic feature of the family Papaveraceae is the accumulation of morphine and benzophenantridine alkaloids (Kutchan et al. 1986). Laticifers of the poppy, *Papaver somniferum*, contain 25 alkaloids: morphine, codeine, papaverine, thebaine, etc. However, morphine prevails: 52% of total alkaloids (Roberts 1986). Alkaloids of latex from fruits of the periwinckle, *Catharanthus roseus*, are represented mainly by vindoline and 19-epivindoline, which are seen under the electron microscope as drops in vacuoles (Eilert et al. 1985). Due to the presence of alkaloids, latex often has a bitter taste and is a toxic. The death-bearing milky sap is known from the South American plant *Maquira coriacea*. Such toxic latex components are found in the upas tree, *Antiaris toxicaria*, whose poisonous substances also includes the cardiac glycoside, antiarin. Aborigenes of Central Malaya use the milky sap of this plant for smearing war-arrow heads (Grudzinskaya 1980). Both the above latex-bearing species belong to the family Moraceae. However, within this taxon there are species such as *Bromium alicastrum*, *B. utile*, *B. potabile*, etc. that have a white latex without toxic alkaloids. The inhabitants of Latin America drink it instead of milk.

The site of alkaloid biosynthesis, as shown for *Chelidonium majus*, is the parenchyma cells which surround the laticifers (Wiermann 1981). The alkaloids thus formed are transported to laticifers, where they are stored either in small vacuoles as in *Chelidonium majus*, or in large central vacuoles, as in roots of *Sanguinaria canadensis* (Wiermann 1981).

Besides alkaloids, amines are frequently found in laticifers. Catecholamines, mainly dopamine, are components of the latex from fruits of the genus *Papaver* (Kutchan et al. 1986; Roberts 1986). Dopamine and its derivatives are produced in the biosynthetic pathway from amino acid phenylalanine via step-by-step hydroxylation to tyrosine and dihydroxyphenylalanine. The latter is transformed into dopamine by decarboxylation. Another metabolic pathway cannot be excluded: first via decarboxylation of tyrosine to tyramine, which is converted to dopamine by hydroxylation. The biosynthesis of dopamine can occur in various cell compartments, including chloroplasts. Finally, amines are accumulated in laticifers (Kutchan et al. 1986). Thus, the formation of biogenic amines takes place in dissimilation of amino acids. As has been supposed (Grosse 1982), This process plays an important role in the detoxification of ammonium and nitrogen-containing compounds as well as in the regulation of growth. Catecholamines being in latex appear to carry out a protectory function when the

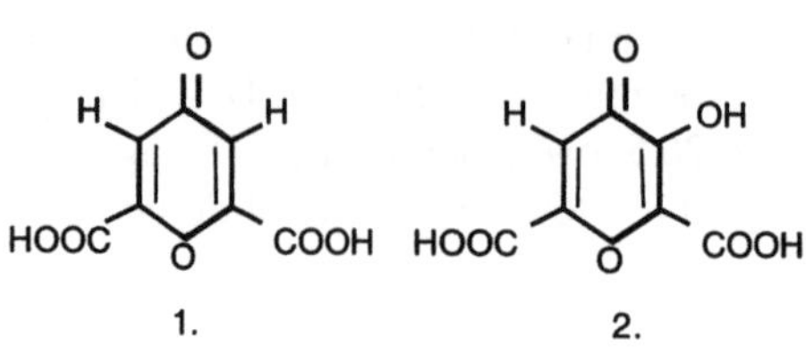

Fig. 22. Chelidonic (**1**) and meconic (**2**) acids in latex

plant is wounded. In particular, their concentration increases in callus forming around the wound in cactus *Carneginea gigantea* (Steelink et al. 1967). Moreover, exogenous catecholamines stimulate the growth of tobacco tissue culture (Protacio et al. 1992).

Along with organic substances, latex contains ions of inorganic salts. In particular K^+, Na^+, Mg^{2+}, Ca^{2+}, Cl^-, NO_3^-, SO_4^{2-}, and PO_4^{3-} are found. The major of these are K^+, Mg^{2+}, Cl^-, NO_3^-, their amount varying from 1 to 6.14 $mol\,m^{-3}$. The pH of latex is usually maintained near 6–6.2. Heinrich et al. (1986) studied the ionic composition of latex in 25 species of milky plants. In comparison with other plant tissues, the latex had more K^+ (17-fold), Mg^{2+} (7-fold), and Ca^{2+} (2-fold), whereas PO_2^- and PO_3^- were lacking in most species studied.

The components of latex are formed in the laticifers themselves, the cytoplasm, or the plastids. The original matter for latex synthesis is sugars and products of their transformation. Saccharides flow from leaves to the sites of latex formation. Sometimes, there are many sugars in the latex, as is the case in *Hevea brasiliens* whose milky sap contains up to 45.6 mM sucrose (Eschbach et al. 1986). The synthesis of latex components occurs at a certain stage of tissue differentiation. A nondifferentiated tissue culture does not form latex but begins to produce this product as soon as laticifer-like structures appear (Dell and McComb 1978a).

According to the opinion of some researchers (Esau 1965; Vasilyev 1977, etc.), laticifers are excretory structures, since it is unknown whether terpenes, once deposited in the cells, are reused by the plant (Vasilyev 1977). At the same time, the variability in the composition of latex and in the proportion of its components in various species permits the supposition that laticifers may fulfill some additive functions (Esau 1965). Indeed, the milky sap of *Taraxacum*, *Cinchorium*, and other plants belonging to the Asteraceae plays a certain role in repelling herbivorous animals, since among the latex components toxins are found (Harborne 1988). Usually, such metabolic products are stored in a closed intracellular compartment and can be liberated only as a result of external mechanical damage.

2.3.3 Secretion of Gum and Essential Oils

Intratissular secretion of gum and essential oils has been poorly studied and there are only few works devoted to this question.

Gum. Gum is usually is amorphous material, soluble (as in the genus *Acacia*) or swelling in water (as in the genus *Sterculia*), due to the presence of polysaccharides, proteins, and other compounds. There are various types of gums in different species, such as the well-known gummiarabic (water-soluble) from the gum arabic or babul tree, *Acacia arabica*, as gum karaya from *Sterculia urens* Roxb (Sterculiaceae). Gum karaya consists of insoluble particles which absorb large amounts of water and form a mucilage-like mixture (Shah and Setia 1976).

Usually, gums are stored in special gum ducts which are found in both tree stems (Esau 1965) and fruits (Aajera 1974); sometimes, in particular in the Brazilian tree *Stryphodendron barbatiman* (Barbatimao), in seeds (Reicher et al. 1991). Gum ducts are well known in plants belonging to the genera *Prunus*, *Brachychiton*, *Citrus*, *Acacia*, etc. Gum has been found in 40 genera from 30 families of angiosperms and in some species belonging to the gymnosperms. However, plants such as *Ailanthus* (Babu et al. 1987), *Bombax* (Babu and Shah 1987), or *Citrus* (Gedalovich and Fahn 1985) do not form gum ducts under normal conditions, but develop traumatic canals or ducts in the secondary xylem after infection. Gum-bearing channels are formed by different modes in different species. They may be either schizogenous or lysigenous. There are cases when gum ducts initially develop schizogenously and at later stages become lysigenous.

Gum that fills up ducts represents complicated mixtures of various chemical compositions. Mostly, it is a complex of polysaccharides and terpenes or a mixture of flavonoids, terpenes, and fatty acids (Babu and Menon 1990). Up to 90% of the carbohydrates in gum arabic from *Acacia senegal* are glycoproteins (Wu Qi et al. 1991).

In most plant species, gums are suggested to be formed in response to physiological disturbances or various diseases when cell walls are damaged (mainly starch and other carbohydrates located in the cell wall) and cell content is exuded. This phenomenon is also called 'gummosis" (Esau 1965). Gum is accumulated not only in gum ducts but also in the cells of the xylem. When in contact with the air, gums thicken and fill up the wound, preventing dehydration and penetration by pathogens. One illustration of this is the formation of gummiarabic (gum arabic) from *Acacia senegal* and *A. verek*. It consists of a high-branched polymer D-galactose, L-arabinose, D-glucuronic acid, and L-rhamnose. The molecular mass of sugar residues reaches 100, whereas the minimum molecular weight of resin is 4000–6000 daltons (Stephen 1980). Generally, the polymer occurs as potassium, calcium, and magnesium salts (Duffus and Duffus 1984). Gummiarabic is found in the cambial zone of *Acacia senegal*, where it is believed to be synthesized (Joseleau and Ullmann 1990) However, no gradient of transport of the resin to the site of exudation was revealed. The wound exudate accounts for 90% of the world yield of this glue, mainly in the Sudan. The resin ghati is an exudate of *Anogeissus litifolia*, which grows in India and Sri Lanka. This is a polymer of L-arabinose,, D-galactose, D-mannose, D-glucuronic acid, and D-xylose, and occurs usually as Ca salts (Duffus and Duffus 1984). The resin glue from the cherry tree contains polysaccharides involving the residues of galactose, mannose, arabinose, D-glucoronic acid, and a negligible amount of xylose (Kretovich 1971). Gum from seeds of *Stryphnodendron barbatiman* consists of galactomannan, which can be recommended as an additive in food and cosmetics (Reicher et al. 1991). The proteinaceous part of gums of the genera *Acacia*, *Araucaria*, *Azadirachta*, and *Lannea* is rich in amino acids such as serine, threonine, and also proline and leucine, especially for *Lannea* (Anderson et al. 1972). Moreover, these gums also contain aspartic acid and glucosamine.

Several types of polymers have been distinguished in the resinous product, depending on the composition of the sugar residues (Stephen 1980). The most widespread type (type A), arabinoglucan, is found in 22 families. It involves sugar residues such as arabinose, rhamnose, xylose, fucose, glucoronic and galacturonic acids. Galacturonorhamnan (type B) was found in the Sterculiaceae family (*Sterculia urens, S. setigera, S. acerifolia, Brachychiton diversifolium*). Its sugar residues are arabinose, galactose, fucose, and glucoronic acid. Xylan (type C) involves residues of arabinose, galactose, and glucuronic acid. It occurs very seldom, mainly in the Liliaceae (*Phormium tenax*) and Iridaceae (*Watsonia pyramidata*) families.

The chemical composition of resin is often very complex. Besides polysaccharides, it contains phenol and protein components. About 20 flavonoids have been isolated from the resin of *Gutierrezia sarothrae* (Hradetzky and Wollenweber 1987) and the structure of six flavonoids has been established: 5,7,4′-trihydroxy-3,6,8,3′-tetramethoxyflavone; 5,7,3′-trihydroxy-3,6,8,4′-tetramethoxyflavone, etc. Protein polysaccharides were found in resin from *Acacia* species (Anderson and McDougall 1985). The protein component in these plants varied from 0.2 to 45%. The amino acid composition of the proteins varied significantly, especially in hydroxyproline, depending on the species. For example, in the resin of *Acacia aestivalis*, per 55 hydroxyproline residues were 1000 other amino acid residues, whereas in the resin of *Acacia saliciformis* this ratio was 287 per 1000. The quantitative ratios of alanine, aspargic acid, proline, and serine were also varied. The resin from *Acacia hebeclada* is arabinogalactan bound to the protein. The protein content in this compound was relatively low (12%) and the amount of arabinose was high (Churms et al. 1986). The secretion of glycoprotens is a complex process (Peterson and Vermeer 1984). The protein component is synthesized by ribosomes attached to the membranes of the endoplasmic reticulum. The synthesized proteins penetrate the cisternae is the endoplasmic reticulum and the Golgi apparatus, where they bind to carbohydrates, thus forming glycoproteins. The produced secretion is eliminated via a granulocrine type of secretion (Peterson and Vermeer 1984). Perhaps ethylene plays a regulatory role in the production of gum. It was shown, for instance, that treatment of almond, *Prunus dulcis*, fruits by ethylene enhances gum formation (Morrison et al. 1987b). It is also known that in normal wood of *Sterculia urens* (family Sterculiaceae) there are no gum ducts. However, after ethylene treatment, these ducts are formed from traumatic parenchyma (Menon and Babu 1989).

Essential Oils. Oil ducts can be found inthe families Calycantaceae, Lauraceae, Magnoliaceae, Simarubaceae, Winteraceae (Esau 1977), and Umbelliferae (Senalik and Simon 1986). They develop in leaves, for example in *Eremophilla*, where they can occupy up to 13% of the leaf cross-section. Oil cavities and ducts of *Hypericum perforatum* (Fig. 23) are seen on the leaf surface as dark spots. A cross-section of the leaf demonstrates the cavity filling with dark secretion,

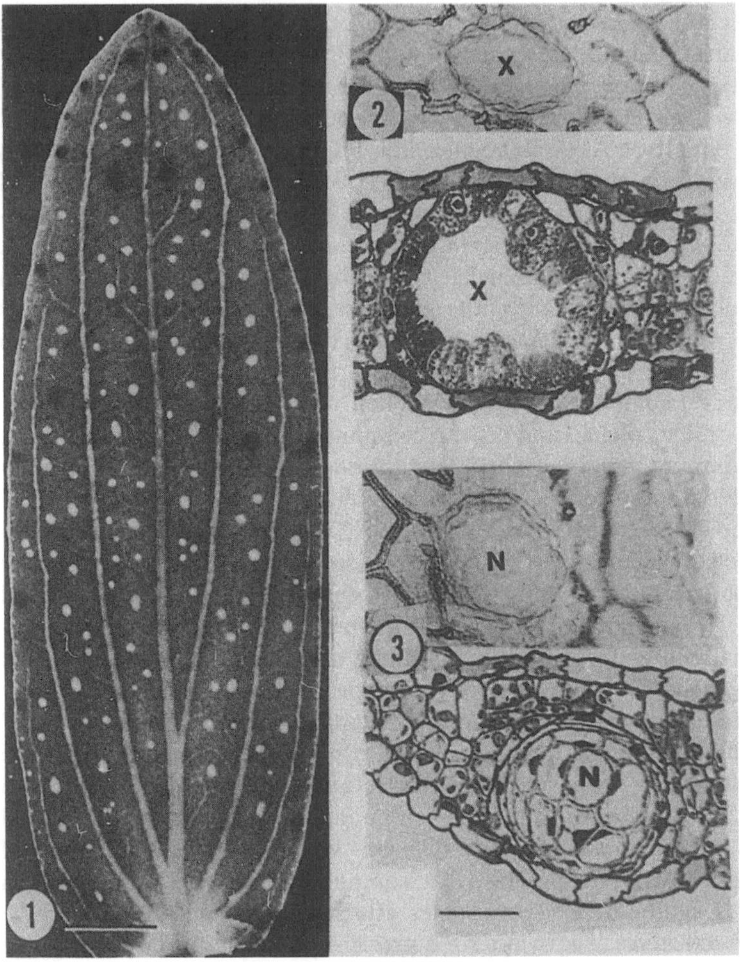

Fig. 23, 1–3. Mature cavities and nodules in leaf of *Hypericum perforatum* **1** Intact leaf. *Light spots* are translucent cavities and *black spots* are nodules. *Bar* 2.0 mm. **2** Cavity (*X*) is seen (*above*) in cleared leaf, surrounded by minor veins and vein ending (*bar* 100 μm) and in cross-section (*bar* 20 μm) **3** Nodule (*N*) is seen in cleared leaf (*below*), surrounded by minor vein and vein ending (*bar* = 100 μm) and in cross-section (*bar* = 20 μm). (After Curtis and Lersten 1990)

perhaps containing hypericin (Curtis and Lersten 1990). The cavities and ducts containing oils in *Tagetes erecta* are located in the parenchyma of the leaf lamina, and protrude above the upper surface of the leaf (Fig. 24). In roots of carrot, *Daucus carota*, the oil ducts are in the phloem, where they form a network connected by concentric rings and radial lines (Senalik and Simon 1986).

Interestingly, the genetically different lines of carrot have a different number of oil ducts. The secretion of the ducts is composed mainly of terpenoids,

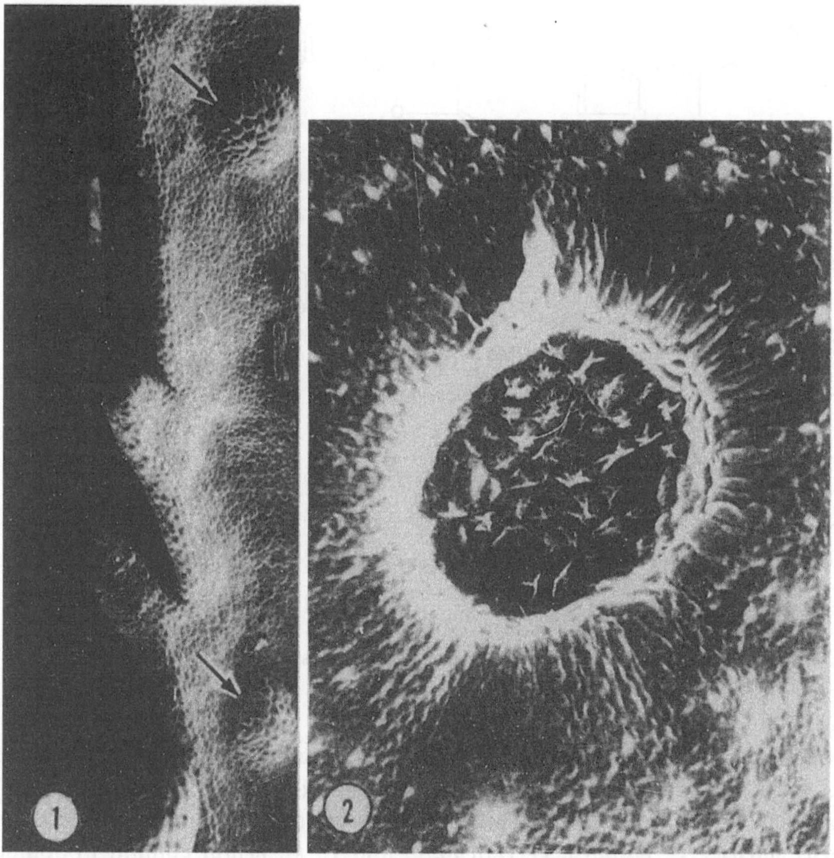

Fig. 24. 1,2. Scanning electron micrographs (SEM) of *Tagetes erecta* leaves. **1** Secretory cavities. (*arrows*) protruding from lower leaf surface. × 100. **2** Collapsed secretory cavity. Collapse occurs at flowering. × 180. (After Russin et al. 1988)

involving a α- and β-pinenes, myrcene, α-phellandrene, α-terpinene, limonene, γ-terpinene, bornyl acetate, β-caryophyllene, and E-γ-bisabolene (Senalik and Simon 1986). Among the terpenoids of essential oils of species belonging to the Lauraceae, furanosesquiterpenes ans sesquiterpene lactones are met (Fig. 25). There is a close correlation between the amounts of volatile terpenoids and nonvolatile products. Formation and accumulation of the secretion proceeds in the plastid thylakoids of secretory cells (Schnepf 1974) (see Sect. 3.6). Then oil drops appear in the cytoplasm and are removed to oil ducts. Lysigenous oil-containing cavities are well developed in the coat of *Citrus* fruits: lemon, *Citrus limon*, mandarin orange, *Citrus nobilis*, orange, *Citrus sinensis*, and are filled with essential oils which consist of terpenes. Essential oil ducts in fruits of cow-parsnip, *Heracleum*, contain oils in which more than 30 individual substances

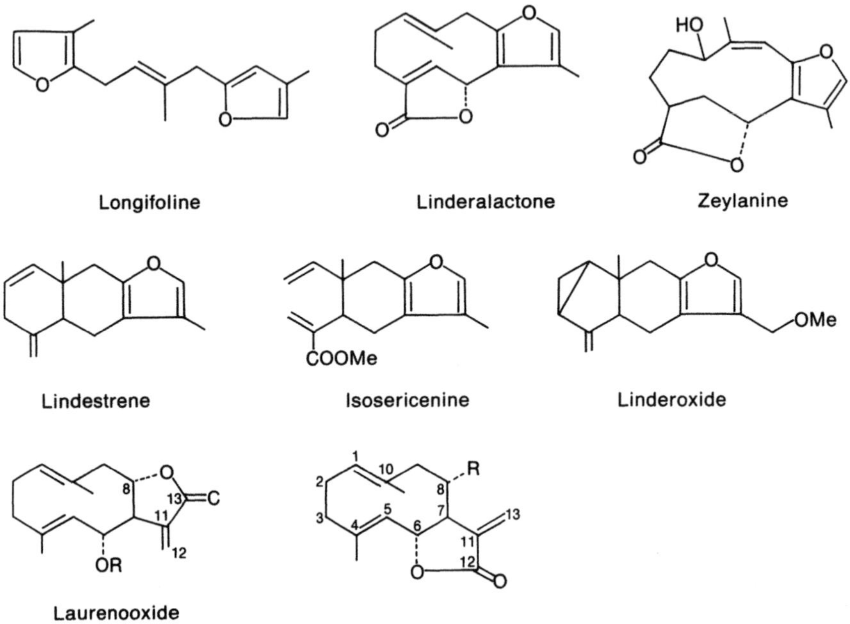

Fig. 25. Furanosesquiterpenes and sesquiterpene lactones in essential oils of Lauraceae

have been identified. The most important of them are esters, where octyl acetate and octyl butyrate prevail. Among the terpenes, α- and β-pinenes were found (Tkachenko 1987). Along with essential oils, secretory ducts in the fruit coat of *Citrus sinensis*, *C. paradisi*, *C. reticulata*, and *C. klementin* contain lipophilic derivatives of flavonoids (Fig. 26) such as poncanetin (1), nobiletin (2), and

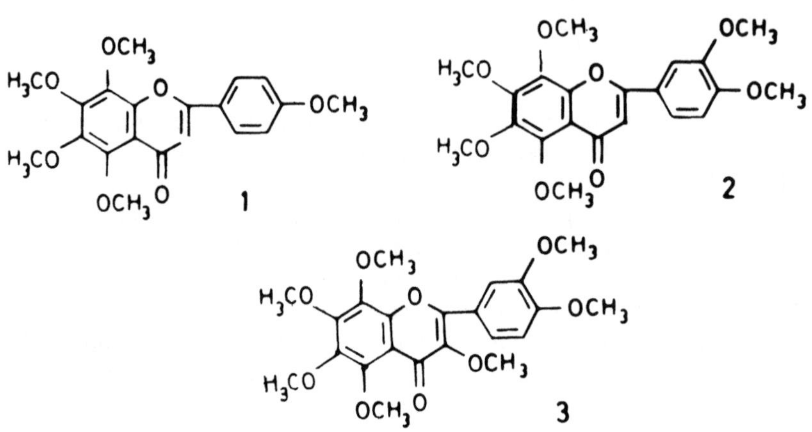

Fig. 26, 1–3. Flavonoids in oil ducts. Poncanetin (**1**), nobiletin (**2**), 3,5,6,7,8,3′,4′-heptamethoxyflavone (**3**)

3,5,6,7,8,3′,4′-heptamethoxyflavone (3) (Walther et al. 1966). Unlike these species, *C. limon* contains only hydrophilic flavonoids; however, they were not found in the secreta.

Very often, essential oil ducts include coumarins, which are dissolved in essential oils. Eleven compounds of coumarinic nature have been found in

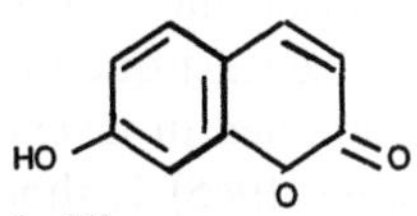

Umbelliferone

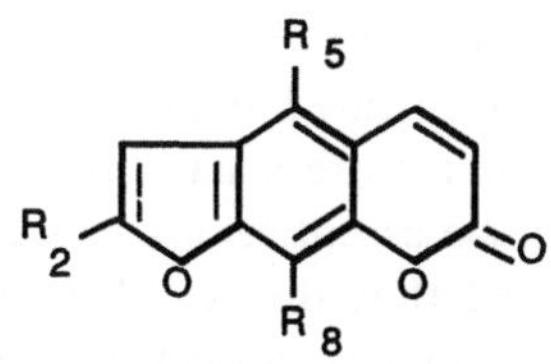

Furanocoumarins

	R_5	R_8	R_2
Xanthotoxin	H	$-OCH_3$	H
Psoralen	H	H	H
Bergapten	$-OCH_3$	H	H
Imperatorin	H	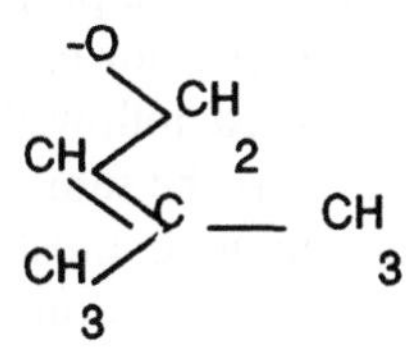	H
Isoimperatorin	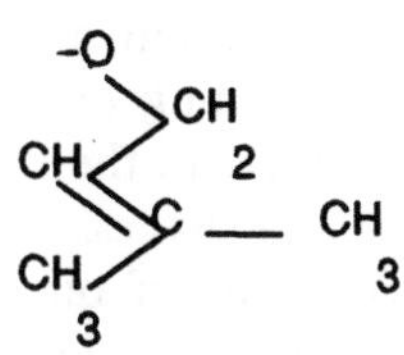	H	H
Rutamarin	H	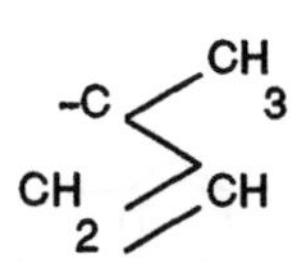	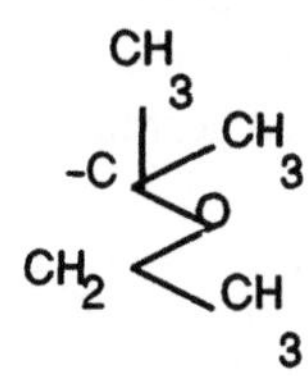

Fig. 27. Coumarins in schizo-lysigenous cavities of rue, *Ruta graveolens*

schizogenous cavities of rue, *Ruta graveolens* (Denisova 1976), six of which have been identified. These are umbelliferone, xanthotoxin, psoralen, bergapten,. isoimperatorin, and rutamarin (Fig. 27). In secretory cavities of *Dictamnus gymnostylis*, six components of coumarinic derivatives have been identified: glycoside of umbelliferone, psoralen, xanthotoxin, auropten, and some other nonidentified compoúnds (Denisova 1976). The compounds of the coumarinic series accumulated in fruits of *Archangelica decurrens* are also dissolved in essential oils and are localized in essential oil ducts (Denisova and Dranitsina 1962). Coumarins as glycosides occur in almost 70 plant species. The amount of compounds is 1.5% in the tonka bean, *Dipteryx* and 0.7% in woodruff, *Asperula*, and concentrated, mainly in oil bearing internal structures. The least quantities of coumarin are usually in glandular hairs. Coumarinis contained in lavender and cassia essential oils, as well as in Peruvian balsam (Bonner 1965). It should be noted that the set of coumarins and their derivatives in the secretion of essential oil cavities is variable in different organs, and depends on the phase of development. the composition of coumarins in different oil reservoirs of the same organ is also variable.

Although secretory products are stored in ducts and cavities, they are parts a of dynamic system which is changed during plant development. This is well seen by the example of Aztec marigold, *Tagetes erecta.* Using high-performance liquid chromatography (HPLC) with identification by mass-spectroscopy, the content of such ducts was studied (Russin et al. 1988) The secretion consists of three components: piperitone, piperitenone, and indole (Fig. 28), whose amount is changed during plant development. The presence of indole in the secretion of oil ducts in marigold was demonstrated earlier by Bicchi et al. (1985), who believed that essential oils with substances dissolved in them have an allelopatic effect and suppress the germination of seeds of other plant species, thus increasing the competitive potential for essential oil-bearing plants. According to Russin et al. (1988), prior to the stage of flowering, the quantity of indole and piperitone is increased, followed by a decrease, and increasing again at the final stage of flowering. At the same time, the amount of piperitone decreased in the process of development, most likely due to the conversion of unsaturated piperitone into the more saturated piperitenone.

α-, β-, γ-Asarones, as well as new compound of acorenone (Fig. 29), were found in the essential oil and oleoresin filling oil ducts of calamus, *Acorus calamus*, roots (Lander and Schreiner 1990). All these chemicals have a characteristic smell which is toxic for insects and snakes.

The biological significance of secretions of oil ducts is not yet known. Essential oil components are thought to have an allelopathic action and depress the seed germination of other plant species, increasing the seed competitiveness

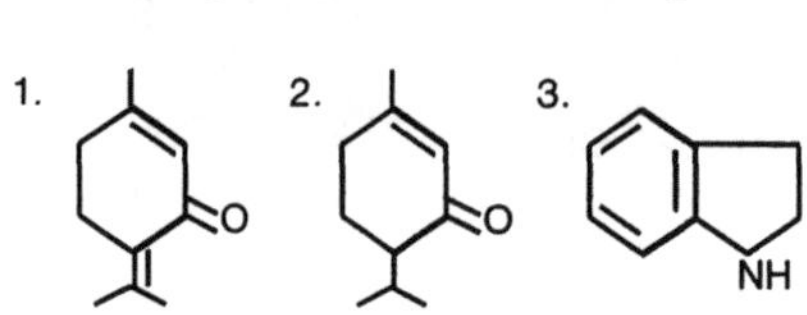

Fig. 28, 1–3. Components of the oil ducts secretion of marigold, *Tagetes erecta.* Piperitenone (**1**), piperitone (**2**), indole (**3**)

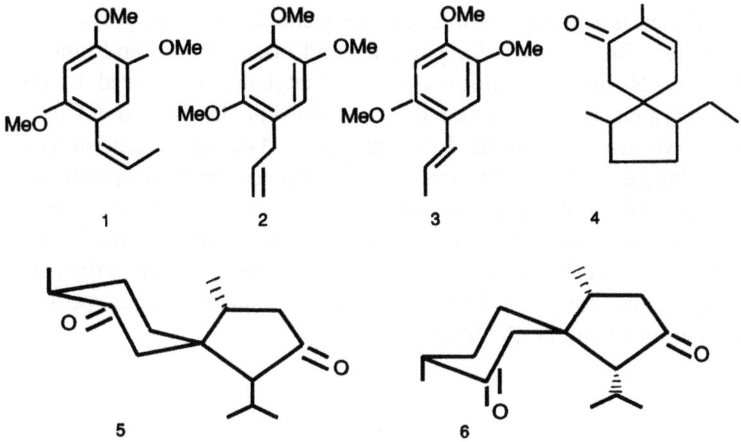

Fig. 29, 1–6. Asarones in essential oils and oleoresins. β-Asarone (**1**), acorenone (**2**), γ-asarone (**3**), α-asarone (**4**), acorone (**5**), isoacorone (**6**)

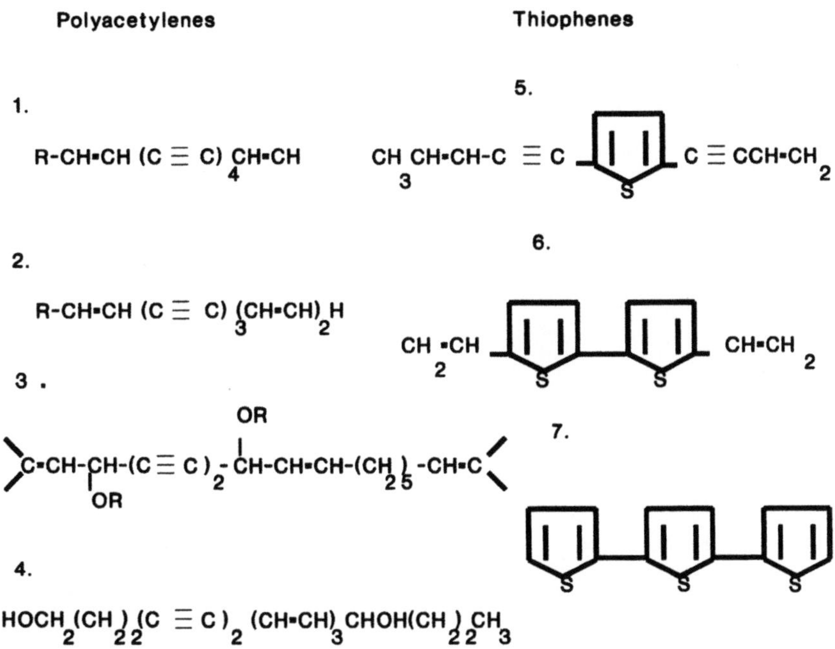

Fig. 30, 1–7. Main structures of polyacetylenes and thiophenes of plant secretory oil ducts and reservoirs. Tetraindienes (**1**) and triindienes (**2**) from *Dahlia tubulata* (Asteraceae); heptadeca-1,9-(*Z*),16-trien-4,6-diin-3,8-diol (**3**) from *Artemisia borealis* (Asteraceae); cicutotoxin (**4**) from *Cicuta virosa* (*Umbelliferae*); one, two and three heterocyclic thiophenes (**5–7**) from *Tagetes patula* (Asteraceae)

of the volatile-oil-bearing plants. Moreover, acetylene derivatives, polyacetylenes and thiophenes (Fig. 30), toxic for parasites and insects, are often found in secretory oil ducts. Noticeable amounts of polyacetylenes are stored in the secretory polyacetylenic reservoirs and ducts of flowers (Lersten and Curtis 1989) and vegetative organs of plants from the family Asteraceae (Christensen and Lam 1990; Wang et al. 1990). Among these substances are some specific for the taxon. For example, polyacetylene cicutoxin has been isolated from root secretory reservoirs of *Cicuta virosa* L. belonging to the family Umbelliferae (Anet et al. 1953). Sulfur derivatives of acetylene thiophenes peculiar to the root secretions of secretory ducts of French marigold, *Tagetes patula*, possess insecticidal properties (Jente et al. 1988).

CHAPTER 3

External Secretion

External secretion is accomplished by specialized secretory cells which are located on the surface of the aboveground parts of the plant (stem, leaves, bud scales, flowers). Among them are glandular hairs or trichomes, glands such as salt glands and glands of carnivorous plants, nectaries, osmophores, and hydathodes. Glandular hairs and glands perform the biosynthesis of many substances such as essential oils, terpenoids, flavonoids, fatty acids, slimes, etc., and often excrete their mixtures (Wiermann 1981). Besides, they can release different aqueous solutions. The morphology of glandular cells depends on the products which they accumulate (Vasilyev 1977; Wiermann 1981).

3.1 Guttation

Guttation, a release of water droplets on the surface of plant leaves, is observed at relatively high air humidity and good water supply. It is believed that guttation is due to excessive water supply of aboveground parts of plants under conditions of strongly decreased transpiration. However, the observations showed that guttation under natural conditions is a complex phenomenon which depends ambiguously on environmental conditions (Henkel 1976; Zlobina 1987). In lady's mantle, *Alchemilla vulgaris* L., and some other plants, the guttation begins after insolation of leaves by direct sunlight and is preceded by a strong jump in transpiration. As a result, the amount of water vapor in intracellular spaces increases, the water is condensed, and then released in the form of droplets. During the course of the day, water deficit increases, water evacuation in the form of vapors is enhanced, and guttation stops (Henkel 1976). Similar observations were obtained for greater burdock, *Arctium lappa* L., and coltsfoot, *Tussilago farfara* L., as well as lateral sprouts of shady horsetail, *Equisetum pratense* Ehrh. It was found under laboratory conditions on wheat seedlings that there is a positive correlation between the increase in soil humidity and light intensity, on the one hand, and between the increase in transpiration and guttation, on the other. Thus, the initial phase of guttation may be associated with an increase in transpiration rather than its reduction. Guttation can be active and passive (Esau 1977; Vasilyev 1977, etc.). Passive release of the guttation liquid is due to root pressure and stops when the root system is removed. Water supplied by roots enters through the ends of the tracheids of conducting bundles to the specialized tissue, the epithem, which consists of parenchymal cells lacking in chloroplasts. Through the epithem the water penetrates the induviate tissue epidermis and effluxes via hydathodes (Fig. 31). Hydathodes resemble the usual stomata, but unlike them, are unable to open and close (Fahn 1979). In some plants there is no epithem, and water moves to these modified stomata through the mesophyll (Esau 1965).

Some plants exhibit active guttation which is supported by the epithem actively secreting water (Esau 1977). As an example, we present an anatomic description of hydathodes in *Potentilla palustris* (Fig. 32) made by Curtis and Lersten (1986). The hydathodes of this plant are placed on leaf teeth and actively secrete water. On each tooth there are 50 and more water pores which are similar in size to normal stomata (14–22 μm long). Unlike stomata, cells surrounding the water pore have a dense cytoplasm and evenly thickened walls. Three conducting bundles end near the hydathode. There are two or three layers of cells of the epithem with a dense cytoplasm and chloroplasts between the xylem and epidermis. The intercellular spaces are small and numerous. A typical hydathode occurs along the leaf margin, but there are three families Crassulaceae (Rosales), Moraceae, Urticaceae in which hydathodes are distributed over the entire leaf surface (Rost 1969; Fahn 1979; Lersten and Curtis 1991). The latter have been called "laminar hydathodes" (Belin-Depoux 1983).

Fertilization needs liquid for its initiation and guttation takes part in the process. Guttation of water in the fern *Asplenium trichomanes* and musci *Bryum pseudotriquetrum* induced by the action potential, which is an electrical stimulus generated in response to the irritation at fertilization (Sinyukhin 1973). When the action potential is blocked, guttation does not take place. Guttation caused by the action potential is also an active process.

Hydathodes may differentiate into glandular trichomes (Esau 1977), whose structure is similar to that of salt glands. As is observed during active guttation,

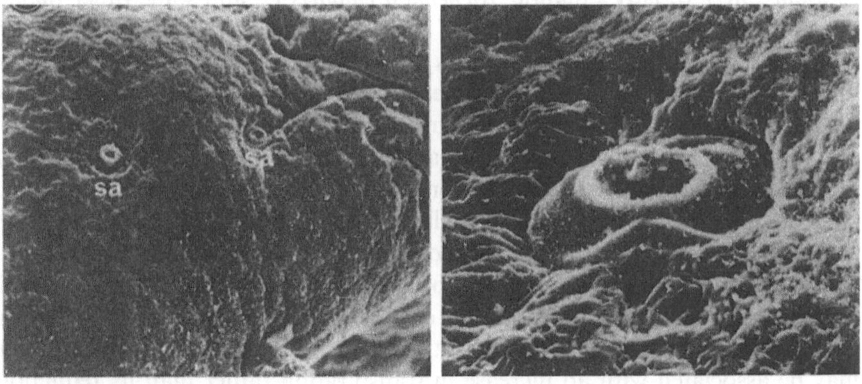

Fig. 31. Scanning electron microscopy micrographs of hydathodes on leaf surface of *Begonia coccinea* ×200 (Sa) and ×1000. (After Brouillet et al. 1987)

Fig. 32, 1–3. The leaf hydathodes of *Potentilla palustris* L (**1,2**) and *Physocarpus opulifolius* (**3**) **1** Adaxial view of guttating leaf of *Potentilla palustris*. **2** Scanning electron microscopy (SEM) of adaxial tooth tip surface showing hydathode pad with sunken water pores (*arrowheads*), line scale 50 μm. Opened hydathodes are seen *above*. **3** Adaxial leaf surface of *Physocarpus opulifolius* showing marginal tooth arrow (*left*), SEM of marginal tooth surface where line represents 50 μm (*middle*) and cleared hydathode with focus on tracheary elements of bundle ending (*right*). Water pores appear as *light circles* superimposed on tracheary elements. (After Lersten and Curtis 1982; Curtis and Lersten 1986)

the removal of the root system in this case does not prevent the release of water solutions. However, if epidermal cells are killed, guttation stops. There is no evidence (Lauter and Munns 1986) that water may be released through glandular hairs. For instance, chick pea, *Cicer arietinum*, grown in a greenhouse under optimal conditions, evolves through glandular hairs an exudate that contains ions of chlorine and hydrogen, malate, and oxalate. Lauter and Munns (1986) believe that exudation plays a role in maintaining the temperature balance. The mechanism of water elimination by active hydathodes and glandular hairs is poorly understood. As a possible explanation, the authors proposed filtration under pressure. Cytological analysis indicated that secretion by hydathodes of *Ficus diversifolia* (Belin-Depoux 1987) is of the ecrine type.

Guttation is inherent in a great variety of plants. According to Fahn (1979), guttation occurs in 350 genera belonging to 115 families. The most intensive guttation is observed in tropic and subtropic zones. One leaf of *Colocasia antiquorum* can evolve 10–100 ml of water per night. Each leaf tip of this plant forms 200 droplets per minute (Kramer 1956). The trichomes of *Monarda fistulosa* exude up to 450 μm^3 of water per day (Heinrich 1973). In the zone of temperate climate, intensive guttation is observed in grasses. Grodzinskii (1965) reported that guttation is often observed in 30 genera and 54 species of grasses, and the process is very intensive.

The composition of guttation moisture varies from almost pure water to a diluted solution of organic and inorganic salts. Analysis of the guttation sap showed that the most abundant cation in different plant species is Ca^{2+}, though K^+, Mg^{2+} and other elements have also been identified (Fahn 1979). The guttation water from leaves of vegetable pumpkin, *Cucurbita pepo*, was found to contain 0.9 mM Ca^{2+}, 0.5 mM K^+, 0.7 mM Na^+, 0.2 mM $H_2PO_4^-$, and 0.4 mM NO_3^- (Clarkson 1974). The total concentration of salts is generally somewhat lower than in the xylem sap of the same plants, indicating the removal of the substances from the xylem sap during its movements through the epithem.

Organic compounds were also found in the guttation liquid (Scheffer et al. 1965). Guttation water collected by decalicified filters for 10 days after the appearance of shoots in six cultivated and two species of wild wheat, *Triticum dicoccoides*, *Aegilopus cylindrica*, was found to contain 15 amino acids, 9 simple sugars, and some nonidentified compounds (amines, organic acids, oligosaccharides, peptides). The qualitative composition and the amounts of evolved substances varied depending on the cultivar and species. The gutta drops from the surface of rice, *Oryza sativa*, seedlings grown on nutrient solution were shown to contain peptides of different molecular mass and the amino acid glutamine (Horiguchi 1987). Presumably, the evolved nitrogen-containing substances are formed by degradation of reserve proteins present in the endosperm. In the form of peptides they are transported along the xylem and enter the secreta of hydathodes, although their main function is to serve as a nitrogen source for developing seedlings. According to Perrin and Karlin (1966), the guttation drops collected from the surface of coleoptiles and leaves of *Zea mays* and *Solanum lycopersicum* contain some vitamins.

The physiological role of guttation is probably in liberating the plant from excessive water and salts. It is also likely that guttation is necessary for thermoregulation. However, it should be noted that salts evolved by guttation, when accumulated and concentrated on the leaf surface, may be harmful for the plant and induce necrosis of tissues.

3.2 Salt Glands and Secretion of Inorganic Salts

Salt glands are inherent in many plant species adapted to living on saline soils, in particular in the representatives of Plumbaginaceae, Tamaricaceae, Verbenaceae, Gramineae, and other families. A great quantity of salts which is constantly released on the surface of the plant leaves, is then blown away by wind or washed off by rain. In *Avicennia marina* (Fig. 33), salt forms a crust on the leaf surface. The shape of salt crystals in this plant depends on the weather. In cold and windless nights, large lamellar crystals are formed, on windy and dewy days the crystals are small and prism-shaped.

Calcium oxalate crystals are often met in secretory trichomes, especially in head cells at the time of resin secretion. Early stages of crystal deposition take place within a small membrane-bound vacuole (Dell and McComb 1978a). Calcium carbonate as crystals is observed in cystolith trichomes of *Cannabis* and glandular hairs of *Nicotiana*. Crystalline calcite and magnesium carbonate are secreted by chalk glands of *Plumbago capensis* (Sakai 1974).

Salt glands originate from the epidermis and are located on the surface of both sides of the leaf and young stems. Generally, they are embedded in the epidermis. More rarely, they rise above the epidermal cells as in leaves of *Cressa cretica* or the spear-leaved orache, *Atriplex halimus*. In Fig. 34 are shown the photos of salt glandular hairs presented in the work of Weiglin and Winter (1988), who studied the structure of trichomes in *Cressa cretica*. The number of salt glands per 1 mm^2 varies between 500 and 1000, depending on the species. In Siberian sea-lavender, *Limonium gmelini*, there are 700 glands per 1 mm^2 and they can release about 1 mm^3 of solution, i.e., approx. 0.05 g of salt per h (Satcliffe 1962).

The anatomy of gland cells varies depending on the genus. Salt glands were described for developed leaves of *Limonium platyphyllum* (Stepanova 1983a,b; Vasilyev and Stepanova 1990) and *Atriplex* (Osmond et al. 1969; Buvat 1989 and others). They have a pedicel and a bulb-like globular head (Fig. 35). The stalk, which connects the gland with the surface of the leaf, is filled with a dense cytoplasm where cell organelles and vesicles are located. The great part of the bulb-like cell is occupied by a large central vacuole in which salts concentrate, and a thin layer of near-wall cytoplasm containing chloroplasts, mitochondria, Golgi bodies, numerous vesicles, and endoplasmic reticulum. The cell wall between the pedicel and the bulb-like cell is crossed by numerous plasmodesmata, which form a symplastic connection between cells. The structure of bulb-

A

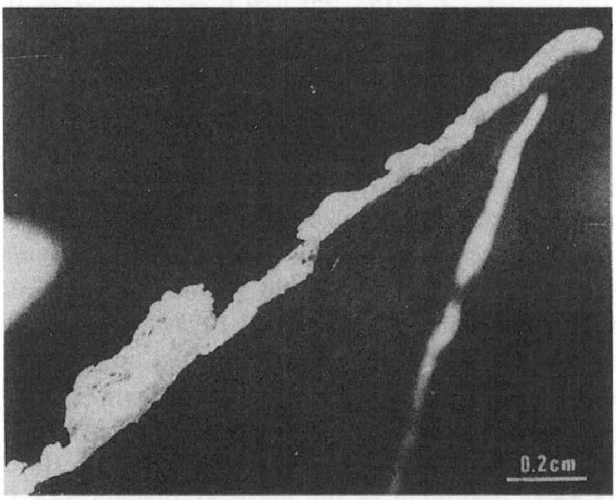

B

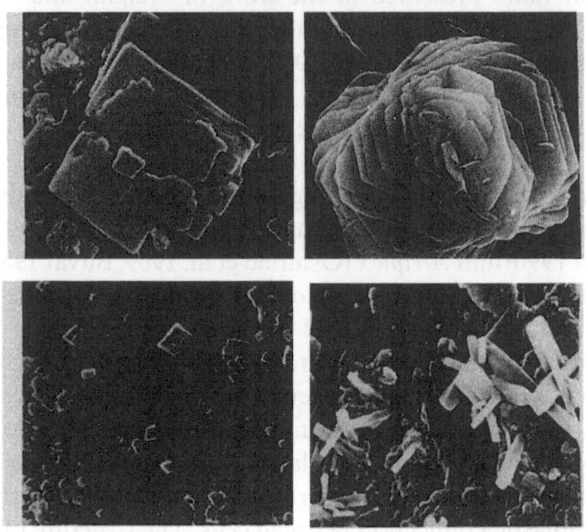

C

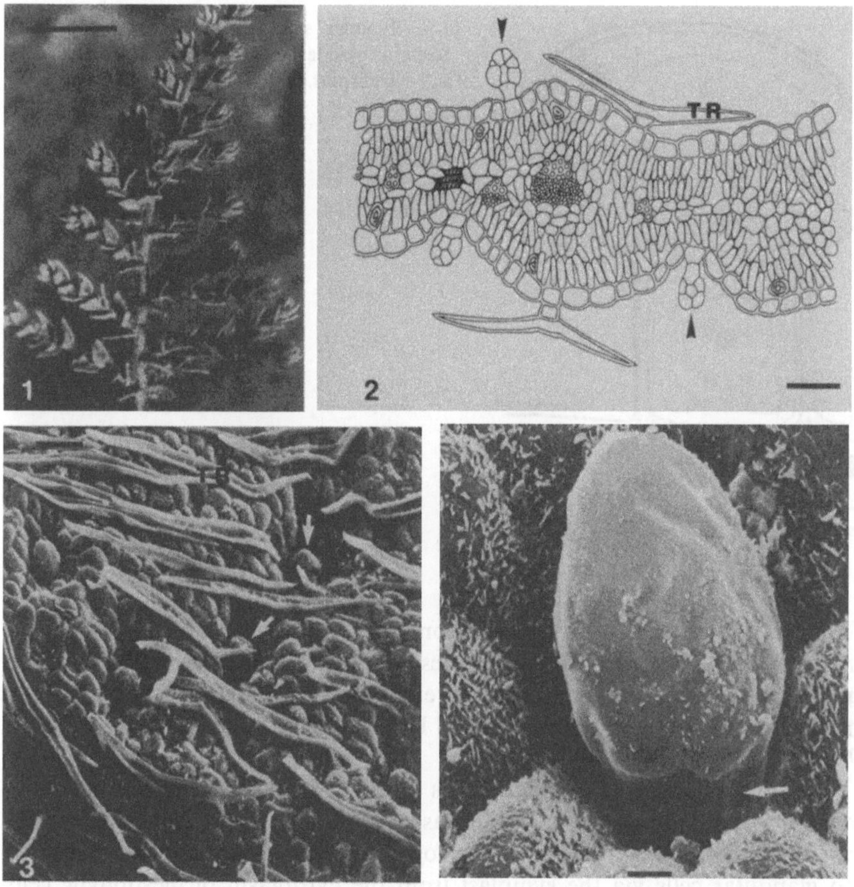

Fig. 34, 1–4. Trichomes of *Cressa cretica* **1** Branch of *Cressa cretica*. *Bar* 0.5 cm. **2** Schematic drawing of leaf cross-section with dead trichomes (*TR*) and glandular trichomes (*arrowheads*). *Bar* 50 μm. **3** Adaxial leaf surface with divided dead trichomes (*TR*) and glandular trichomes (*arrows*). Scanning electron micrograph *marker bar* 50 μm. **4** Glandular trichome of *Cressa cretica*, consisting of the stalk (*arrow*) and the pluricellular glandular head. Epidermis covered with wax crystals. Scanning electron micrograph *marker bar* 10 μm. (After Weiglin and Winter 1988)

like hairs of *Atriplex nummularia* is similar to that of *Atriplex halimus*; the size of the head, however, is very large, up to 100–200 μm (Uchiyama and Sugimura 1985). On the contrary, the salt gland of *Cynodon* has an expanded emidermal basal cell and a small apical cell (Thomson and Healey 1984). Salt glands of *Avicennia* have one basal and eight to ten radially arranged secretory cells

Fig. 33, A–C. Salt excretion by salt glands of *Avicennia marina*. Leaf covered by salt crust, especially along the margins (**A–B**). Salt crystals (**C**) are varied in size and form. Pure large crystals formed under undisturbed conditions (cold night without wind and dew), and small impure crystals under disturbed conditions (dewy or windy night). (After Ish-Shalom-Gordon and Dubinsky 1990)

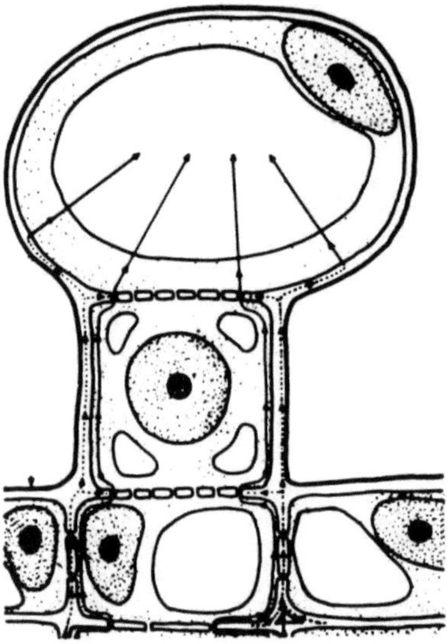

Fig. 35. Bladder salt hair of *Atriplex halimus* L. *1* Stalk; *2* bladder cell: *arrows* indicate pathways of active transport. (After Fahn 1979 and Buvat 1989)

(Thomson and Healey 1984). In cell contents, along with the common ultrastructural features, there are differences in the degree of vacuolization, the amount and structure of endoplasmic reticulum, and the activity of the Golgi apparatus (Vasilyev 1977; Thomson and Healey 1984; Uchiyama and Sugimura 1985).

The salts migrate into glandular cells together with the transpiration stream from the xylem endings of conducting vascular bundles. Their movement occurs predominantly through the apoplast. However, some ions may be translocated into glandular cells via the symplast from the cytoplasm of mesophylic cells through plasmodesmata. Then from the cytoplasm of the bladder-like cell of the gland, ions pass through the plasmalemma by the mechanism of active transport (active ecrine secretion) into extraplasmic (free space) to be further excreted through the pores in the cuticle (Thomson and Healey 1984). Another mode of salt excretion is granulocrine secretion (Vasilyev 1977). This is evidenced by the electron-microscopic studies (Platt-Aloia et al. 1983), indicating that in the halophyte *Tamarix aphyllia*, the ions to be secreted are accumulated in microvacuoles and then, as their inclusions penetrate via exocytosis through plasmalemma out onto the surface of the glands. In the bladder-like cell, salts are also concentrated in a large central vacuole, where they pass from the cytoplasm through the inner membrane, the tonoplast. This penetration through the tonoplast occurs by active mechanisms. For this process, ATP is supplied by the mitochondria of the bubble-like cell. Energy can also be delivered by chloroplasts, since they are present in the cells of the stalk and head cells (Osmond et al. 1969). According to data of other authors (Lüttge and Higinbotham 1979), these plastids are photosynthetically inactive and thus cannot participate in the energy supply of cells. Accumulation of salts in the bladder-like cells persists until the cell is ruptured and the salts are released to the surface of the leaf (Thomson and Healey 1984; Uchiyama and Sugimura 1985). In this case, the secretion is supposed to be of the holocrine type.

According to Ish-Shalom-Gordon and Dubinsky (1990), the excretion of salts by salt glands in the salt-tolerant *Avicennia marina* (family Verbenaceae) occurs by two modes, i.e., merocrine and holocrine secretions (Fig. 36). In the former case (Fig. 36, I), the salt solution is mixed with pectins in the subcellular space from where it passes through the pores of the cuticle, forming a balloon-like vesicle which grows in volume till the bubble is ruptured. In the latter case (Fig. 36, II), the salt solution accumulates in the subcellular space, creating a high pressure on the cuticle which is ruptured releasing this solution. The first mechanism is regular in plants while the second is only occasionally found. The sequence of the principle steps of secretion is shown in Fig. 36.

According to Stepanova's observations (1983b), the vacuolar membrane of the salt gland cells of *Limonium platyphyllum* leaf is in close contact with the plasmalemma, and probably a passive release of ions into the cell wall (free space) occurs through this contact rather than via exocytosis. Yet it remains unclear by which mechanism the plasmalemma-tonoplast barrier is overcome by ions. The glands of *Diplachne fusca* have pores through which the salts are excreted outside. On saline soil, the glands excrete Cl^-, Na^+ as well as K^+, Ca^{2+}, and Mg^{2+} to the surface of the leaf (Breauv et al. 1983). Being an active physiological process, the secretion depends on the temperature and the presence of oxygen, and is suppressed by metabolic inhibitors (Arisz et al. 1955; Gorham 1987). The composition and concentration of secreted salts vary according to the composition and concentration of the solution surrounding the

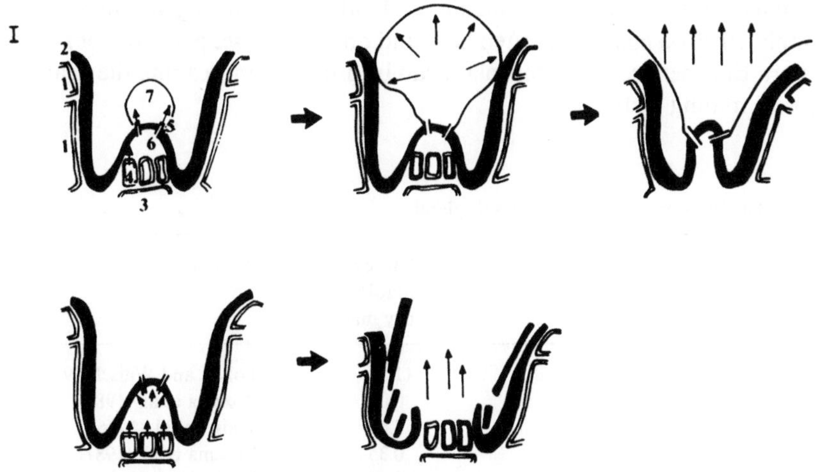

Fig. 36, I,II. Scheme of two modes of salt secretion from salt glands of *Avicennia marina*. **I** Merocrine secretion. **II** Holocrine secretion. *1* Epidermal cell; *2* cuticle; *3* stalk cell; *4* secretory cell; *5* pores in cuticle; *6* subcellular space; *7* vesicle. (After Ish-Shalom-Gordon and Dubinsky 1990)

roots. The concentration of salts in the secretion product is usually by two orders of magnitude higher than in the xylem sap and is rarely the same.

The secretion product was found to involve different cations such as Na^+, K^+, Mg^{2+}, Ca^{2+}, and anions such as Cl^-, NO_3^-, HCO_3^-, SO_4^{2-} as well as organic substances (Thomson 1975). The secretion of salts is most likely selective, and not all ions are easily released. Wieneke et al. (1987) were able to show that at an equal quantitative salt composition of a solution in which cuttings of *Leptochloa fusca* L. leaves were placed, their secreta contained more Na^+ ions than K^+ ions and more Cl^- than Na^+ ions. In the presence of Ca^{2+} ions, the secretion of Cl^- decreased, a greater quantity being observed in the tissue of the leaf, while the secretion of Na^+ and K^+ ions was slightly stimulated.

The intensity of salt secretion was very high (Table 3). For example, the weight of the fluid elaborated by the glands of a *Limonium latifolium* leaf per day is half the weight of the leaf per se (Arisz et al. 1955). The rate of NaCl secretion by the salt glands of *Cynodon* plants was 2.14×10^{-6} nmol per day per gland, or 5.25×10^{-9} nmol $s^{-1} m^{-2}$ (Oross et al. 1985).

The rate data for other plants are summarized in Table 3. The ability of glands to excrete salts depends on the peculiarities of the species: it is the highest in the inhabitants of mangrove thickets as well as of dry, arid, and saline regions. At low salt concentration in the surrounding solution, the salt glands secrete almost pure water and thus function as typical hydathodes (Esau 1965; Vasilyev 1977). It should be noted that some investigators do not separate salt glands into a special group of secretory structures but consider them as water glands, hydathodes, having similar structure and function.

The secretory function of salt glands is considered to be an adaptation to excess salinization allowing the plant to maintain the salt balance. Owing to this peculiarity, some representatives of the Plumbaginaceae tolerate the presence of high concentrations of heavy metals and other harmful agents in soils. For example, *Armeria halleri* can often be found on old slag heaps of copper mines in Western Europe. Probably, this plant can be used for landscaping the near-mine heaps in mining districts.

Table 3. The rates of salt excretion by salt glands

Species	External salinity, mol m^{-3}	Salt excretion, μmol kg^{-1} of dry mass s^{-1}	Reference
Aeluropus litoralis	300	11.5	Pollak and Waisel (1979)
Armeria maritima	200	0.09	Rozema et al. (1981)
Avicennia marina	700	3.49	Waisel et al. (1986)
Glaux maritana	200	0.35	Rozema et al. (1981)
Leptochloa fusca	250	0.45	Gorham (1987)
Limonium vulgare	200	0.94	Rozema et al. (1981)
Spartina anglica	200	0.96	Rozema et al. (1981)

3.3 Secretion of Nectar

Nectar is elaborated by specialized structures called nectaries which are found on flowers (floral nectaries) and vegetating organs (extrafloral nectaries). Floral nectaries may occupy different positions on the flower and have different shapes. Ring-like, disk, and other nectaries occur at the base of stamens, below the androecium, at the base of the ovary, between the androecium and the gynoecium, and sometimes as glandular multicellular hairs on petals. The flowers of the Australian bush *Thryptomene calycina* (Myrtaceae) contain two types of secretory tissue (Fig. 37) (Beardsell et al. 1989): the nectaries of the floral tube which secrete nectar rich in fructose and galactose, possibly through numerous stomata in the epidermis, while another type of anther glands containing lipids and probably phenols releases nectar through the pore at the end of the gland, where it is mixed with the pollen, thus serving as a food source for insects. This double secretion of nutritious material is necessary in order to attract as many various insect pollinators as possible. In addition to the nectaries of the floral tube and anthers, there are also dispersed oil glands in the subepidermal layer. The exterior of the ovary is also covered by dispersed oil glands (Beardsell et al. 1989). A good example of poisoning and functioning of nectaries is provided by *Asclepias syriaca* (Asclepiadaceae), whose nectaries are located in special cavities (sacs) of the pistil stigma (Kevan et al. 1989). The nectar accumulated in these stigma chambers is a good medium for the pollen to develop, because it would

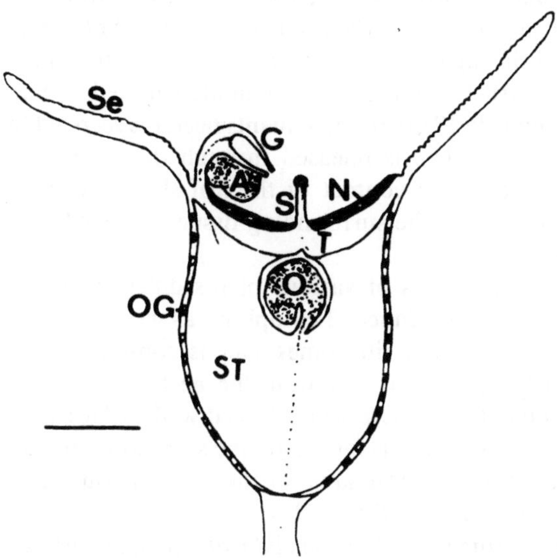

Fig. 37. Longitudinal section of a flower of *Thryptomene calycina* at anthesis *G* anther gland; *N* nectariferous tissue of the floral tube (*T*); *S* style. The ovule (*O*) is surrounded by spongy tissue (*ST*) and a subepidermal layer of scattered oil glands (*OG*). *Se* sepal; *A* anther. *Dotted lines* indicate vascular traces. Scale: 1 mm. (Beardsell et al. 1989)

not occur in a dry stigma. For the pollen to germinate, the nectar should contain 5–30% sugars essential for pollen development as well as for feeding the insect pollinators. On warm dry days the nectar is evaporated and the concentration of sugars is more than 40%, which may inhibit pollen germination. However, as soon as the flower is covered with a night moisture, the pollen resumes growth (Kevan et al. 1989). Extrafloral nectaries are found on the stems, leaves, stipules, and pedicels of flowers. The exterior of the nectary of the distal part of the calyx *Campsis radicans* is shown in Fig. 38. Most commonly, the nectaries develop from epidermal and subepidermal cells which become meristematic, and are divided many times, forming glands of different shape, the so-called nectary emergences. Nectaries of certain plants have no special morphological characteristics. In this case, epidermal and subepidermal cells of particular parts of the flower can perform the functions of nectar production and secretion. In such a manner, for example, nectar is excreted by the stamen bases and petals of *Thea*, and the ovary walls of *Jasminum*. The nectaries of monocotyledonous plants consist of only one layer of glandular epidermis.

The structure of nectaries is often varied and sometimes very complex. A schematic representation of extrafloral nectaries is given in Fig. 39. The basis of the nectary is the secretory tissue whose cells have a high metabolic activity. The secretory tissue underlies the epidermis, which is covered by the cuticle. Sometimes the epidermis may have stomata. Nectaries are closely connected with conducting vascular bundles providing a continuous inflow of organic substances which are needed for nectar production. Nutrients pass through the parenchyma cells into the conducting bundles, which may either penetrate into the secretory tissues or have their ending near the nectary surface. In both cases, the metabolites are transported through the parenchyma cells. The representatives of Marantaceae have nonstructured nectaries (nondifferentiated nectar-bearing tissue) (Kirchoff and Kennedy 1985). In these plants nectar is secreted by the nectar-bearing parenchyma and then is released through the pore. Such nectaries are often located at the site of contact of the leaf sheath with the petiole. They can be distinguished from the surrounding tissue by their lighter color and the absence of hairs.

During secretion, the cells of nectaries of various plants differ from each other (Vasilyev 1977). However, the abundance of cytoplasm as well as cellular organelles such as plastids, mitochondria, ribosomes, etc., is common to the ultrastructure of nectaries. In the period of secretion, the nectaries show an extension of the plasmalemma due to the increase of the cell wall surface. Such changes may result in secretion via exocytosis (granulocrine secretion). Besides, a large surface of plasmalemma provides more area for the membranous pump systems of the carriers (Lüttge and Higinbotham 1979).

The secretion of nectar varies from 0.5–2.1 (*Lavandula latifolia* and *Rosmarinus officinalis*) up to 8.3–10.9 μl per flower (*Lonicera periclymenum*) (Herrera 1985). Among 23 families studied, flowers of species belonging to the Labiatae, Cistaceae, and Fabaceae excreted the maximum amount of nectar. Highest sugar concentration (3.58–4.98 mg/flower per day) is observed for the family Scrophullariaceae (Herrera 1985).

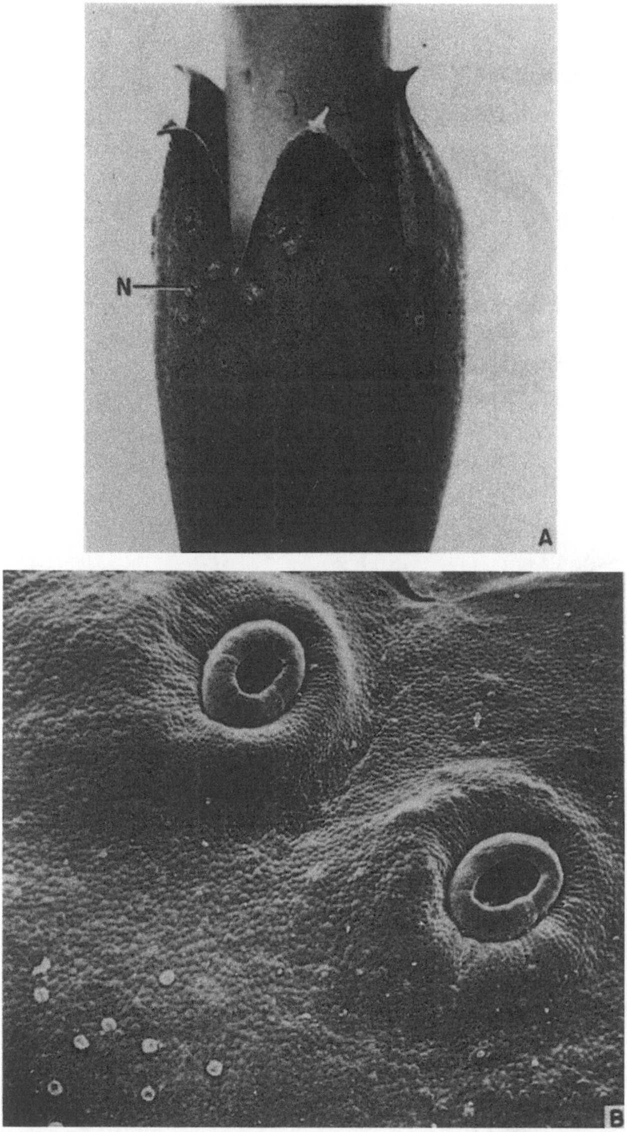

Fig. 38, A, B. Extrafloral nectaries (*N*) of *Campsis radicans* on the outside of the distal part of the calyx. **A** Calyx and part of the carolla tube × 5, **B** Scanning electron micrograph showing two nectaries × 50. (Fahn 1979)

Nectar produced by secretory structures is a complex combination of substances among which sugars such as sucrose, glucose, fructose, maltose, raffinose, etc. prevail. However, among the carbohydrates the first three sugars are predominant. According to sugar content, the nectars of angiosperms are

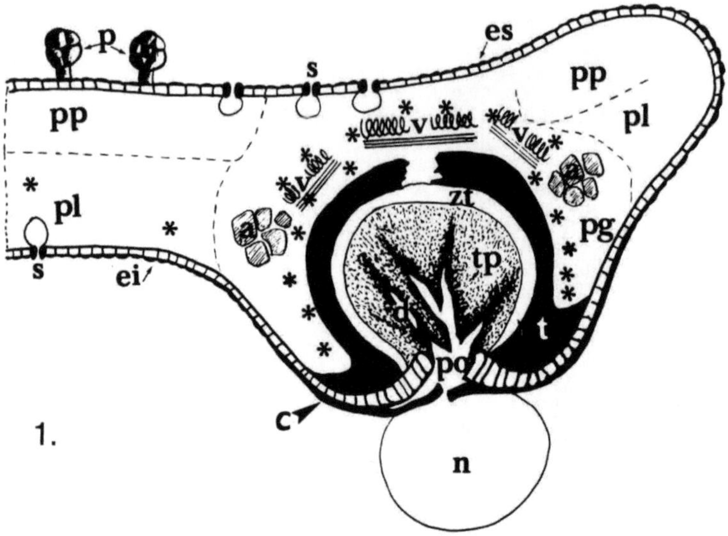

1.

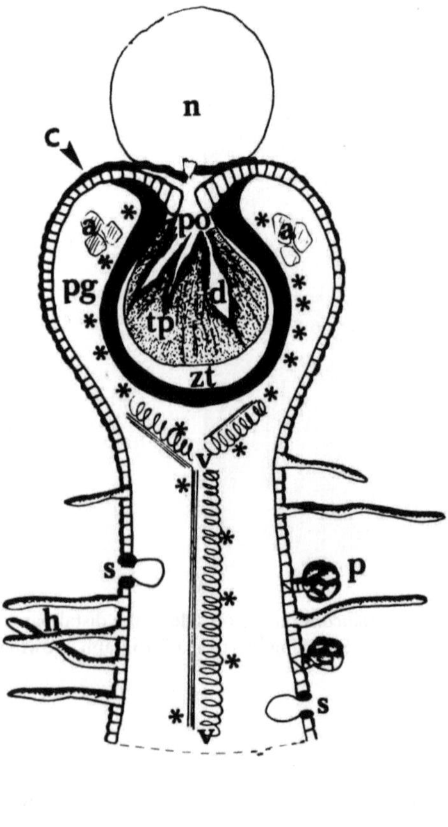

2.

Fig. 39. Extrafloral nectaries of *Ailanthus glandulosa* 1. Nectary of the middle part of leaf (×45); 2. Nectary of the stem stipule (×210); *e* Upper and lower epidermis; *c* cuticle; *s* stomata; *g* glandular hairs; *h* nonglandular hairs; *pp* palisade parenchyma; *pl* spongy parenchyma; *tp* tissue of prismatic nectary; *zt* zone of nectar transport; *v* veins; *p* pore; *n* nectar; *a* alkaloids; *T* tannins; *- crystals of calcium oxalate. (After Clair-Maczulajtys and Bory 1983) usually divided into three groups: nectars in which sucrose is prevalent, those in which glucose and fructose are predominant, and those with an equal amount of the three monosaccharides. Floral and extrafloral nectaries of *Sansevieria* have a varying disaccharide/monosaccharide ratio. The prevalence of sucrose in the former, and glucose and fructose in the latter can be observed (Tanowitz and Koehler 1986). The disaccharide/monosaccharide ratio varies in the floral nectar from 1 to 4.4, in the extrafloral nectar from 0.16 to 0.55. As a whole, sugars make up 12–87%of the dry matter of the secretion product (Fahn 1979). In addition, nectar involves minor quantities of amino acids. Moreover, it was found to contain all the proteinogenous amino acids, among which arginine, histidine, lysine, tryptophan, phenylalanine, methionine, threonine, leucine, isoleucine and valine are the most widespread. The nectar of Australian *Acacia* involves 18 amino acids, mainly glutamine, phenylalanine, and tyrosine, together reading 6.7 μmol ml^{-1} (Knox et al. 1986). Glutamine and asparagine are also frequent. Probably secretion of sugars and amino acids proceeds independently, as is inferred from the analysis of floral nectar of *Hibiscus rosa-cinensis*, *Malvaviscus arbreus*, *Abutilon pictum*, and *Fushsia speciose* (Gottsberger et al. 1989). Other components found in nectar are proteins, organic acids, vitamins, enzymes, and mineral salts (Baker and Baker 1975; Vasilyev 1977). Sometimes floral nectars also contain toxins, most often terpenoids, alkaloids, and glycosides. In other cases, the chemical nature is not always well understood. Nectar of *Ruta graveolens* contains a set of coumarinic derivatives (Andon and Denisova 1974). Glycoside arbutin determines the toxicity of *Ledum palustre* (family Ericaceae) nectar, and saponine the toxicity of *Aesculus hippocastanum* (Baker and Baker 1983). Alkaloids were found in the nectar of the pistil stigma of *Nicotiana sylvestris* (Baker and Baker 1983). In some cases, as in 20 Argentinian species belonging to the family Bromeliaceae, there are no or very few nitrogenous compounds in nectar, and saccharides predominate (Bernardello et al. 1991).

Using laser mass-spectroscopic microanalysis, Heinrich (1989) studied the cation content of nectar from 20 plant species. The predominant cation of nectar was K^+ (35–74% of total cation amount). Concentrations of other cations were less (in %, on the average): Na – 17.9; Ca – 12.8; Mg – 5.9; Al – 4.6; Fe – 1.2; Mn – 0.8. Ions play an essential role in the secretion of nectar (Kronestedt-Robards et al. 1989). The amount of Ca^{2+} and Mg^{2+} was found to increase at the late stages of secretion: from 18 to 26.46 μg g^{-1}(w/v) of nectar (by 47%) and from 8 to 12 μg g^{-1} (w/v) (by 56%), respectively. The amount of released K^+ showed no variations for different stages and that of Na^+ was insufficiently increased at the late stages from 30 to 38 μg g^{-1} nectar (Kronestedt-Robards et al. 1989). The pH of nectar may vary from strongly acidic, pH = 3, in *Silene alba* (family Caryophyllaceae) to strongly alkaline, pH 10, in *Viburnum costaricanum* (family Caprifoliaceae) (Baker and Baker 1983).

The composition of the secreted product may vary substantially, depending on the position and type of the nectary. This may be best shown by the example of *Vigna unguiculata*; in which nectaries are located on leaf stipules and flowers

(Pate et al. 1985):

	Fruit phloem sap	Extrafloral nectar	
		Of leaf stipules	Of inflorescence
Sucrose	180–210	127–244	100–271
Glucose	5.7–6.6	170–352	145–252
Fructose	3.4–4.3	170–354	145–252
Amino acids	16.6–33.2	0.05–0.1	12.5–17.0
Organic acids	9–60	10–57	36–188

Particularly great differences are found in the balance of sugars and other compounds. The nectaries of leaf stipules have a lower amino compounds/sugars ratio than the nectaries of flowers. The organic acid content is also much lower in these structures (Pate et al. 1985). The sugar composition in nectar, for example in *Acacia*, may vary substantially depending on the weather (Knox et al. 1986). The source of sugars for the production of nectar is, to a minor extent, the reserves of the secreting cell per se, while the bulk is provided by the flow of assimilates moving through the sieve tubes. This migration occurs essentially via the symplastic route, along the cytoplasm and plasmodesmata (Vasilyev 1977; Zauralov 1985). However, it is generally believed (Kursanov 1976) that sugar is transported through the apoplast, i.e., the intermicellar and interfibrillar spaces of cellular walls.

Experiments on the incorporation on ^{14}C-sugars into the plant have shown that the main transported sugar is sucrose, sometimes hydrolyzed to a minor extent at the final secretion stage (Nichol and Hall 1988). However, in representatives of Rosaceae, the main transported carbohydrate during nectar secretion is sorbitol (Bieleski and Redgwell 1980).

In nectaries, carbohydrate is converted into nectar, whose composition is species-dependent (Schnepf 1974; Zauralov 1985). The released nectar is not similar to phloem sap in composition. The comparison between phloem sap and nectar (Pate et al. 1985) shows that nectar contains many more sugars (500–650 $mg\,ml^{-1}$) and less nitrogen-containing substances such as amino acids (0.05–17.0 $mg\,ml^{-1}$) than phloem sap. Thus phloem sap is translocated to nectaries and turned into nectar. There occurs a selective retention of nitrogen-containing substances and metabolic conversion of sugars (Pate et al. 1985).

The hypothetical model of transport and secretion of nectar by secretory hairs of *Abutilon striatum* (family Malvaceae) is presented in Fig. 40. Each trichome is covered with epidermis and consists of the stalk cell (Fig. 40,5), intermediate cells transporting prenectar and head cell (Fig. 40,2). The first stage of nectar transport is the efflux of phloem sap from the sieve tubes. Further transport occurs either via the symplastic mode, i.e., through the plasmodesmata (Fig. 40, 7) or through the apoplastic compartment sealed on the outside by the impermeable cuticle of the cell wall and on the inside by the plasmalemma (Fig.

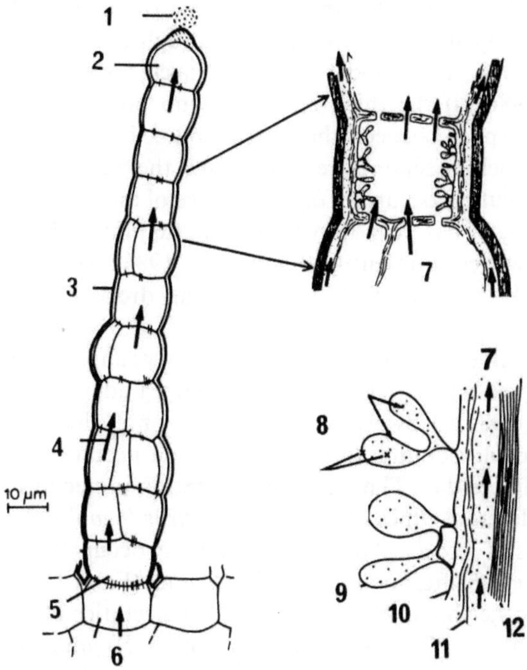

Fig. 40. Nectar transport and excretion by hairs of *Abutilon.* (After Robards and Stark 1988)

40, 7). Using electron microscopy and freeze-fracture replication, Robards and Stark (1988) were able to show the existence in the secreting hair cells of a stable intracellular membranous system called the secretory reticulum (Fig. 40, 9). This membranous system is closely connected with the plasmalemma. Filling the secretory reticulum with the secreted product results in an increase in hydrostatic pressure. It is believed that this is due to the hydrolysis of sucrose, the main transported sugar, into glucose and fructose. In turn, the increase in hydrostatic pressure within the compartment leads to the opening of "sphincters" connected with the cisternal space of the secretory reticulum to the outside of the plasmalemma. A pulse of nectar is forced into the apoplast. The nectar is eliminated from the tip cell of the nectary trichomes via the transient pores passing across the plasmalemma and the cell wall.

The mechanism of membranous transport during secretion is far from being understood. It is likely that the secreted sugars are transported in a phosphorylated form. The ecrine mode of secretion is suggested for nectar elimination from the apical cell of the trichome. For *Abutilon,* the co-transport of H^+-sugar is proposed (Robards and Stark 1988). Holocrine secretion can also take place. The nectar is first accumulated in the endoplasmic reticulum and then is transported in detached vesicles through the outer cytoplasmic membrane by exocytosis (Fahn 1979; Zauralov 1985; Robards and Stark 1988). Data are available indicating that nectar release by various glands occurs by different mechanisms. The nectar passes across the plasmalemma to be excreted through the cell wall, which is believed to be easily permeable. The surface of the cell wall has a cuticle. Dissolved sugars may penetrate the cuticle through the pores if there are any (Fahn 1979). In *Citrus*, *Ruta*, and *Vinca rosea*, nectar is excreted through the stomata (Jeffree 1986). If the thick cuticle has no pores, nectar accumulates in the subcuticular space between the cuticle and cell wall. After that, by the rupture of the cuticle, nectar is released onto the surface. In Australian species of *Acacia*, the cuticle occurring on the surface of the nectaries has no visible pores and breaks, but is very thin (less than 1.6 μm) and has discrete regions of high- and low-electron density (Knox et al. 1986). Possibly, nectar can diffuse through such semipermeable sites, forming drops on the surface of the gland.

The time of secretion may be different. Extrafloral nectaries of young trees of Australian *Acacia* may secrete the whole year round, especially under greenhouse conditions (Knox et al. 1986). However, adult trees strongly increase nectar excretion during autumn blossoming. In this period each nectary may release from 3.5 to 15 μm nectar per day. The maximum secretion of nectar in different plants is observed at different times of the day. In *Citrus limon* (Fahn 1979) and the Australian species of *Acacia* (Knox et al. 1986) this occurs in the morning, in *Tropaeolum majus* in the afternoon and in the evening (Fahn 1979). The rate of nectar secretion is rather high. Within 2-3 days, the nectaries of the banana, *Musa*, can release 1–2 ml $\sim$ of 32% sugar solution, or 25–50 μmol per flower. For comparison, we present the excretion rate of glucose from the store epidermal cells of *Allium* L. into a sugar-free solution. According to Lüttge and Higinbotham (1979), it is only 10 nmol/h cm^2. In *Vaccinum myrtillus* (Pahlke 1989), the rate of nectar release by the floral nectaries is from 1.34 to 0.68 $g\,m^{-2}$ per day, as calculated per unit of the area, and from 0.66 to 2.7 mg per day, depending on the age of the flower, as calculated per flower. The nectaries of flowers excrete from 6.45 to 3.2 $g\,m^{-2}$ nectar within the whole period (14 days) of blossoming. In flowering plants of *Acacia terminalis*, the maximum diurnal secretion of nectar is 15 μl per nectary. In rain, the amount of released nectar is increased, while on foggy days this is not the case (Knox et al. 1985). The concentration and composition of sugars in secreted nectar may vary during excretion. For example, at 20 °C, the nectaries of flowers of *Strelitzia reginae* release nectar at the rate of $\sim$5 $mg\,h^{-1}$ sugar within 7 days. At the beginning of secretion, the concentration of sugars reached 25%, but at the end of secretion less than 10% (Kronested-Robards et al. 1989). Moreover, at the first stage of secretion, glucose and fructose are eliminated to a lesser degree than sucrose, and the other way round at the end of the secretory process. Besides, the stabile level of K ($\sim$30 $\mu g\,g^{-1}$ fresh weight) and Na ($\sim$150 $\mu g\,g^{-1}$ fresh mass) and the increase in Ca ($\sim$ 47 $\mu g\,g^{-1}$ fresh mass) and Mg (56 $\mu g\,g^{-1}$ fresh weight) at the late stages of secretory process was noticed (Kronestedt-Robards et al. 1989). The intensity of nectar release was found to depend on the concentration of sugars flowing to the nectaries, as shown by model experiments with extrafloral nectaries of *Ricinus communis* (Nichol and Hall 1988). Isolated nectaries can secrete nectar only when the same sugars are present in the artificial medium as in the nectar (sucrose, fructose, glucose). Other sugars do not maintain this process. The nectary yield depends on the sugar concentration in the culture medium.

Both synthesis of nectar and its secretion, are active processes which require the energy of cellular metabolism. Energy in the form of ATP is provided by oxidative phosphorylation. This is greatly favored by the availability of a great number of mitochondria in secretory cells with well-developed crists. In this respect, the communication of Zauralov et al. (1986) is of interest, showing that the nectaries of *Cucurbita pepo* contain more phosphorus compounds than other plant tissues. Besides, the maximal phosphate content coincides with the period of enhanced release of nectar, which may be accounted for by the utilization of phosphates in the processes of respiration and sugar transport through the membranes. ATPase, the enzyme necessary for ATP splitting during active secretion, is localized in the epidermis, i.e., at the final stage of secretion (Nichol and Hall 1988). Being an energy-dependent process, the secretion is suppressed by 30–95% under low temperature, anaerobiosis, or respiration inhibitors such as 2,4-dinitrophenol, sodium fluoride, and phloridzin (Nichol and Hall 1988).

Generally, nectar release is considered to be an ecological adaptation of the plant to attract insect pollinators. Hence, nectar is an attractant for these insects. The nectar, secreted by the insectivorous trap plants, *Nepenthes*, *Cephalotus*, *Sarracenia*, *Darlingtonia*, is a most effective lure for certain insects (Joel 1986). Nectar may have also other functions. It stimulates germination of pollen on the stigma of the pistil (Kartashova 1965) and may contain repellents scaring away undesirable insects (Baker and Baker 1982). The function of plant protection against herbivorous insects, e.g., aphids, may also be performed by extrafloral nectaries. They attract ants which collect sweet nectar and scare away insect pests (Harborne 1982, 1988). Since nectaries are also located on leaves of nonfloral plants (e.g., ferns), their physiological role should be considered more closely (Kursanov 1976). It is suggested (Kursanov 1976) that the plant rids itself of excessive sugars through nectaries.

However, the significance of nectaries is not limited to the above functions. Isotope incorporation experiments show that the flower per se utilizes nectar, since it is absorbed by the cells of the flower. At the moment of pollination ^{14}C-labeled nectar sugars were observed in the stigma, and after fertilization in the region of ovules where the embryo begins to develop. Moreover, the nectar components also move into other parts of the plant, i.e., leaves and roots. It is believed that, like essential oil glands, nectaries can synthesize some hormonal substances needed for the processes of fertilization and development of the ovary and fruit. It is likely that such substances are the steroid hormones observed in higher plants.

The evolution of the nectaries is a complex and so far poorly studied problem. An attempt to analyze the problem is made by Jeffree (1986), who suggested that extrafloral nectaries appear in the plant kingdom only beginning from the Pteridophytes. Darwin first described them in *Pteridium aquilinum* in

1877 (cited from Jeffree 1986). In ferns, nectaries probably have the function of protection, mainly against ants and other Arthropoda. However, such a defensive mechanism is characteristic mostly for tropical species, while in temperate zone, ferns do not secrete nectar. This phenomenon is observed in many cereals, except *Campsis*, which, though growing in the temperate zone, originates from the tropical group of the family Bignoniaceae. Thus, extrafloral nectaries are the most ancient type. Transition to a floral type of nectaries began from the Gymnospermae, and is related to their appearance on reproductive organs in the genera *Ephedra* and *Welwitschia*. However, floral nectaries are not peculiar to anemophilous plants. Naturally, they are infrequent in Monocotyledoneae plants as compared to Dicotyledoneae. The appearance of floral nectaries is associated with cross-pollination.

3.4 Secretion of Polysaccharides

At a certain stage of development, the ability to secrete polysaccharides is characteristic of any plant cell, especially when the cell wall is forming (see Chap. 1.2.). However, in some cases, this function becomes basic in differentiated secretory cells. Both single cells, called the slime idioblasts, and the mucilage glands, which are multicellular external secretory structures, are attributed to them (Chap. 1.5). The slime glands are located in various organs: leaves, stems, roots, flower buds, and ripening fruits. Peripherical cells of the root cap which differ from others in.their structure can also be considered as secretory (Rougier 1981). They contain the developed Golgi apparatus with normal and greatly elongated hypertrophied cisternae from which small vesicles (from nonhypertrophied cisternae) and large spherical vesicles (from hypertrophied cisternae) ~ 0.3 μm have gone. Membranes of these vesicles are included in the plasmalemma, liberating their secretions into the free space of the cell, the secretion then passing through the cell wall.

Gel-like mucilage around the root cap is observed in many species. The dry weight of slime released by the root cap from 3-day-old seedlings of maize is about 0.1%. Mucilage, like any gel, possesses a weak water-retaining power (Guinel and McCully 1986). Under water deficit, mucilage is not eliminated through the cell wall and is accumulated in the periplasmic space. Guinel and McCully (1986) have shown that when such a root is dipped into a cuvette with water, a drop of mucilage appears very quickly (after 3–4 min).

The main part of slime polysaccharides consists of β-(1 $\rightarrow$ 4)-glucan chain with a number of free hydroxylic groups. The total molecular mass of mucilage from maize root cap has been estimated as 9×10^7 (Floyd and Ohlrogge 1971) or 2×10^6 daltons (Paull et al. 1975).

Monosaccharides have also been found in slime (Morre et al. 1967). Besides mono- and polysaccharides, mucilage contains ferulic acid (Harris and Hartley 1976) whose residues bind with hydrophylic sugars, forming diferulic bridges between the polymers (Smith and Hartley 1983). In mucilage of the root cap there are sugars such as fucose (up to 20%), arabinose, xylose, mannose, galactose, and glucose. The total amount of polysaccharides excreted by the root is from 15 to 20 μg per 100 ml of extract prepared by washing 200–300 root tops in 100 ml for 20 h (Chaboud and Rougier 1984). According to Paull et al. (1975), the rate of mucilage release by the root cap of maize, *Zea mays*, is 0.79 $\mu g\,h^{-1}$. The composition of mucilage excreted by roots of wheat, *Triticum*, and cowpea, *Vigna*, has been studied better than that of other plants (Moody et al. 1988). The water-soluble component of the slimes consisted of carbohydrates (respectively 95.5 and 97.5%) and proteins (5 and 3%). The main monosaccharides in wheat were arabinose, xylose, glucose, and galactose, and in cowpea arabinose, galactose, and glucose. In slime of both species there is a small quantity of fucose. Mucilage of cowpea root contained more uronic acids (11.5%) than mucilage of wheat root (4%). Arabinogalactanic protein was identified in root slimes of both species (Moody et al. 1988).

The release of slime is also observed in ferns. Young gametophytes of *Botrychium dissectum* (family Ophioglossaceae) produce a mucilage containing mucopolysaccharides which coated the external surface of proximal cells like a cover (Fig. 41). Histochemical study of the secretion has shown that in mucilage composition there are sugars with neighbor hydroxylic groups, carboxylated sugars, and a few sulfated sugars, proteins, and phenols. The chemical composition of fern slime is essentially different to that released by the root system. For instance, the mucilage of maize roots contains chains of β-1-1-glucan and hydrophylic chains of fucose and galacturonic acid (Miki et al. 1980) and most important in rice are sugars such as glucose (30 ~ 38%), fucose (20%), and

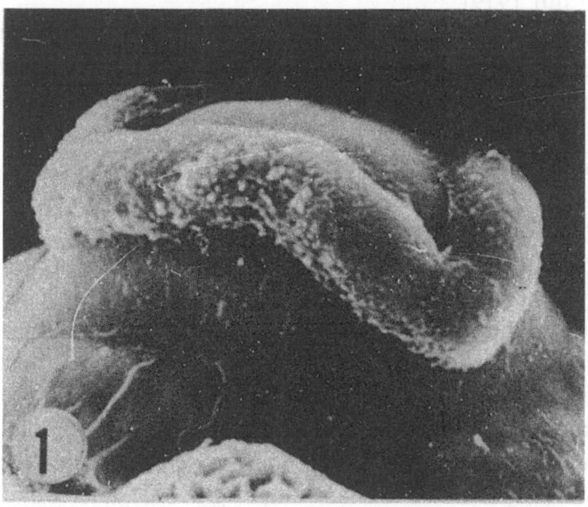

Fig. 41. Scanning electron microscopy of young gametophyte of fern *Botrychium dissectum* (Ophioglossaceae) with thin triangular ring of mucilage on proximal cell (×1460). (After Melan and Whittier 1989)

xylose (18%) (Chaboud and Rougier 1984). Mucilage of the fern *Botrychium* is similar to the root slime in physical features because fibrillar elements can be observed after freezing, but it includes neither the main component of maize root slime, β-1-1-glucan, nor sulfated sugars (Melan and Whittier 1989). The mucilage of fern protects gametophytes from drying and acts as a water reservoir. In contrast, Guinel and McCully (1986) doubt that it is a defense against root dehydration. It is more realistic that mucilage helps mycorrhizal fungi to infect the root systems, and also has chemotactic attractivity for symbiotic microorganisms (Rougier and Chaboud 1985). Slime formation and mechanisms of uncoating, i.e., undressing root cap cells of *Raphanus sativus* was reported by Barmicheva (1982). Excretion of mucilage via exocytosis leads to an accumulation of secretion out of the plasmalemma that induces rupture of the plasmodesmata and thrombosis of the plasmadesmata's channels by mucilage. Enzymes which induce a swelling of the middle lamina of the cell wall as well as changes in the cell wall structure, were transported with mucilage. This leads to the breaking of contact between cells and the isolation of cells from each other. Mucilage of the root cap facilitates root movement in soil.

Traps of carnivorous plants such as *Utricularia vulgaris* and *Drosera rotundifolia* have slime glands which release a trap mucilage that sticks an insect to the leaf surface (Fahn 1979; Fineran 1985; Joel 1986; Juniper 1986). Recently, it was successfully shown that mucilage contains toxic substances paralyzing insects. Among them is the alkaloid coniine, which is met in *Drosera.*

Slime-like excreta are also characteristic for the glands of the pistil stigma and serve to adhere pollen grains to the stigma (Cresti et al. 1986). Outer cells of the seed coat of *Lilium cydonia* also release a sticky slime which anchors seeds onto the soil surface and, by its own hygroscopity, improves the water regime and prevents dessication (Fahn 1979).

Thus, polysaccharides are excreted in slimes of plant cells located in the root cap, traps of carnivorous species, the stigma or pistil (see Sect. 3.5), the seed coat etc. Furthermore, these substances are released in the composition of gums eliminated when the plant is wounded (see Chap. 2.3.3). This phenomenon is observed when the trunks and branches of trees belonging to the genera *Cerasus, Prunus*, etc., are damaged.

The chemical composition of mucilage, consisting of high molecular mass polysaccharides, varies in plants. If mucilage from leaves of yew, *Taxus baccata, Taxus arborvitae*, and *Thuya occidentalis* undergo hydrolysis, galactose, rhamnose, glucose, arabinose, xylose, and small amounts (less than 5%) of uronic acid are found in the reaction mixture (Distelbarth and Kull 1985). Similar treatment of the root cap slime of *Zea mays* leads to the liberation of glucose, galactose, galacturonic acid, xylose, arabinose, fructose, and mannose (Fahn 1979). The trap slime released by glands of the carnivorous plants *Drosera, Drosophyllum, Pinguicula*, etc. contains acid polysaccharide consisting of gluconic acid, galactose, arabinose, xylose, rhamnose, and other substances (Fahn 1979). Besides polysaccharides, the stigma mucilage of pistil is rich in free monosaccharides, proteins, amino acids, phenols (Knox 1984), and lipids (Cresti et al. 1986). The chemical nature of the mucilage polysaccharides and the mode of their excretion are similar to the pectin material of the cell wall matrix.

Mucilage-forming cells have a structure particular to any secretory cells (see Chap. 3.3; 3.6; 3.7). Synthesis of polysaccharides takes place in their Golgi apparatus, where sugars flowing to secretory cells from other organs are used. The release of secreta occurs via Golgi vesicles according to the granulocrine type ((Vasilyev 1977; Fahn 1979). The kinetics of slime formation has been illustrated by Peterson and Vermeer (1984) for glands of *Mimulus tillingii* (family Scrophulariaceae). They established that Golgi vesicles occupy about 18.2% of total volume of every secreting cell and form 34.4 μm^4 mucilage per minute, i.e., nearly 530 vesicles can be anchored to the plasmalemma every minute. Slime can be accumulated between the plasmalemma and the cell wall or squeezed out of the cell wall into the environment. In root cap cells of *Zea mays*, mucilage is secreted with participation of glucose-6-phosphatase, which is localized in both the plasmalemma and the cell wall (Moore and McCleven 1985). The same enzyme also plays a role in the formation and secretion of slime by animal cells.

Mucilage which is released on the plant surface can play the role of a lubricant for root growth, and flood the wound surface on mechanical damage. It is supposed that the slime then plays an important role in plant defense against drying, and increases frost resistance. This conclusion is based on the following experiments and facts. Isolated and dried slimes from leaves of *Taxus baccata* and *Thuya occidentalis* are able to bind great amounts of water, up to 300% of dry plant mass (Distelbarth and Kull 1985). More mucilage is accumulated in frost-resistant plants. The amount of slime increases in the winter, and is minimal in summer (Distelbarth and Kull 1985).

Slime eliminated by the roots is able to bind heavy metals and thus prevent damages which can be induced by toxic metals: aluminum (Horst et al. 1982), lead, copper, cadmium, and zinc (Mench et al. 1987). However, mucilage also binds the necessary microelements such as iron and manganese (Jauregui and Reisenauer 1982) and therefore makes the mineral nutrition of plants more difficult.

The ability of slimes to bind mineral elements is also evident from the experiments of Mench et al. (1987) where a high molecular fraction > 1000 daltons, containing mainly protein and glycoproteins, was isolated from root exudate of *Zea mays*. This fraction was able to bind metals in the following sequence: Pb > Cu > Cd > Zn. It is possible that there are several mechanisms (modes) of the metals binding: absorption, formation of complexes, chelation (Mench et al. 1987).

One of the functions peculiar to the root cap is a sense of gravitation which depends on the slime secretion. Miller and Moore (1990) showed that non-efficient secretion of mucilage primary maize roots of an ageotropic cultivar causes a lack of sensibility to gravitation due to the dictyosomes lacking the ability to secrete slime. Otherwise, gravisensible maize cultivar secretory root cells released a large amount of slime, which fills the spaces between the root cap and the main body of the root. Using electron microscopy in the peripheric cells of the root cap, the authors demonstrated dictyosomes with vesicles which migrated to plasmalemma and fused with its membranes; in this way mucilage is excreted.

The biological role and main characteristics of mucilage secretion by roots are described in great detail in the review of Rougier and Chaboud (1985).

3.5 Secretion of Proteins

The protein secretion is widespread in the plant world. The secretion of enzymes by digestive glands of carnivorous plants is usually considered as the most brilliant example of this process. This accommodation for additional nutrition by food of animal origin is known for 400 plant species (Fahn 1979; Juniper et al. 1989). The structure of the plants of the family Utriculariaceae has been more closely studied than others (Fineran 1985), and one example is shown in Fig. 42. Tamarin (!986) described the structure of small trap bladders of *Utricularia vulgaris*. Nearly 60–70 bladders, as accommodations to entrapping and diges-

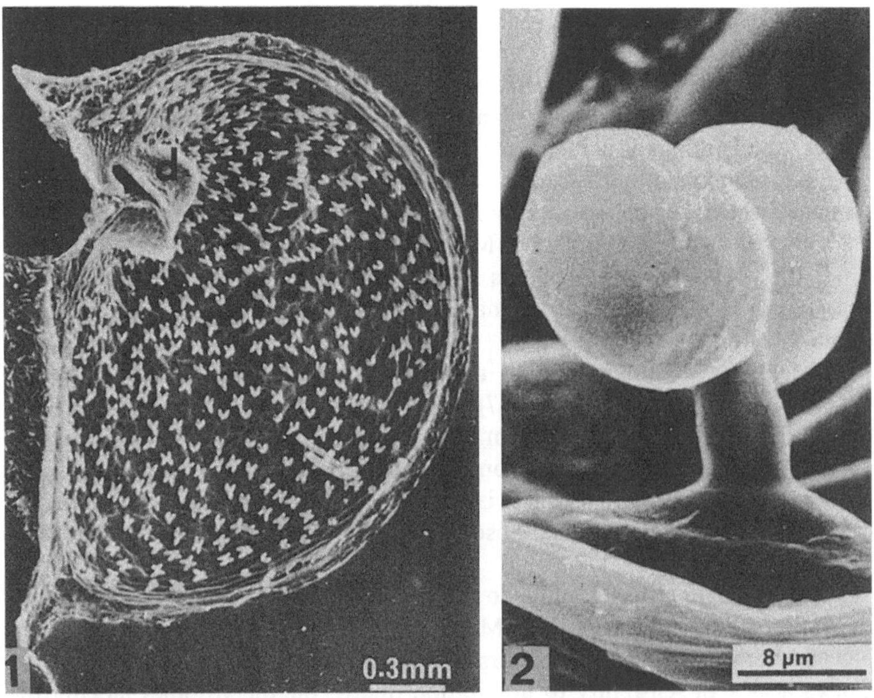

Fig. 42, 1,2. Digestive glands in trap of bladderwort *Utricularia monanthos* Hook. **1** Interior of sagittaly cut trap with numerous quadrifid glandular hairs on inside walls, × 30. **2** Electron microscopy of separate glandular hair in oblique view, × 1000. (After Fineran 1985)

tion, are located on leaves dissected to the milliform (filliform) parts. Within trap bladders there are approximately ten types of trichomes, among them sensory (tentacle) secreting, or glandular and soaking.

The morphology of digestive glands as a whole is similar to the essential oil-bearing secretory cells of plants (see Sect. 3.6), and their anatomy to the protein-secreting animal cells. They release ribonuclease, lipase, esterase, acid phosphatase, and large amounts of protease (Fahn 1979). Hydrolytic enzymes are eliminated on the leaf surface just as in insect settles on it. Before this, the enzymes are synthesized in the secretory cell or head of the digestive gland and then are mainly accumulated in the periplasmic space and vacuoles. The appearance of the insect in the trap is the signal triggering the excretion of the protein slime from the digestive glands. In this situation, a mechanical irritation is not so important (because rain does not induce excretion) as a chemical one, which appears when nitrogen-containing organic substances are released by the insects themselves. This signal is received by special sensory cells or sensor hairs (tentacles), where the action potential is generated and then transferred from cell to cell. A similar action potential induces tigmonastic movements: a twisting or closing leaf-trap in *Dionaea* and *Drosera* (Williams and Spanswick 1976). The mechanism of the event resembles the excitation of an action potential in animals, and consists of a depolarization of the membrane (with an amplitude up to 120 mV for *Dionea*) due to a change in ion permeability. The formation of the action potential lasts for 2–3 min and then it spreads along the sensory and triggering hairs. Carnivorous plants also have nectar-secreting glands which attract insects and are usually developed before the enzyme secretion (Bentrup 1979).

Electron microscopy of leaf glands of *Pinguicula vulgaris*, also belonging to the family Utriculariaceae, shows that cytological changes in secretory cells of digestive glands are observed when transition to synthesis and secretion of hydrolytic enzymes occurs (Vasilyev and Muravnik 1988). The volume of the granular endoplasmic reticulum is increased, whereas the density of the distribution of free ribosomes is not. The synthesis of hydrolases is presumed to occur in the membrane-bound ribosomes. Then the enzymes are transported to the storage compartments either via the Golgi apparatus or bypass. More details about the structure and function of glandular hairs connected with trap bladders can be found in the basic review of Fineran (1985).

As another example of cells secreting enzymes, cells of the aleurone layer of the caryopsis should be considered. When seeds are germinating, such cells synthesize hydrolytic enzyme amylase, protease, phosphatase, glycosidase, RNAase, etc. and then secrete them into the storage cells of the endosperm. Synthesis and secretion of α-amylase have been studied in greater detail. Histochemical experiments of Akazawa and Hara-Nishimura (1985) demonstrated that isoenzymes of α-amylase are formed not only in the aleurone layer but also in the scutellum, where they are identified at an earlier stage of development. However, only approximately 5–10% of the total amylase is synthesized in the scutellum. Using gold-labeled antibodies to α-amylase,

Gubler et al. (1986), showed that the Golgi apparatus participates in the secretion of enzymes from the aleurone cells. As a result of enzyme elimination into the endosperm, the stored substances of the seed are transformed into a form which can be assimilated. Soluble products of hydrolysis are taken up by the scutellum of the embryo and expended on its growth (Vasilyev 1977).

Phytohormones perhaps regulate the secretion of hydrolytic enzymes. Incubation of the isolated aleuronic layer with gibberellic acid (10^{-7}M) induces synthesis of hydrolytic enzymes and their secretion. It is presumed (Brown and Brodl 1988) that gibberellin can regulate synthesis and secretion of the proteins in barley seeds at the level of transcription. Abscisic acid is an antagonist of gibberellin. Heat shock stops the synthesis of secretory proteins and their secretion from the aleuronic layer via the unstacking of the lamellae of the endoplasmic reticulum, and a blockade of the association of the α-amylase mRNA in polysomes bound with the endoplasmic reticulum (Brown and Brodl 1988).

Besides these cases of protein secretion, the pistil stigma can also release proteins. The surface of the stigma is covered by glandular epidermal cells forming papillae (Fig. 43). Such cells are rich in cytoplasm and have a cuticle (Fig. 43). The top of the papilla is permeable for secreta, and the numerous ruptures in the cuticle perhaps also serve the release of the surface secretion and water in order to moisten the pollen (Reger 1989). Basing on the exudate composition, one distinguishes lipophylic and hydrophylic secretions. The first occurs according to the holocrine type as in *Forsythia* (family Oleaceae). The granulocrine type of secretion is more characteristic of elimination of hydrophylic excreta as in *Aptenia cordifolia*.

In spite of the variety in the chemical composition of stigma secretion among plant species, the major substances contained in it are proteins which may differ from the usual intracellular proteins. In any case, some essential distinctions for both types of protein mentioned have been shown in *Pennicetum americanum* using methods of electrofocusing and electrophoresis in polyacrylamide gel. The fraction of protein from the pellicula, a so-called protein film on the stigma surface, was a glycoprotein with molecular mass 180–200 kilodaltons. This major protein plays an important role in an early recognition of pollen, when the latter contacts with the pellicula. In a review by Dumas et al. (1988) the conceptions are developed according to which glandular cells of the pistil stigma have specialized receptors which are capable of recognizing pollen that is untrue to the same plant genotype. During interbreeding, the pistil stigma of angiosperms recognizes the pollen of other species and seizes it, maintaining the stability of the species. In contrast, during interspecific crossing, its own pollen is torn away, which compels the plant to the unrelated interbreeding characteristic of pollen self-incompatibility. Receptory proteins are S-glycoproteins, which are found only on the surface of the stigma, not in the style or vegetative tissue (Nasrallah et al. 1985). The molecular interaction with identical or complementary S-proteins located in pollen grains is perhaps, a basis for pollen recognition.

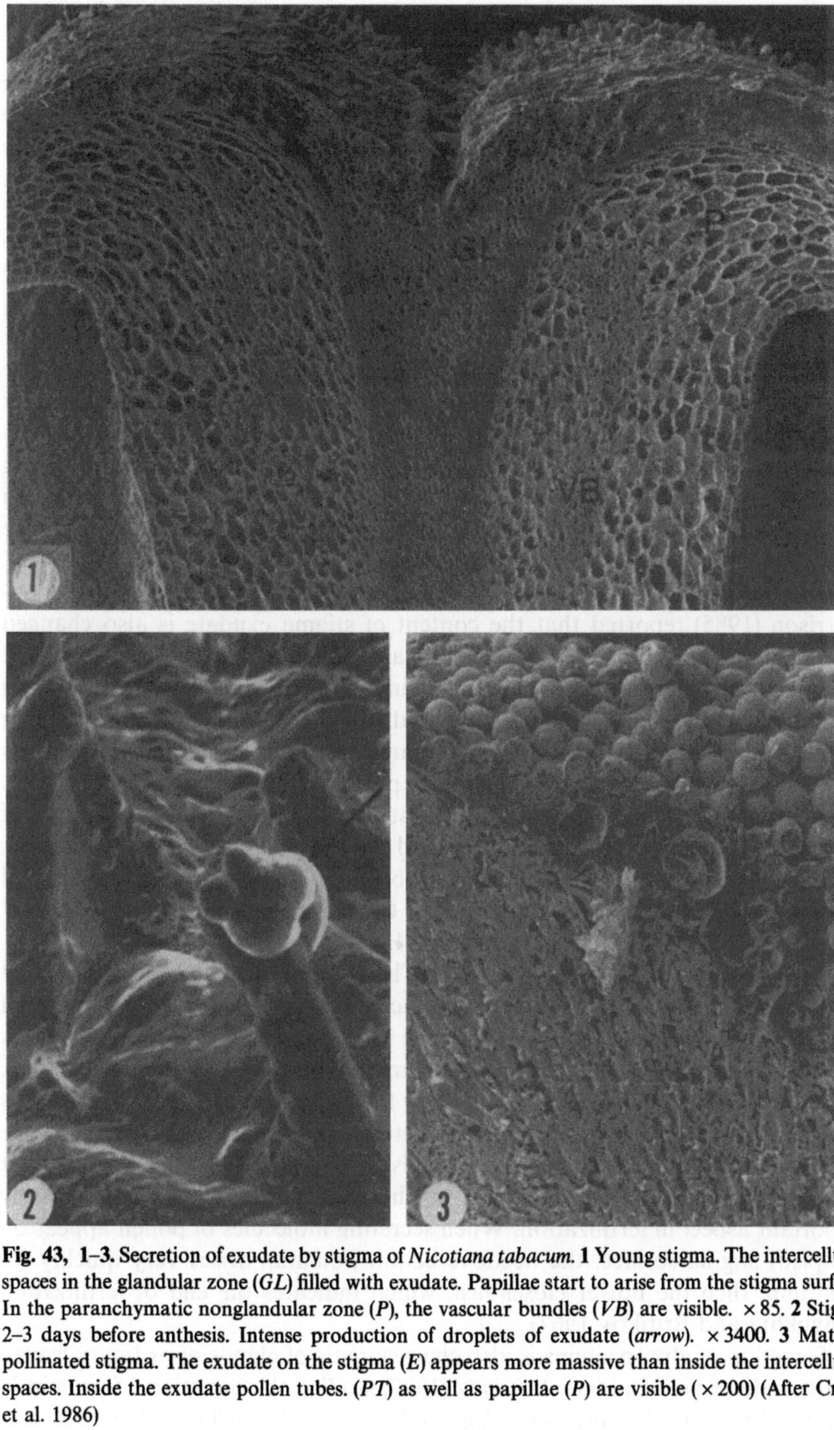

Fig. 43, 1–3. Secretion of exudate by stigma of *Nicotiana tabacum*. **1** Young stigma. The intercellular spaces in the glandular zone (*GL*) filled with exudate. Papillae start to arise from the stigma surface. In the paranchymatic nonglandular zone (*P*), the vascular bundles (*VB*) are visible. × 85. **2** Stigma 2–3 days before anthesis. Intense production of droplets of exudate (*arrow*). × 3400. **3** Mature, pollinated stigma. The exudate on the stigma (*E*) appears more massive than inside the intercellular spaces. Inside the exudate pollen tubes. (*PT*) as well as papillae (*P*) are visible (× 200) (After Cresti et al. 1986)

The exudate of the pistil stigma possesses extremely high biological activity. Here esterase (mainly cholinesterase) activity is observed (Bednarska and Tretyn 1989). Protein with features of esterase and glycoprotein with molecular mass 180–200 kdaltons have been found (Reger 1989) in the pistil of *Pennicetum americanum*. In the excreta of the pistil ATPases, and nucleases, the main components of the messenger system of cyclic AMP, have also been found. Sometimes, for instance in maize, the release of nuclease is observed in the first minutes of pollen contact with the liquid medium, and it induces complete degradation of exogenous DNA. Nucleases can diffuse not only from the pistil but also from pollen. In maize this sometimes occurs simultaneously from both the pistil stigma and pollen (Dumas et al. 1988).

Glandular tissues of the pistil in other cases release a secretion of more complicated composition. Besides proteins, in the stigma, exudates of *Citrus limon* lipids, polysaccharides, glycoproteins, amino acids, phenols, and alkaloids have been found (Tiezzi et al. 1983). Kandasamy (1986) showed that during anthesis of *Cyamopsis tetragonoloba* the amount of total soluble proteins in stigma exudate was 2–2.5-fold greater than in soluble saccharides. In the same plant, lipids, free fatty acids, amino acids, and phenols were also identified. Among the phenols, p-coumaric acid prevailed. Heslop-Harrison and Heslop-Harrison (1985) reported that the content of stigma exudate is also changed during stigma ripening. The mature stigma of *Oenothera organensis* is covered by a secretion which is an emulsion consisting of continual lipid phase (68–75%) and dispersal phase in the form of water and aqueous polysaccharide-containing globules (Heslop-Harrison and Heslop-Harrison 1985). The secretory products are rich in glucose and sucrose. The minor protein fraction has esterase activity. At the beginning of the development, the stigma has a short phase secretion of protein. Then secretion of lipids occurs and finally polysaccharides are released, after which the stigma becomes susceptible to pollen. The average rates of emission have been measured: one cell of the stigma surface releases 109 μm^3 secretion per hour (Heslop-Harrison and Heslop-Harrison 1985).

Pollen grains also excrete proteins. The hydration of the pollen when it contacts the papillae of the stigma, is accompanied by pollen excretion of various proteins during the first hour. The peak in protein release in *Nicotiana alata* is observed already after 15–20 min (Shpilevaya and Il'chenko 1984). Secretion of protein also continues during the growth of the pollen tubes. This process is important for interrelations between pollen and pistil. Due to the protein secretion, a specific micromedium is created around the pollen tube that promotes its growth. The emission of chemicals by pollen is an extremely important aspect in fertilization. When secreting molecules of pollen appear on the pistil stigma surface, the negative action potential arises very quickly for 3–17 min, then the flower closes fast, which indicates the end of fertilization (Sinyukhin and Britikov 1967).

Secretion of glycoproteins is also characteristic of glands of ordinary leaves, and is widespread in plants. Vasilyev (1977) studied the secretion of protein in glands of poplar, *Populus suaveolens* and *P. deltoides*, willow, *Salix* sp., and

Scopolia stramonifolia. Highly specialized structures were characteristic of the glandular cells of these plants. There are no large vacuoles in them, as a result of which the cytoplasm was seen as dense. The granular endoplasmic reticulum, and the Golgi apparatus were very developed. Glandular cells of the plants studied developed structures characteristic of the cells secreting proteins only at a certain stage of leaf development. Before that the same cells appeared terpenogenous. A change in secretory function of the glands from terpenoid secretion to protein secretion is presumed to occur (Vasilyev 1977).

Thus, protein secretion can have many functions, from digestive purposes in carnivorous plants and hydrolyzing seed storage substances, transforming them to a more suitable form for embryo nutrition, to participation in the recognition and orientation of pollen on the pistil stigma. It should be noted that the capacity to secrete proteins is extremely high in plants. Under some conditions, protein can be secreted by any cells. Experiments with tissue cultures have shown that high active peroxidase is released by cells of cowpea, *Vigna* sp., (Moreno et al. 1990), and peanut, *Arachis hypogaea* (Hystee and Turcon 1973), whereas cells of callus of tobacco, *Nicotiana tabacum*, excrete superoxide dismutase (Nicolova et al. 1990).

3.6 Secretion of Essential Oils

The glandular epidermis and various glandular hairs are attributed to the external secretory structures where synthesis and secretion of essential oils occur. Moreover, production of essential oil components can also take place in cells which do not belong to specialized structures.

Glandular hairs are located only on the above ground part of plants. Often they cover both the abaxial and adoxial sides of leaf. Figure 44 shows one such glandular hair on the tomato leaf at the moment of excretion of a drop of essential oil. The essential oil-bearing structures have been studied by electron scanning microscopy. Two types of glandular hairs of *Diplopeltis petiolaris* and *Newcastelia viscida* secrete the same terpenoids and have heads of spheric form seated on the stalk (Fig. 45a). Trichomes of *Newcastelia viscida* (Fig. 45b) are characterized by either branched (left) or spheric tops, which are connected with the epidermis by a short pedicel (right).

Essential oil-bearing glands vary in their structure. The glandular hair of peppermint, *Mentha piperita*, consists of a multicellular pedicel and a unicellular head. For sage, *Salvia glutinosa*, the stalk is, in contrast represented by one cell, while the head is multicellular (Fahn 1979). Glandular trichomes of *Artemisia* are formed by ten cells, two of which, being apical, have no chlorophyll and the remaining eight (four subapical, two pedicel, two basal cells) include chloroplasts (Ascensao and Pais 1985). The amount of essential oil-bearing glands located on the leaf is hereditary. When the leaf trichomes once arise their number is not changed throughout differentiation. However, some factors,

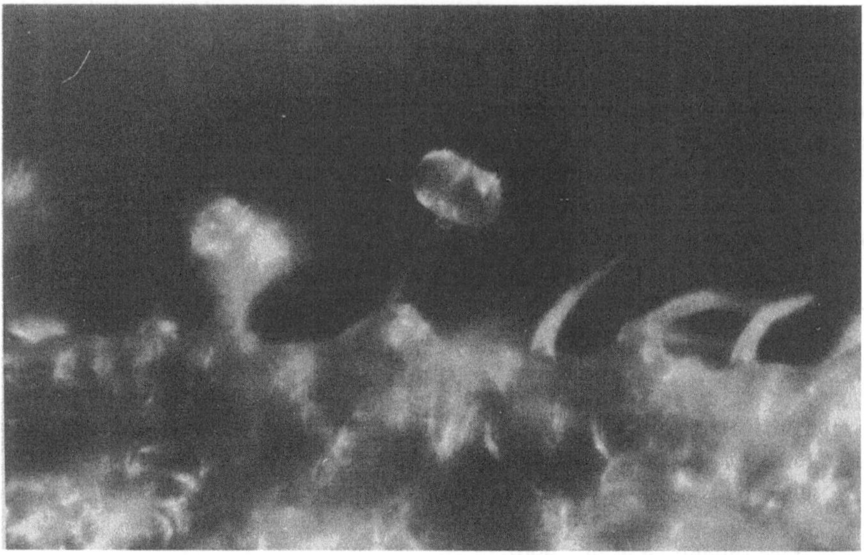

Fig. 44. Glandular hair on cell surface of *Lycopersicon esculentum* in the moment of secretion of the essential oil drop. (Photo courtesy Professor J. Lovett, Australia)

including the balance of hormones, influence trichome differentiation. In connection with this problem are the experiments of Bhaumic and Datta (1989), performed on shoots of *Mentha arvensis* treated with lanolin paste including hormones. It was found that indole acetic acid and kinetin ($1 \times 10^{-6} - 6 \times 10^{-6}$ M) did not influence the number of the glands, whereas naphthalene acetic acid (1×10^{-6} M), gibberellic (2×10^{-6} M) and 2,4-D (2,4-dichlorophenoxyacetic) 4×10^{-6} M) acids noticeably increased it. Moreover, Berger et al. (1990) showed that treating *Coleonema album* plants with growth regulators such as 2,4-D, p-chlorophenoxyacetic, or naphthalene acetic acids in small concentrations and kinetin in higher ones stimulates the formation of essential oil. The specialized glands of some plants which produce the essential oils that are characteristic of the flower aroma are called osmophores. They originate from different parts of the flower in the form of wings, celiates, or villuses (Esau 1965). Osmophores of *Scaphocepalum* (family Orchidaceae), a plant of the wet mountain forest, are themselves glandular pulvini located on the adaxial surface of the sepal tops (Pridgeon and Stern 1985). In flowers of *Restrepia*, functions of osmophores belong to the layer of epidermal cells with small excrescences–papillas (Pridgeon and Stern 1985). The same tissue parts are observed in osmophores of *Stanhopea* (Stern et al. 1987) (Fig. 46). Osmophores release the odor substances which attract the pollinators.

The structure of all glandular cells during the secretory period has similar features. They are rich in cytoplasm with strongly developed agranular endoplasmic reticulum which often becomes a dominating cell organelle. The

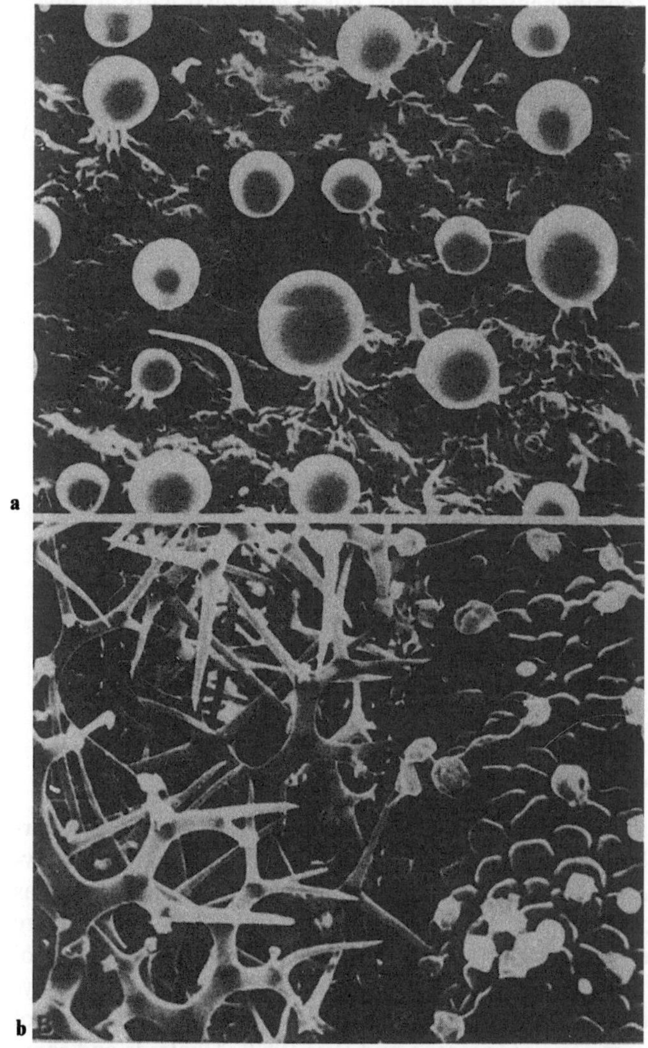

Fig. 45a,b. Scanning electron micrographs of stipitate leaf glandular hairs ($\times 250$), **a** Hairs of *Diplopeltis petiolaris* with secretion droplets, **b** *Newcastelia viscida*. Note the branched trichomes with scattered glandular tips on the left and short stalked glands on the *right*. Both types secrete the same array of terpenoid compounds. (Dell and McComb 1978a)

plastidic apparatus, consisting of a number of leucoplasts, is also well developed. Plastids are closely connected with the elements of the agranular endoplasmic reticulum, which forms some kind of sheath around the leucoplasts (Vasilyev 1977; Fahn 1979; Bernard-Dagan et al. 1982).

Essential oil-bearing secreta produced by glandular cells are a mixture of substances. Almost 1000 organic compounds have been isolated from essential oils. Some examples of components of essential oils in glandular hairs are given below:

Species	Prevailing component (%)	Reference
Calamintha methifolia	Piperitone oxide (62.9–67.6)	Hanlidou et al. (1991)
Coridothymus capitatus	Thymol (38.8)	Werker et al. (1985)
Majorana syriaca	p-Cymene (22.2), carvacrol (33)	–
Melissa officinalis	Geraniol (31.6), neral (19.5)	–
Mentha piperita	Menthone (35.21), Isomenthone (12.52)	Maffei et al. (1989)
Micromeria fruticosa	Pulegone (61.4)	Werker et al. (1985)
Phellodendron amurense	Myrcene (90)	Heinrich and Schultze (1985)
Rosmarinus officinalis	α-Pinene (17.0), eucalyptol (23.2)	Werker et al. (1985)
Salvia dominica	α-Thujone (24.8) linalool (15.3–24.6) Camphor (19)	–
S. fruticosa	Eucalyptol (23.5), camphor (17.7)	–
S. sclarea	Sesquiterpenes (70), linalool (32), lynalyl acetate (44)	–
Satureja thymbra	γ-Terpinene (50.9)	–

Among these are saturated and unsaturated aromatic terpenes, sesquiterpene mono- and bicyclic hydrocarbons and their oxygenated derivatives–alcohols, aldehydes, ketones, ethers, esters, acids, and lactones as well as heterocyclic substances (Bonner 1965). In some essential oils phenols are present. For example, oils from buds of *Populus nigra* contain a range of methylated flavonoid aglycones (sect. 3.7). In addition to sesquiterpenes, polyacetylenes are also observed in the oil composition of glandular structures of wormwood *Artemisia campestris* (Ascensao and Pais 1987). Alkaloids and some other components are also found. However, terpenes, mainly monoterpenes (C_{10}) and sesquiterpenes (C_{15}), are most characteristic for essential oils (Figs 47–49). Monoterpenes such as geraniol, limonene, camphor, and α-pinene present in volatile essential oils are included in important plant odor components, in fact, due to them the flower flavor of many plants is known. The determining matter of the aroma can be caused either by one individual compound in the flower glands or more often by a complicated mixture of chemicals. The subtle odor of linden, *Tilia*, comes from the aliphatic sesquiterpene alcohol farnesol (Denisova 1989). The perfume of rose is associated with primary alcohols such as geraniol, nerol, and citronellol, and the scent of lily of the valley *Convallaria* L. (Liliaceae) with the tertiary alcohol linalool (Harborne

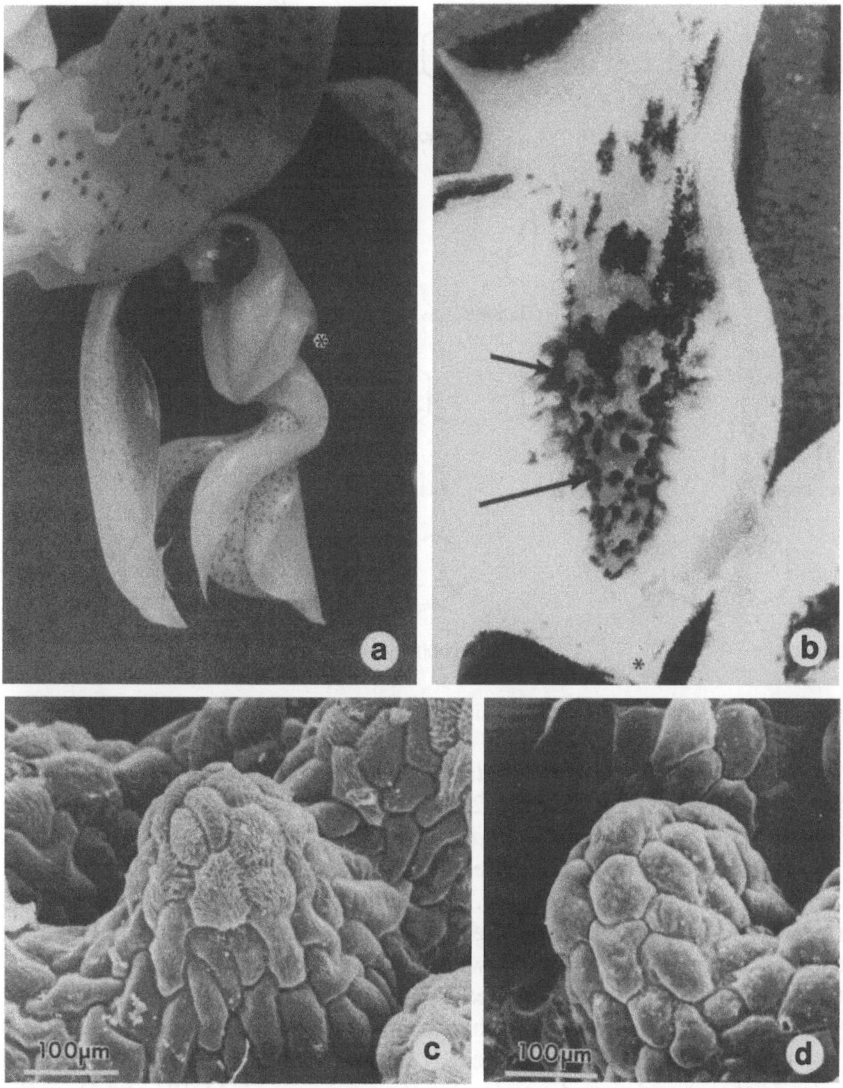

Fig. 46a–d. Osmophores of *Stanhopea* belonging to the Orchidaceae. **a** Flower of *Stanhopea oculata*. An *asterisk* at the tip of hypochile marks the osmophore. **b** Bisected hypochile of *S. oculata* showing the papillate osmophore surface (*arrows*). **c**, **d** Scanning electron micrographs of osmophores of *S. wardi*. (After Stern et al. 1987)

1988). The main source of the aroma is usually the glandular epidermis of the petals, but the fragrant compounds can be emitted by the stamens, nectaries, and many other glands as well. A normally developed trichome produces abundant secreta. Up to 0.07–0.1 μg of essential oil is contained in the head of a gland of *Mentha piperta*. The highest yield (18–43 ml m^{-2}) of essential oil is observed for

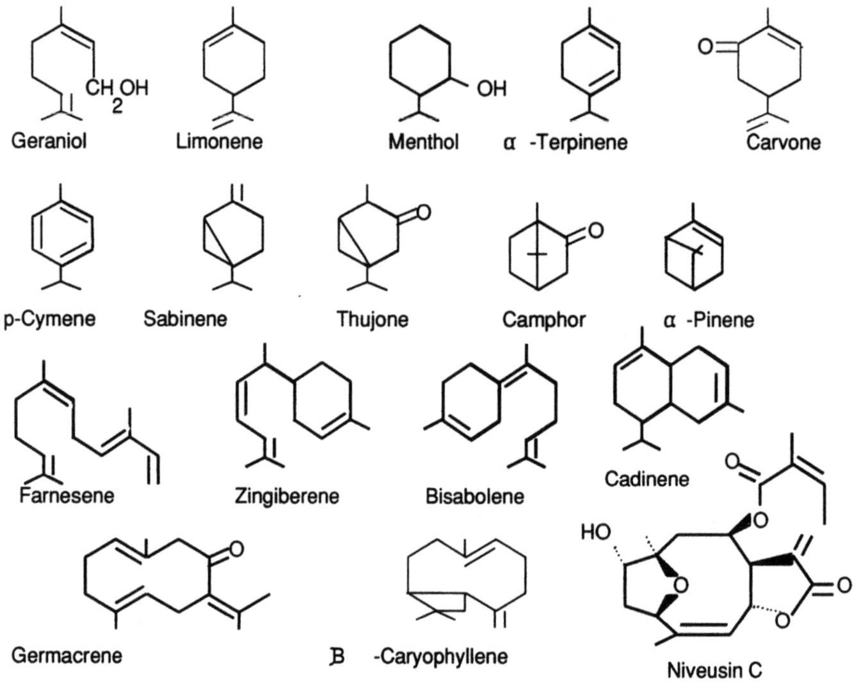

Fig. 47. Terpenes in secreta of glandular hairs

Satuteja thymbra, Micromeria fruticosa, Majorana syriaca, Salvia fruticosa, Rosmarinus officinalis, with only 2–14 ml m^{-2} for *Salvia officinalis, Melissa officinalis, and Coridothymus capitatus* (Werker et al. 1985). The oxygenated derivatives of monoterpenes are present in the composition of volatile oils and their study is of great importance. Karp et al. (1987) analyzed in detail the transformation of (+)-sabinene, the major olefin component of volatile oil from the leaves of sage, *Salvia officinalis*, wormwood, *Artemisia absinthium*, and tansy, *Tanacetum vulgare* into oxygenated products: ketones ( )-3-iso-thujone and (+)-3-thujone. It has been shown that the process of oxidation of (+)-sabinene to (+)-*cis*-sabinol etc. was catalyzed by the microsomal fraction of epidermal oil glands of sage and were dependent on the presence of NADPH and O_2 in the medium. Evidence has been given that epidermal oil glands of sage, wormwood, and tansy contain cytochrome P-450-dependent hydroxylase-oxygenase, highly specific to (+)-sabinene and transforming olefins to oxygenated products in the microsomal fraction. Sesquiterpenes play a lesser role in aroma formation, although this is the greatest group among the known terpenes. To the sesquiterpene belong some rare compounds, monocyclic germacrene, bicyclic caryophyllene, and others released from the yellow centaurea, *Centaurea solstitialis* (Buttery et al. 1986). Zingiberene, isolated from *Zingiber officinalis* (Lundgren et al. 1985) is also found in exudates of glandular trichomes of tomato, *Lycopersicon*, species (Lin et al. 1987). The concentration of zingiberene in young leaf of

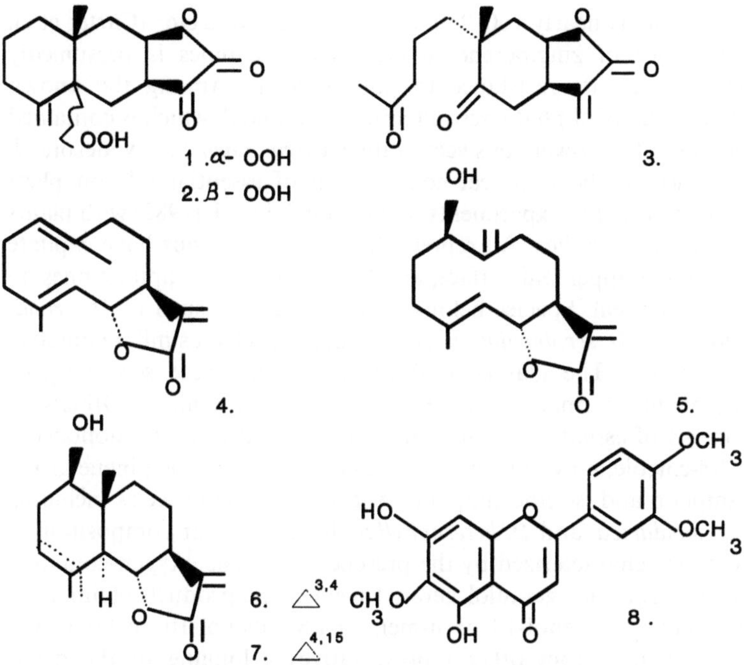

Fig. 48, 1–8. Sesquiterpene lactones in glandular hairs of *Artemisia umbelliformis*. **1** and **2** Hydroperoxyeudesmanolides. **3** 4,5-Secoeudesmanolide. **4–7** derivatives related to costunolide. **8** Eupatilin

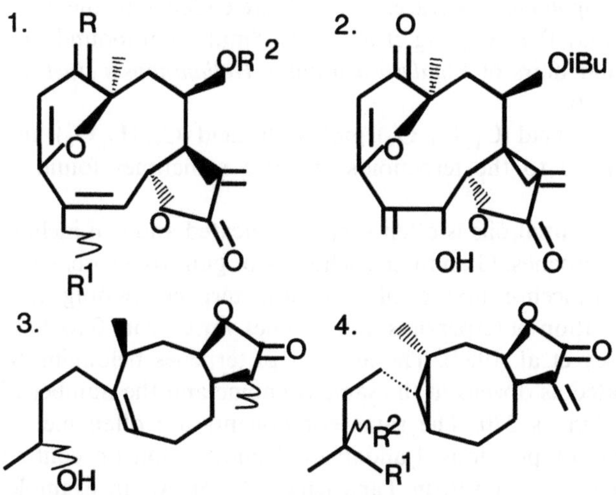

Fig. 49, 1–4. Sesquiterpene lactones from glandular hairs of three species, *Pappobolus*, *Viguiera ocutifolia*, *Helianthopsis stuebelii*; *H. sagasteguii*, *H. lehmannii*. (Asteraceae). **1** and **2** Heliangolides budlein A and isobudlein isobutyrate R′=OH, R_2=Ang. **3** and **4** Xanthanolides 11,13-dihydroxanthanolide and carabrone R′=R_2=O

Lycopersicon hirsutum is nearly 160–250 μg over an area of 2 cm^2 (Carter et al. 1989b). The presence of zingiberene in glandular trichomes is presumedly connected with the plants resistance to potato beetle. Among the known oxygenated derivatives of aliphatic sesquiterpenes, nerolidol, which is contained in the essential oil of the flowers of sweet orange, *Citrus sinensis*, is widespread.

The dependence of the chemical composition of essential oils on plant species was demonstrated in experiments by Dembitskii et al. (1982) with plants from the genus *Dracocephalum*. All representatives of this genus have capitate glands, mainly on the upper leaf surface, and both peltate and capitate ones on the lower side of the leaf. Species belonging to this genus, such as *D. grandiflorum*, *D. nutans*, and *D. scrobiculatum*, are characterized by a similar composition of the secretions. The glands of these species eliminate sesquiterpene hydrocarbons, mainly germacrene D (Fig. 47) and gratissimene. Otherwise, major components of essential oils in glands of *D. nodulosum* are monocylcic terpenes like 1,8-cineole, n-cymol, and bicyclic compounds of the pinene group like E-pinocamphon and isopinocamphon. Another group of species including *D. foetidum*, *D. moldavica*, and *D. heterophyllum* have a similar composition of glandular secretions characterized by the presence of acyclic oxygen-containin compounds such as geranial, geraniol, geranyl acetate, and products of limonene oxidation (n-mentha-1,8-dienol-10 or n-mentha-1,8- -dienol-10 acetate), that distinguish these plants from other representatives belonging to the genus *Dracocephalum*.

In glandular hairs of *Artemisia*, genera especially *Artemisia umbelliformis*, there are sesquiterpene lactones (Fig.48). New sesquiterpene lactones, such as furanohelioangolides and xanthanolides, are found in glandular trichomes of plants from the genus *Pappobolus* (Asteraceae), which are endemic to the Ande in Columbia, Ecuador, and Peru (Spring et al. 1991). Similar compounds are contained in the glandular hairs of *Viguiera acutifolia*, *Helianthopsis stuebelii*, and *H. sagasteguii* (Fig. 49).

Plant hormone: abscisic acid ($C_{15}H_{20}O_4$), gibberellic acid ($C_{19}H_{20}O_6$), and steroids (C_{28-30}) belonging to the terpenoids are also sometimes found in essential oils.

The composition of essential oils is often very complicated and can include some dozens of various terpenes. Glandular trichomes of gum rockrose, *Citus ladaniferus*, synthesize an incense mixture of 70 various terpenes (Königs and Gülz 1974). The concentration of terpenoids in trichomes varies from 0 to 20% of dry plant mass (Kelsey et al. 1984). The amount of terpenes belonging to various groups is estimated as dozens in the same trichome and the number of individuals can be as high as 270. The same components are often met in essential oils in different compositions. Fundamental information on content and quantity of essential oils is given in Tanasienko (1985). As an example, major components of essential oil extracted from flowers and pollen of five plant species are given in Table 4. The essential oils of *Lippia alba* and *Aloysia chamaedrifolia* differ in the unusual composition of their prevailing components. The essential oil of pollen is especially varied in terpenoid content. Essential oils

Table 4. Some components of essential oils from flowers and pollen

Plant	No. of identified components	Prevailing components	Reference
Rosa davurica (flowers)	152	Eugenol (23.8 %), benzyl alcohol (24 %), β-ionone (7 %), linalool (6.0 %), dihydro-β-ionone (3.3 %), benzaldehyde (2.5 %)	Ueyama et al. (1990b)
Lippia alba		Camphor (18.2),1,8-cineole (16.5 %), β-cubebene (6.9 %)	Dellacassa et al. (1990)
Aloysia maedrifolia		Globulol (15.6 %), spatulenol (9.3 %), γ-elemene (9.9 %)	–
Chimonanthus praecox (flowers)		Trans-ocimene, nonal, linalool, bornyl, acetate,3,7-dimethylocta-1,5-(E) triene-3-ol, δ-cadinene, benzyl alcohol, elemol. All these components constitute 32.53 % of the total essential oil	Ueyama et al. (1990a)
Vitis L. var. muscat (pollen)		Ethanol, isobutanol, isopentanol, pentanol, hexanol, *cis*-hexen-3-ol-l, isopropyl acetate, ethyl capronate, hexyl acetate, isobutyl capronate, hexyl capronate, benzyl acetate, acetate phenylethanol, acetic and valeric acids β-Ionone	Egorov and Egofarova (1971)
		Aldehydes: formaldehyde, acetaldehyde, propionaldehyde, isobutyraldehyde, hexylaldehyde, caprylaldehyde	–

are widespread, being found in 2000 plant species belonging to 60 families (Vasilyev 1977) and most often are met in representatives of the families Labiatae, Umbelliferae, Rutaceae, Myrtaceae, Cruciferae, Compositae, and Cupressaceae.

The composition of essential oils changes in ontogeny and depends on plant species, its contents differing in various plants: peak production is at the time of inflorescense, or sometimes at flower bud formation or even earlier, before the flower buds appear. Changes are observed not only in the quantity but also the quality of the evacuated oils. Their content also depends on soil, climate, season, humidity, and the age of the plant. In most cases, all parts of the plant contain oils of the same composition. However, there are cases where oils of various plant organs differ greatly in their components. The essential oil composition of the same organs, for instance leaves, can vary, depending on their position on the plant, in both the qualitative and quantitative ratio of their chemicals. The highest amount of essential oils, calculated by weight, is 0.075 % in the flower of *Rosa* sp., 0.084 % in petals of *Acacia farnesiana*, and 0.4–0.05 % in *Jasminum grandiflorum* (Fahn 1979).

Pathways of terpene biosynthesis for the production of essential oils are shown in Fig. 50, based on experiments with isolated trichomes, capable of forming monoterpenes (Dell and McComb 1978b). Terpenoids are formed from mevalonic acid synthesized from sugar transported to leucoplasts via phosphorylated intermediates; first, isopentenyl diphosphate, the so-called active "isoprene" (Fig. 50); then, via condensation of the isopentenyl diphosphate molecule, compounds, containing from two to some thousands of "isoprene" units, are formed. In the biosynthesis of secreted terpenoids, ATP is used as the energy source and donor of phosphate.

Besides terpenes themselves, essential oils of glandular hairs also include many of their alcohols, esters, and aldehydes. Sometimes these excreta contain rare components which nevertheless are the main components in the same plant species like esters of sugars and diol-diterpenes in exudates of glandular trichomes of tobacco, *Nicotiana tabacum* (Fig. 51). (Severson et al. 1985; Kandra and Wagner 1988).

Nonbenzoic aromatic compounds contain the condensated system from 5- and-7-carbon cycles named azulenes (Heilbronner 1959). Chamazulene (1,4-dimethyl-7-ethylazulene) was isolated from essential oils of wild camomile, *Matricaria chamomilla*, and prochamazulene from blossoms of the common yarrow, *Achillea millefolium* (Fig. 52). Moreover, by histochemical methods, it has been shown that prochamazulene is located in the glandular trichomes

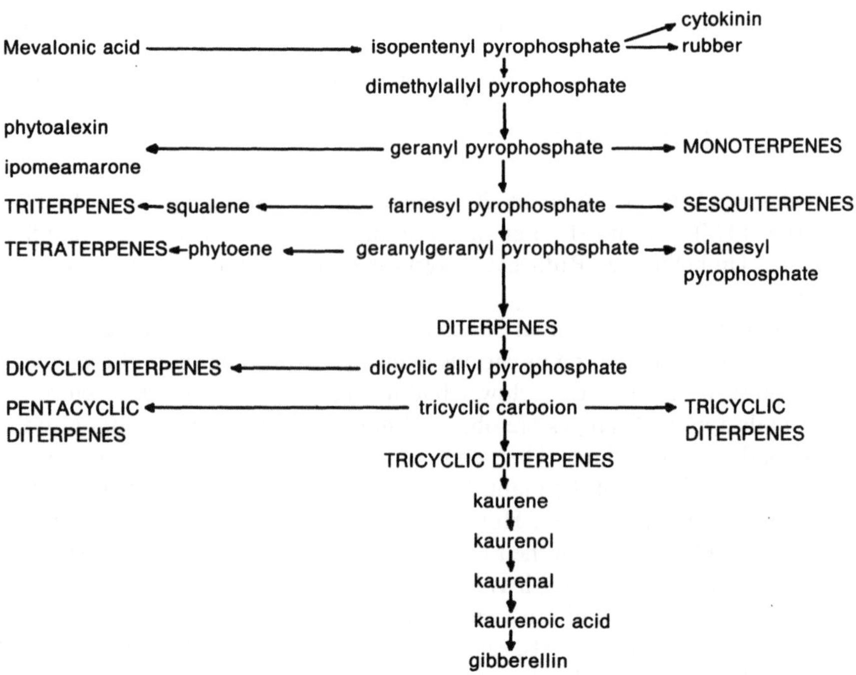

Fig. 50. Scheme of terpene formation from mevalonic acid

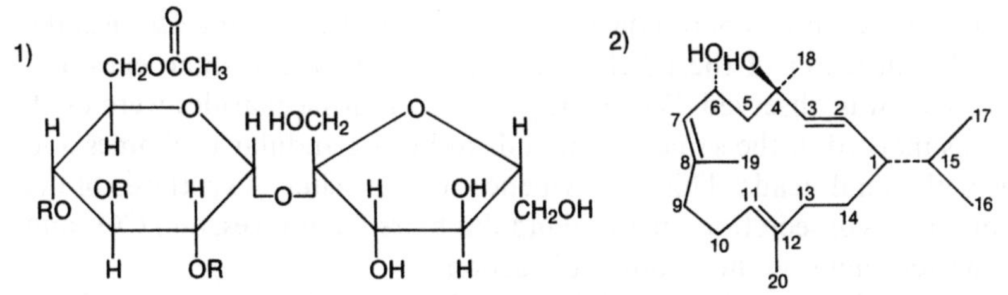

Fig. 51, 1,2. Ester of saccharose (**1**) and duvatriendiol (**2**) in the composition of glandular trichome excreta from *Nicotiana tabacum*

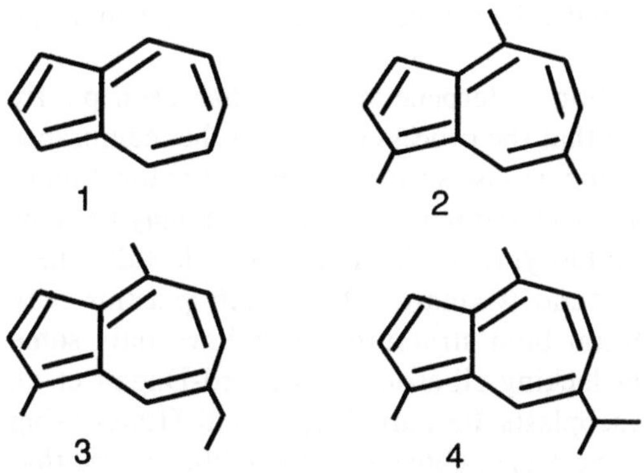

Fig. 52, 1–4. Azulenes of essential oils in glandular hairs. **1** Azulene. **2** Chamazulene. **3** Prochamazulene. **4** Gujazulene

(Stahl 1952, 1957). Most of the azulenes are colored blue or blue-violet (maximums of absorbance near 700, 640, and 580 nm). Many azulenes possess bacteriostatic activity (Heilbronner 1959).

In essential oils of the genera *Asarum*, *Thymus*, and *Mentha*, the benzoic compound asarone or 4-propenyl-1,2,5-trimethoxybenzol has been found (Ozarowski 1956). This compound kills *Paramecia* cells (Tokin 1980a). Most researchers presumed that all the steps of terpene biosynthesis take place in the secretory cells themselves (Vasilyev 1977; Croteau and Johnson 1983). Experimental evidence of this was given in experiments with *Nicotiana tabacum* by Kandra and Wagner (1988), who used isolated glandular capitate hairs and segments of epidermis without glandular trichomes to determine the location of the synthesis of the major components of their exudate (Table 5). After 20

Table 5. Sites of biosynthesis of exudates on the light in tissue of the middle vein of *Nicotiana tabacum*, % of labeling. (Kandra and Wagner 1988)

Tissue	Duvatriene diol	Esters
Isolated glandular heads	0.06	0.11
Epidermis with heads	1.93	0.82
Epidermis without heads	0.0	0.0
Subepidermal tissue	0.0	0.0

exposure in water solutions of radioactively labeled saccharose, glucose, acetate, and CO_2, the inclusion of the label mainly in duvatrienediol diterpenes and esters of sucrose was observed. When CO_2 and/or monosaccharides were used, the label was marked in the same compounds in both glandular trichomes and epidermis with gland heads. This was evidence that the sites of synthesis of the major components of secretion are the glandular hairs themselves, and CO_2 and monosaccharides may be their source of carbon.

Glandular trichomes of sage, *Salvia officinalis*, and *Tanacetum vulgare* contain enzymes of terpene biosynthesis (+) and (−)-bornyl pyrophosphate cyclase,1,8-cineole cyclase, humilene cyclase, (+)-sabinene hydroxylase, and caryophyllene cyclase (Gershenzon et al. 1987). Some details of C_5–C_{20} terpene formation (as components of essential oils) are considered in detail in Beale (1991) and Wagner (1991).

The intracellular compartmentation of terpene synthesis has been poorly studied. It is thought (Vasilyev 1977) that the process occurs on the membranes of the agranular endoplasmic reticulum, whose surface serves as the location of the enzymes participating the hyaloplasm and matrix of organella may perhaps also play a part data show terpene biosynthesis in specialized plastids where enzyme phytoene synthetase was found (Camara 1984). Such plastids may possibly be large leucoplasts with ameboid structure which have only some internal membranes and a stroma lacking ribosomes (Bernard-Dagan et al. 1982). Galndular cells are rich in leucoplasts. Bernard-Dagan et al. (1982), using isolated vesicles of leucoplasts from the peel of *Citrofortunella mitis*, showed that they synthesize monoterpenes from isopentenyl diphosphate. Synthesis of monoterpenes also occurs in cell tissue of grape var. muscat (Dell and McComb 1978b). The formation of terpenes took place in leucoplasts connected with the endoplasmic reticulum. These organella synthesized the c_{10}-monoterpenes linalool, geraniol, citral, and citronellol, whose concentrations change during plastid development. Since the membrane of the leucoplast is connected with the endoplasmic reticulum, terpenoids can pass from the site of their synthesis – the external plastid envelope – to the site of their accumulation – the extraplasmic space between the cellulose wall and cuticle. Elimination through the plasmalemma occurs according to either the ecrine type via monomolecular transport (Vasilyev 1977) or the granulocrine type (Fahn 1979; Pridgeon and Stern 1983). From the periplasmic space, essential oil is emitted via rupture in the cuticle, after which new cuticle can be formed and a new drop of secretion is collected. At favorable temperatures, essential oils diffuse in gaseous form through the cell wall and cuticle. Neither the cellulose cell wall nor the cuticle is a barrier to the emission of volatile low-molecular weight terpenoids. The substances are capable of penetrating the water phase of the cell wall, as is seen from facts such as the emission of the flower aroma and action of the vapors of essential oils on the accompanying organisms around the excreting plant (Kisser 1958; Denisova 1989). Curry (1987) showed the location of enzyme hydroxymethyl-CoA-synthetase, which is important for the metabolic pathway of terpene biosynthesis in osmophores of the tropical plant *Stanhopea anfracta* (Orchidaceae). The enzyme is found in the smooth endoplasmic reticulum, between inner and outer membranes of the mitochondria and in membranes of plastids. Later, the same authors (Curry et al. 1989) observed methyl-glutaryl-CoA-synthetase on studying another object, *Gongora truncata*. It is located in the smooth endoplasmic reticulum, between the inner and outer membranes of amyloplasts and mitochondria. The stage of differentiation of cells is also important for terpenoid synthesis. Brown et al. (1987) have shown that the nondifferentiated tissue culture of *Pelargonium fragrans* produces essential oils, and they are only 3% of the total terpenoids contained in the adult leaves. This problem has also been analyzed by Berger et al. (1990), who found 41 components, mainly monoterpenes and sesquiterpenes, in essential oils of intact plants of *Coleonema album*. The production of terpenes by the photomyxotrophic callus of the same plant grown in the light only began after 48 h. The composition of essential oils included 11 monoterpenes and 5 sesquiterpenes. The heterotrophic callus of plants grown in the dark was lacking in differentiated plastids and the components of essential oils. Thus, the correlation between light-induced differentiation of plastids and synthesis of essential oils is evident. The synthesis of monoterpenes in cell culture depends on the stage of tissue differentiation, as is shown in cell culture of *Artemisia dracunculus* (Cotton et al. 1991a). At, non differentiated tissue produces phenylpropene (allylanisol) as the major component (Fig. 53). Then methyleugenol, anehtol, eugenol, and chavicol arise. In such cultures, allylanisole and methyl eugenol often prevail, and their synthesis is inhibited by α-naphtyl acetic acid and benzylaminopurine (Cotton et al. 1991b).

Terpenoids secreted by glandular hairs are more often considered as secondary compounds which have no clear function in the plant itself, although the number of terpenoids is nearly 5000 (see Fig. 1). Perhaps it is true for certain groups of terpenoids, however, that many of them possess important physiological activity. The components of essential oils are thought to be reutilized in metabolism (Harborne 1988). Many terpenoids show hormonal activity, such as cytokinins, gibberellins, and abscisic acid. Danilova and Kashina (1987a,b) found terpene-containing glandular tissues of *Perilla ocymoides*: glandular trichomes and scales (peltate glands), differ both in their ultrastructure and their

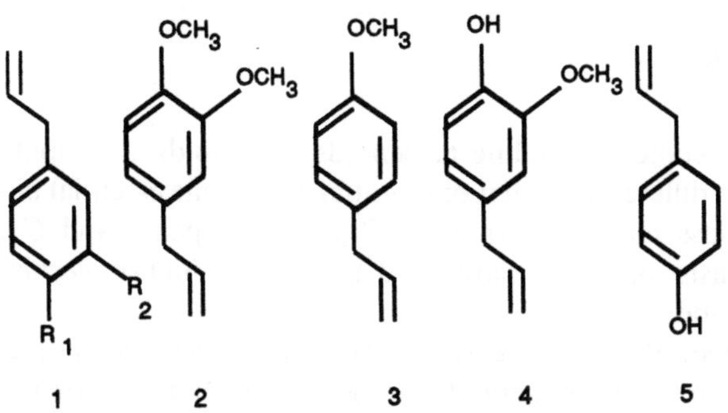

Fig. 53. 1–5 Components of essential oils in plant cell tissue of *Artemisia dracunculus*. **1** Allylanisol. **2** Methyleugenol. **3** Anethol. **4** Eugenol. **5** Chavicol

physiological functions. The ultrastructure of the glandular scale (including the development of the agranular endoplasmic reticulum and other organella) is similar to cells usually secreting terpenes as major components of essential oils. Glandular scales secrete terpenes, and perhaps steroid hormone of flowering. (gibberellin is its possible precursor). In glandular hairs derivatives of shikimic acid auxins and phenols predominate, whereas secretion of terpenes is small.

Mono- and sesquiterpenes are often thought to play the role of allelochemicals in interactions between plants. Brown et al. (1987) showed that components of essential oils, monoterpenes, isolated from tissue culture of *Pelargonium fragrans*, at a concentration of 0.1 gl^{-1} are capable of depressing the growth of tissue culture in this plant. The most active inhibitors were carvone, α-terpinene, and limonene. Studies of the biochemical basis of interaction between plants and insects are also now being undertaken. Low-molecular weight terpenes are easily evaporated and possess a sharp smell. Insects are very sensitive to even small concentrations of them. Secreting terpenes of the flower trichomes can live insect-pollinators as chemical attractants. (Kullenberg and Bergström 1975). Some sesquiterpenes are defensive substances against insects (Gregory et al. 1986).

One interesting discovery is that of alarm pheromones for insects in plant excreta (Jeffree 1986). Trichomes of leaves of *Solanum berthaultii* produce a significant amount of the sesquiterpene E-β-farnesene, which is in exudate drops and acts as an alarm pheromone (Gibson and Pickett 1983). In *Humulus lupulus*, the same substance is located within the hairs in amounts needed for the deterrence of insects. It appears on the leaf surface just after mechanical damage (Pickett and Griffith 1980). Since substances which are known as alarm pheromones are met in plant trichomes and frighten off insects, it is possible that they play a certain role in ecology. Perhaps the compounds are synthesized by plants as a special means of self-defense against insects. However, it should be emphasised that there are few direct experimental data confirming the existence of one or the other physiological role of glandular trichomes.

3.7 Secretion of Resins

Resins are mixtures of substances including terpenoids, flavonoids, and lipids. Their terpene composition differes from that of essential oils. While essential oils contain low-molecular mass terpenoids (mainly C_{10} monoterpenes and C_{15} sesquiterpenes), resins consist of both volatile and nonvolatile high-molecular C_{20} terpenoids (Croteau and Johnson 1983)

Resins are formed in specialized glandular trichomes or on the inner surface of resinous ducts (see Chap. 2.3.1). The number of glandular hairs on the leaf surface is high, estimated per leaf as from 300000 for *Beyeria* to 600000 for *Eremophila* (Dell and McComb 1978b).

The structure of the glandular hairs producing resins can vary. The glandular trichome of *Beyeria* (Fig. 54) consists of two cells, that the stalk of that of the head. The stalk cell includes numerous chloroplasts and rather small vacuoles. On the contrary, cells of the multicellular stalk of *Newcastelia viscida* (family Dicrastylidaceae) contain a few chloroplasts and large vacuoles. The structure of the head cells of both glands is similar and is characterized by numerous ameboid plastids producing the terpenoid components of resins.

Numerous yellow resinous glands of *Azadirachta indica* (fam meliaceal) consist of a short pedicel, subglandular tissue, and one or two layer secretory tissue like the epitelium (Inamdar et al. 1986). The common structure of the gland is shown to have a similarity with extrafloral nectaries. The origin of the gland is like an epidermal cell looking on ordinary papilla. The amount of resins which can be produced by some species of Australian plants is shown in Table 6. From these data, it is seen that some plant species can vary significantly both in resin production per dry weight of leaf and in the quantity of resins in the glandular hairs. The former is highest in *Halgania lavendula* and *Anthocersis viscosa*, and the latter in *Acacia glutinosissima* and *Cyanostegia angustifolia.*

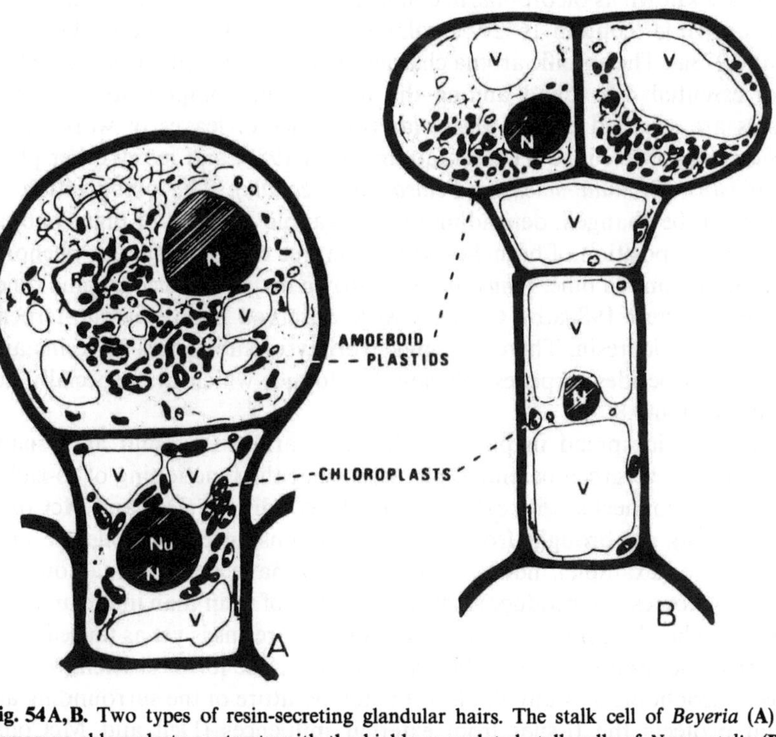

Fig. 54A,B. Two types of resin-secreting glandular hairs. The stalk cell of *Beyeria* (**A**) with its numerous chloroplasts contrasts with the highly vacuolated stalk cells of *Newcastelia* (**B**), which contain only a few chloroplasts. Head cells of both gland types have similar fine structure, and are characterized by numerous amoeboid plastids. Drawn from electron micrographs. *M* mitochondria; *N* nucleus; *Nu* nucleolus; *P* plastid; *R* resin; *W* cell wall; *V* vacuole. (Dell and McComb 1978a)

The sites of resin synthesis are glandular hairs. Here mevalonic acid, which is a precursor of terpenes, is found. The compartments of resin synthesis on the cell level are plastids. In the last few years, using gas chromatrography, some rare components of resins such as six sesquiterpene lactones, niveusins B and C; 4,5-dehydroniveusin; 15-hydroxy-3-dehydrodesoxyfruticin, and argophyllins A and B were identified in resinous excreta of multicellular peltate glandular hairs of young leaves of *Helianthus annuus* seedlings (Fig. 55). Up to 1.67 μg of niveusin C and 13.9 μg of 15-hydroxy-3-dehydrodesoxyfruticin are observed on 1 cm^2 of leaf surface.

Sesquiterpene lactones from *Helianthus annuus* possess very strong antimicrobial activity, acting on DNA, RNA, and protein synthesis. The sesquiterpene lactones cumambrin A and B are also found in glandular trichomes of *Artemisia nova*, as well as monoterpenes (Fig. 56) (Kelsey and Shafizadeh 1980).

In glandular hairs of *Beyeria*, the rate of resin secretion through the head wall was determined (Dell and McComb 1978b) as 6.3×10^{-5} μg μm^{-2} day^{-1} or 2×10^{-2} pmol s^{-1}. The total amount of leaf resins can be up to 30% of dry weight of the mature leaf and 50% of that of the young leaf. As a secretory product, mixtures of resin and oil are often found in plant glandular hair. These secretions are known as oleoresins, in conifers also called balsams. For instance, oleoresins of *Abies* contain 16–20% volatile oils and 70–80% resin (Dell and McComb 1978a). The specific aroma characteristic for some plants is created by drops of essential oils found among the nonvolatile components or resins. Oleoresins are secreted by the glandular trichomes of leaves of wormwood, *Artemisia campestris* (Fig. 57) (Ascensao and Pais 1987) and many other plant species (*Cistus, Primula sinensis, Pelargonium zonale*). The composition of oleoresins can be changed, depending on geographic position of plant habitat and even on the position of branches within the tree crown. A similar phenomenon has been found in pine, *Pinus contorta*, grown in different geographic zones of America (Forrest 1980a, b). Changes were observed mainly in the terpene composition of oleoresin. There is a great variety of substances in resins and oleoresins, and besides terpenes, phenols are found, which are especially described in Section 3.8.

Resins are widespread in plants in habiting arid zones, and are usually characteristic of evergreen perennial plants, so that the functioning of so-called surface resins is connected with extremal climatic conditions (Dell and McComb 1978a). If plants are brought from extremal circumstances to moderate ones (greenhouse, for example), newly forming leaves have a smaller amount of glandular trichomes and produce smaller quantities of resin than in the unfavorable habitat. The physiological function of "surface" resins is yet as unclear. It is possible that the resin flow on to the leaf or bud surface forms a defensive film, which is hydrophobic and unvolatile at the temperature of the surrounding air. This film protects the tissues from external influences (Dell and McComb 1978b). Glandular hairs are also supposed to lower leaf heat, which leads to a decrease in transpiration and staves off depression of photosynthesis. Resins are capable of preventing the invasion of insects, fungi, and bacteria into the leaf.

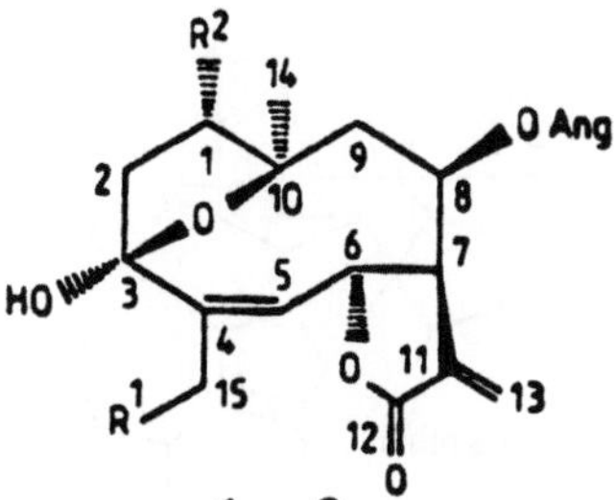

	1	3
R1 -	H	OH
R2 -	OH	H

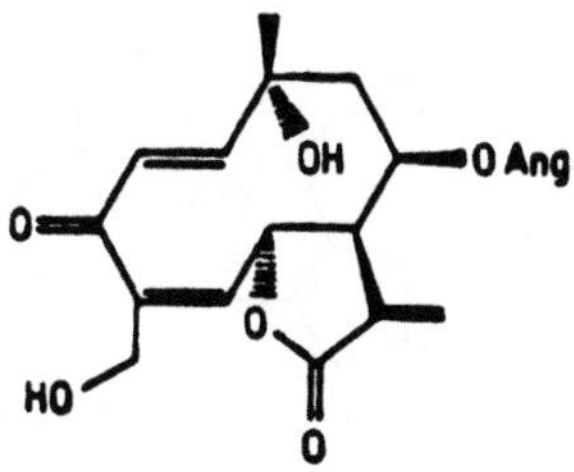

2

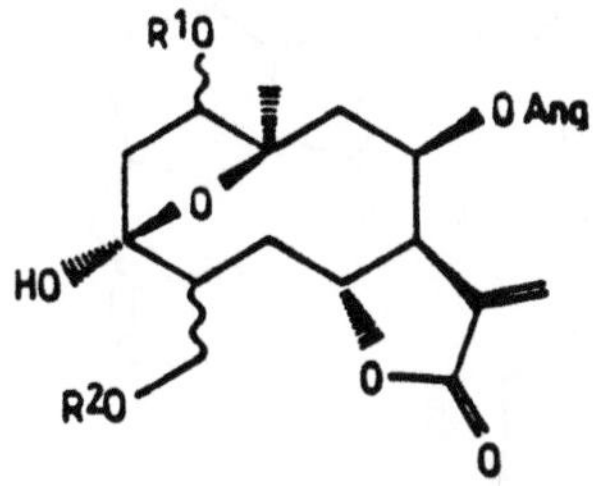

	4	4a	4b
R1 -	H	Me	Me
R2 -	H	H	Ac

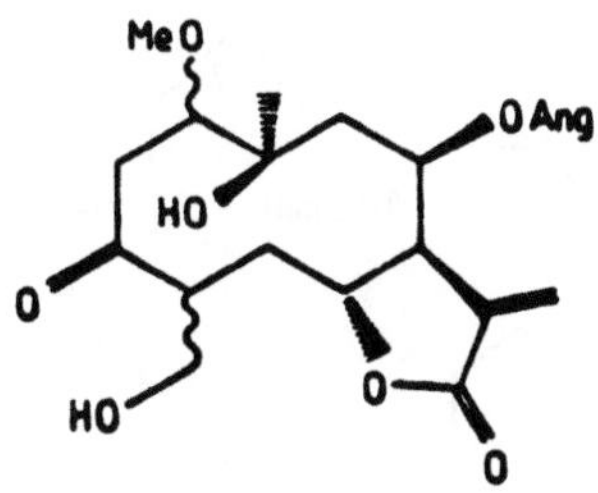

4c

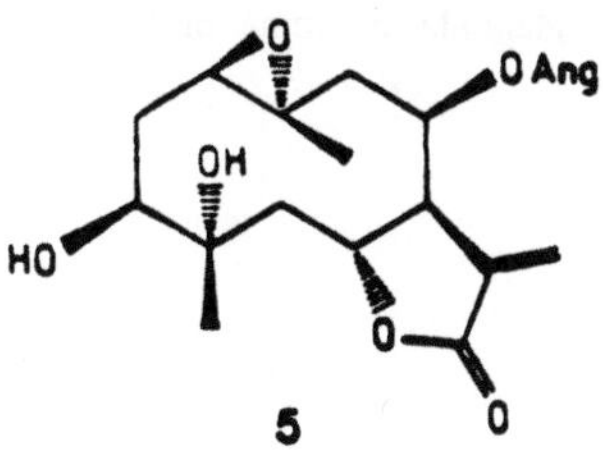

5

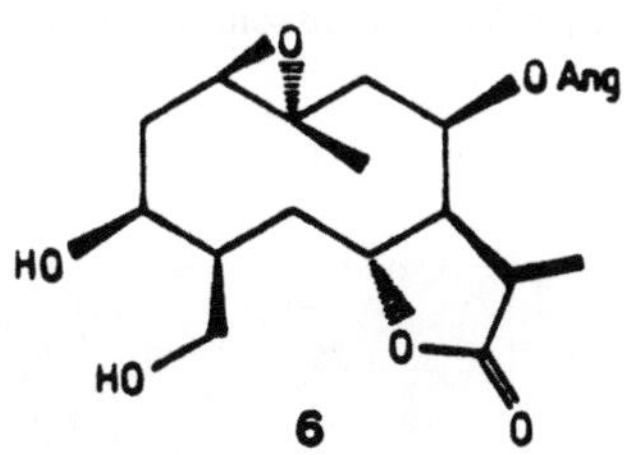

6

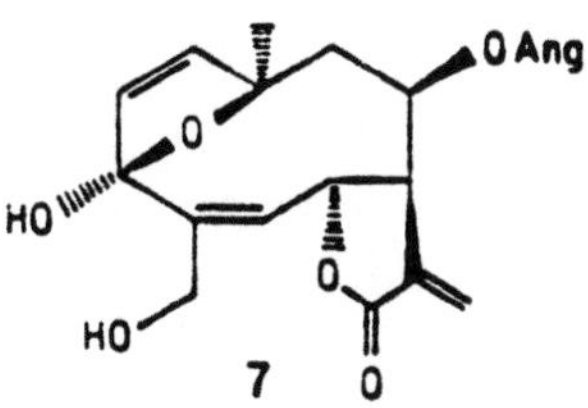

7

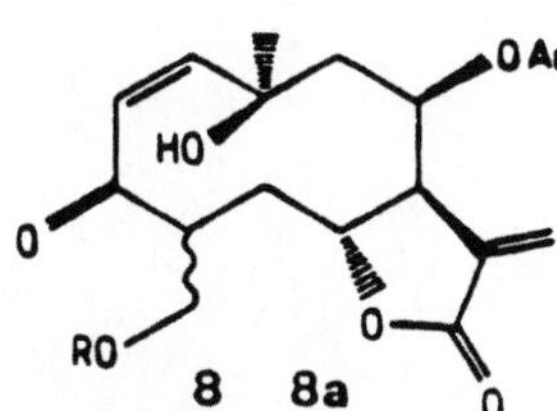

	8	8a
R -	H	Ac

Fig. 55, 1–7. Sesquiterpene lactones from glandular hairs of *Helianthus annuus*. **1** Niveusin C. **2** 15-Hydroxy-3-dehydrodesoxyfruticin. **3** Niveusin B. **4** 4,5-Dehydroniveusin A. **4a** 1-Methoxy-4,5-dehydroniveusin A. **5** Argophyllin A. **6,7,8** Germacranolides. *Ang* 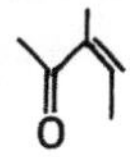 ; *Ac* acyl

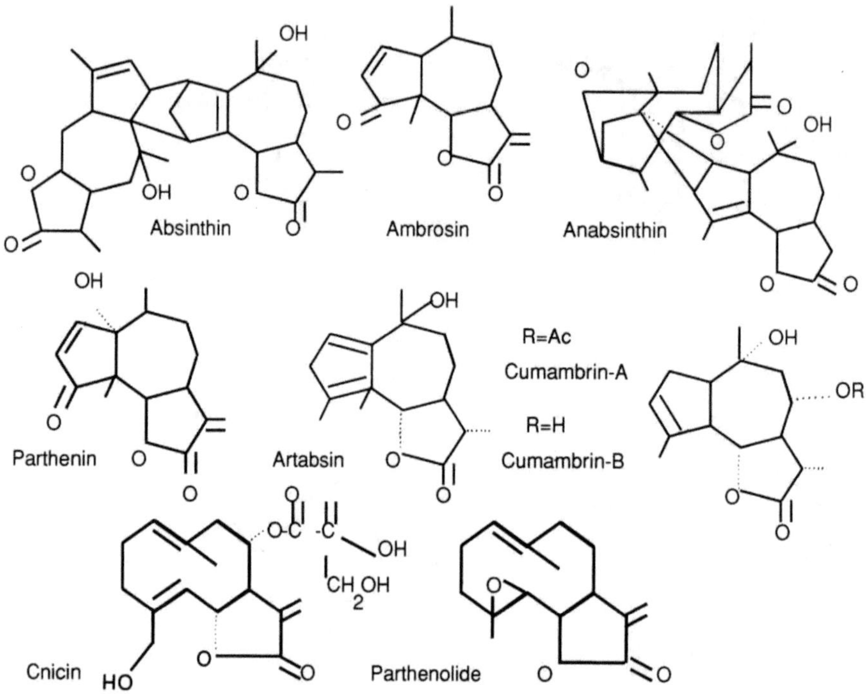

Fig. 56, 1–8. Cumambrines A and B and some other lactones from glandular trichomes of *Artemisia nova*

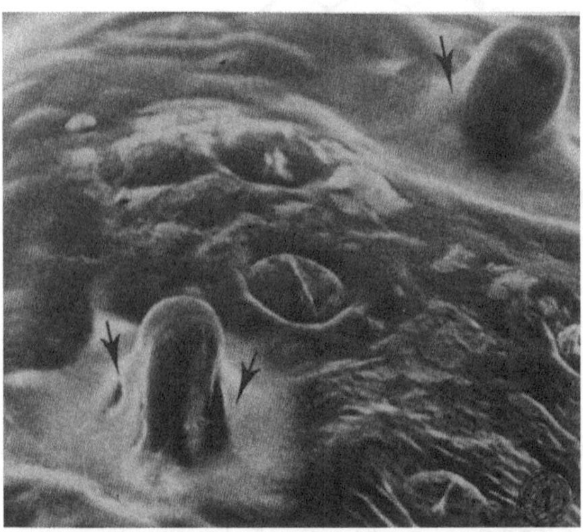

Fig. 57. Scanning electron micrograph of young leaf surface of *Artemisia campestris* (ssp. *maritima*) partly covered by oleoresins secreted by glandular hairs (*arrows*) (Ascensao and Pais 1987)

Table 6. Resin yields for some Western Australian plants. (Dell and McComb 1978a)

Family	Species	Resin (% dry weight)	mg Resin/ mature leaf	μg Resin of glandular head	Diameter of glandular head (μm)
Mimosaceae	*Acacia glutinosissima* Maiden and Blakely	17	17.6	2.4	55–65
Euphorbiaceae	*Beyeria viscosa* (Labill) Miq	20	46.0	0.2	15–25
Sapindaceae	*Diplopeltis petiolaris* (F. Muell) Benth	25	11.6	1.3	50–90
Sapindaceae	*Dodonaea viscosa* Jacq	7	2.2	0.5	95–115
Boraginaceae	*Halgania lavendulacea* Endl	29	8.1	1.5	20–30
Dicrastylidaceae	*Newcastelia viscida* E. Pritzel	15	2.3	0.2	20–45
Dicrastylidaceae	*Cyanostegia angustifolia* Tursz	17	2.8	1.6	35–45
Lamiaceae	*Prostanthera grylloana* F. Muell	12	0.1	0.7	30–75
Solanaceae	*Anthocercis viscosa* R. Br.	28	24.4	0.9	45–60
Myoporaceae	*Eremophila fraseri* F. Muell	17	38.0	0.2	35–50
Asteraceae	*Olearia muelleri* (Sond) Benth	9	0.8	1.0	40–65

Among other isoprenoids, steroids are known as components of plant excreta. Jeffree (1986) reported the release of molting hormones of insects (such as ecdysones, hydroxylated sterols, relative to cholesterol in the chemical nature) structure of conifers from Podocarpaceae, and Taxaceae, mosses, especially of the family Polypodiaceae, and many angiosperms. This phenonenon is of practical significance, as similar compounds induce serious disturbances in insect development. As a result, ugly deformations appear or death of the larvae occurs. It should be noted that gymnosperms and ferns which excrete ecdysones invariably do not suffer from insect attack (Harborne 1988).

3.8 Secretion of Phenols

Although isoprenoids are the major metabolites of the glandular trichomes of flowering plants, phenols are also found in the secretions. Although little information exists on the substances, they are thought to be the ordinary components of such glands (Kelsey et al. 1984). Beckman et al. (1972) showed that glandular trichomes of 32 species, from 39 studied, possessed the ability to synthesize and store phenolic substances. According to Harborne (1988), nearly one-third of the total leaf phenols is concentrated in the trichomes. The total amount of the substances located here has been determined by Beckman et al. (1972), Salatino et al. (1988), and Feucht and Treutter (1990), together with its dependence on the species and genera of the plants. Woody plants contain the highest quantity of flavonoids in trichomes of *Acer platanoides* and *Prunus subhirtella*, next in *P. sargentii*, and lowest in *Aesculus hippocastanum* and *Tilia grandiflora* (Feucht and Treutter 1990).

Phenols of the trichome cell exist in the form of water-soluble glycosides located in the cell sap (see Chap. 1.4), but their lipophilic aglycones can be excreted, together with resinous and oil secretions, by glandular trichomes to the environment. In this case they are observed on the plant surface as either components of leaf resins and small thin crystals (Wollenweber 1984).

Simple phenolic C_6-compounds are seldom found in glandular hairs. However, Duffey (1986) reported that in the periplasm of tomato trichomes, catechols were discovered, and in the cytoplasm were enzymes, oxidizing them. C_6–C_3–C_6 phenols, namely flavonoids of various classes and their glycosidized derivatives, prevail in resinous and oil plant excreta. C_6–C_3 phenols (coumarins and esters of caffeic acid) and C_6–C_2–C_6 phenols (stilbenes and anthraquinones) are found to a lesser degree. Additionally, terpenoids and their phenolic derivatives such as cannabinoids are found in glandular secreta as well as various esters of flavonoids. Some formulas of flavonoids and coumarins are seen in Fig. 58.

Exudates of leaf resins differ from each other by their great variety of flavonoid aglycones. For example, nearly 50 different flavonoids are found in resins of *Cistus* (Vogt et al. 1987), and there is a certain taxonomic specificity in their composition. Leaves and inflorescences of some species belonging to the genus *Primula* are covered by numerous peltate multicellular glands, eliminating an exudate which looks like a farinose bloom consisting of rod-like or needle-shaped crystals of phenols. Wollenweber (1985) has shown that these excreta are composed of unsubstituted flavonols. 5-Hydroxyflavone and 2-hydroxyflavone are found in 18 species and 5,8-dihydroxyflavone in 11 species of this genera. The presence of 5-hydroxy-6-methoxyflavones was characteristic for only one species, *Primula sinensis*. The taxonomic specificity of the flavonoid composition is confirmed by the study of flavonoid aglycones in the genus *Ambrosia* belonging to the family compositae or Asteraceae (Table 7). In resinous excreta of glandular trichomes, 6-methoxy-derivatives of flavonoids dominated, including such rare products as 3′-dimethoxysudachitin, xanthomicrol, sideritiflavone, and oxyajapin B, some of them peculiar to certain species. The major components of the exudates were 3′-dimethoxysudachitin and xanthomicrol for *Ambrosia deltoides* and *A. artemisiifolia*. In the latter species, the coumarins scopoletin and ajapin were also identified, whereas derivatives of quercetagetin, such as scutellarein and hyspidulin, predominated in *A. ambrosioides*. Similar aglycones are found in many families of higher plants, but external flavonoids appear to be most abundant in the Asteraceae and relatively frequent in plants

Flavones

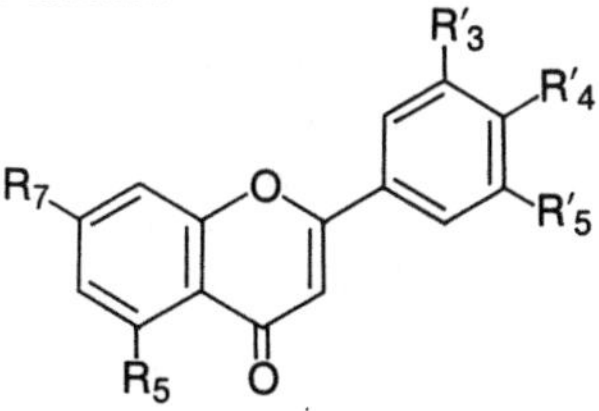

Apigenin R_5, R_7, R'_4=OH; R'_3, R'_5=H
4′,7-Dihydroxyflavone R_7, R'_4=OH; R_5, R'_3=H
Luteolin R_5, R_7, R'_3=OH; R'_5=H
Chrysin R_5, R_7=OH; R_3, R'_4, R'_3=H

Flavanones

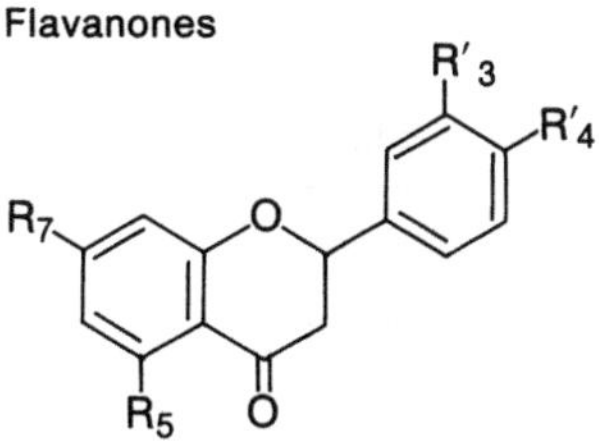

4′,7-Dihydroxyflavanone R_7, R'_4=OH; R_5, R'_6, R'_3=H
Diplacone R_5, R_7, R'_3=OH; R'_4=H; R_6=geranyl
Naringenin R_5, R_7, R'_4=OH; R_6, R'_3=H
Pinocembrin R_5, R_7=OH; R_6, R'_3, R_4=H
Eriodictyol R_5, R_7, R'_3=OH; R'_6, R'_4=H

Flavonols

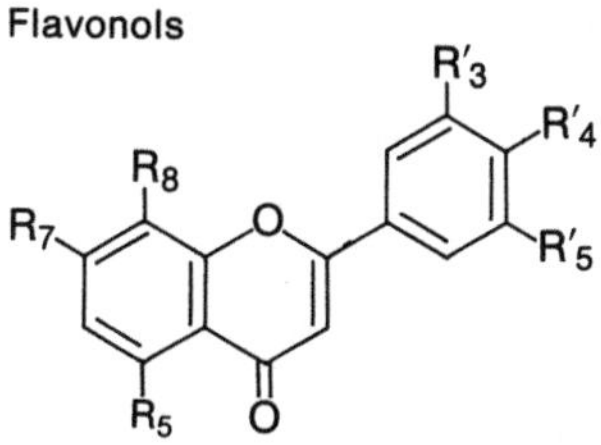

Galangin all R=H
Herbacetin R_5, R_7, R_8, R'_4=OH; R_3, R_6, R'_3, R_5=H
Gossypetin R_5, R_7, R_8, R'_3, R'_4, R_3, R_6, R'_5=H
Izalpin R_3, R_5, R_6, R_8, R'_3, R'_4, R'_5=H; R=CH_3
Izalpinin R_3, R_5, R_6, R_8, R'_3, R'_4, R'_5=H; R_7=OCH_3
Isorhamnetin R_5, R_7, R'_4=OH; R_3, R_6, R_8, R'_5=H; R'_3=OCH_3
Quercetin R_5, R_7, R'_4=OH; R_3, R_6, R_8, R'_5=H
Quercetagetin R_5, R_6, R'_3, R'_4=OH; R_3, R_8, R'_5=H
Kaempferol R_5, R_7, R'_4=OH; R_3, R_6, R_8, R'_3, R'_5=H
Oxyajapin B R_6, R'_3, R_5=OH; R_8, R'_5=H;
R_3, R_7, R'_4=OCH_3

Catechins

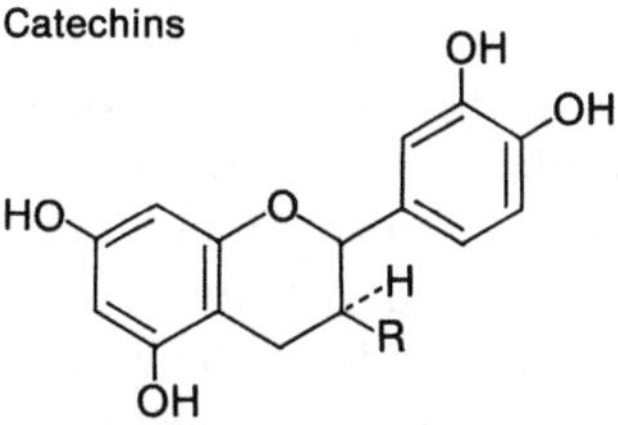

(−)-Epicatechin R=OH; cis-H
(+)-Catechin R=OH - trans-H

Chalcones

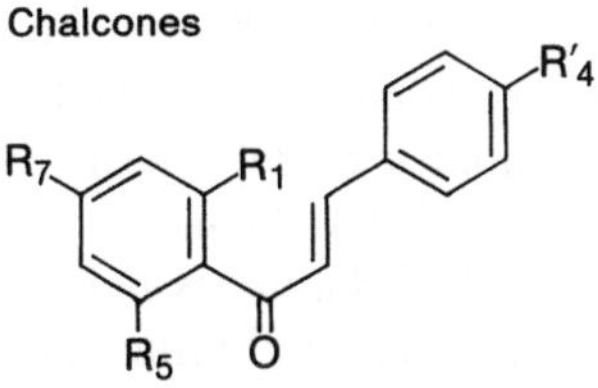

4,4′-Dihydroxy-2-methoxychalcone R_7,
R'_4=OH; R_1=OCH_3; R_5=H
Neosakuranetin R_7=OH; R'_4=OMe; R_1=R_5=H

Coumarins

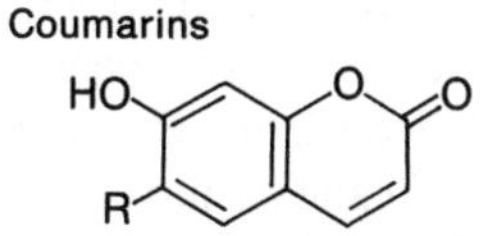

Scopoletin R=OCH_3
Ajapin (methyl ether of scopoletin)

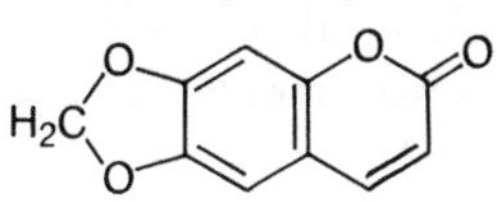

Fig. 58. Formulas of main flavonoids and coumarins in resinous and oil plant excreta

Table 7. Flavonoid aglycones found in resinous exudates of various species of *Ambrosia*. (After Wollenweber et al. 1987b)

Species	Flavonoid aglycone	Major component or traces
A. ambrosioides	Hispidulin	Major
	Apigenin	Traces
	Luteolin	–
	Nepetin (6-methyoxyluteolin)	–
A. artemisiifolia	Quercetagetin-3-,6-dimethyl ester or axillarin	Major
	and its derivatives Tomentin and patuletin	Major
	Nepetin, luteolin, quercetin	Traces
A. camphorata	3,6,4′-Trimethyl ether 6-hydroxykaempferol	Major
	Hispidulin, centaureidin or quercetagetin-3,6,4′-tri-methyl ester and its derivatives (chrysosplenol D and others) and pinocembrin	Traces
A. cordifolia	Oxyajapin B or quercetagetin-3,7,4′-trimethyl ether	Major
	Quercetin, its 3-methyl ether (isorhamnetin), rhamnetin	Traces
A. deltoides	3-Dimethoxysudachitin and xanthomicrol	Major
	Hispidulin, luteolin and its methyl ether, nepetin, sideritiflavone, kaempferol, quercetin, rhamnetin, rhamnasin, ombuin	Traces

of semiarid habitat. The structural diversity of the exudate flavonoids is normally greater than that of the tissue glycosides in the same plant. Many flavonoids (up to 11% dry weight) are found in exudates of the ferns *Pityrogramma chrysophylla* and *P. triangularis*. For instance, galangin (3,5,7-OH flavonol) and izalpinin (7-OMe flavonol) have been identified. A taxon-specific chalcone, neosakuranetin (4′-OMe chalcone), is also known from the excreta (Wollenweber 1984). In the farinose exudate of leaves of an other fern species, *P. triangularis*, new flavonoids, 5,7-dihydroxy-3-methoxy-6,8-dimethylflavone and 5,6,7,8-tetrahydroxyflavone, are found. More new hydroxylated and methylated flavonoid aglycones have been identified in frond exudates of such ferns as *Cheilanthes argentea*, *Northolaena*, and *Platyzona microphylla* (Wollenweber and Roitman 1991), and in resins from leaves and stems of *Baccharis biglovii* (Asteraceae) (see Fig. 59), namely flavonol alnusin, dihydroflavonol alnustinol, 2,5,7-trihydroxyflavanone, and 7-benzoyl chrysin (Arriaga-Giner et al. 1986). Some other lipophilic phenols are also found in leaf resins of the genera *Baccharis*. One of them is 2′,4′β-trihydroxy-6′-methoxychalcon from *B. salicifolia*, while in *B. sarothroides* phenylethyl caffeate and benzyl caffeate have been identified (Wollenweber et al. 1989a). *Gutierrezia sarothrae* also has resin-containing glands whose excreta include flavonoid aglycones (Hradetzky and Wollenweber 1987) such as jaceidin and sudachitin (Fig. 59).

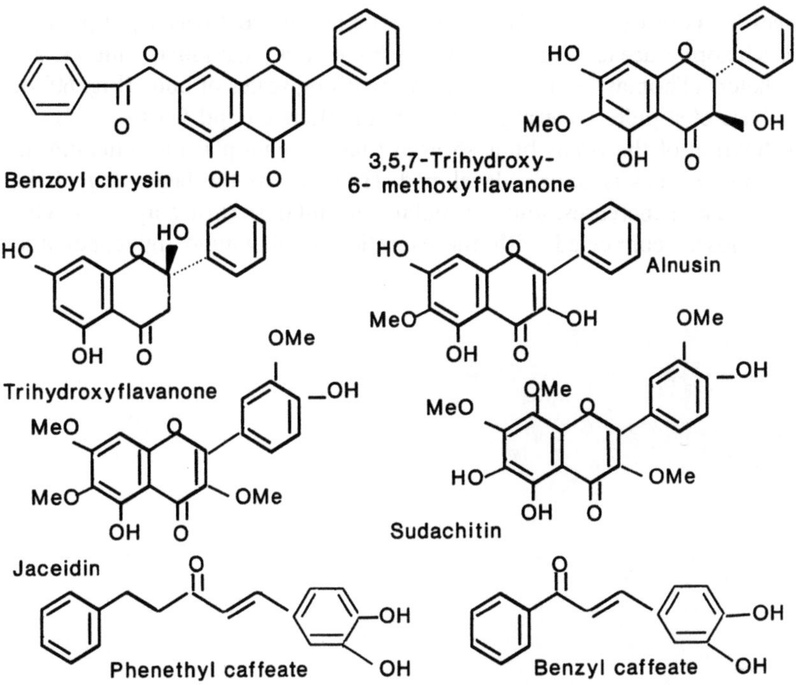

Fig. 59. 1–8 Flavonoids and caffeates from *Baccharis* and *Gutierezia. 2a R* acyl

Some flavonoids are found in isolated glandular trichomes of *Helianthus annuus* (Riesenberg et al. 1987). They were identified by potassium hydroxide treatment, which induced the appearance of the orange color characteristic of a flavonol reaction. Flavonoids were observed in glandular hairs located on both surfaces of the leaf, whereas chalcone aglycones are thought to be only in the waxy layer of the leaf cuticle (Riesenberg et al. 1987). Two new, earlier unknown flavonoids, diplacone and diplacol, were found in resins of glandular trichomes of the shrub *Diplacus aurantiacus* (Scrophulariaceae), which is widespread in California. Diplacone is 6-geranyl eriodictyol, and diplacol is the corresponding dihydroflavone. The resin amounts to 30% of dry leaf weight and is represented mainly by diplacone, diplacol, and their derivatives (Lincoln and Walla 1986; Lincoln 1990). The leaf exudate of the weed *Polanisia trachysperma*, consisting of terpene resins, contains more than 12 methylated flavonoid aglycones including rare compounds. Among the identified substances are 6,8-dimethoxy-substituted derivatives (Wollenweber et al. 1989c). Another group of flavonoid catechins, flavan-3-ols, are located in the leaf hairs of many woody plants, among them catechin and epicatechin (Feucht and Treutter 1990). Rich material dealing with phenol excretion was collected on buds of plants belonging to the family Betulaceae. Winter buds of the trees are covered by scale-shaped multicellular glandular hairs and release a slimy lipophilic secretion containing phenolic derivatives (Asakawa et al. 1977). Plants of the families Betulaceae, Hippocastanceae, Rhizophoraceae, and some others have sticky glandular villuses, so-called colleters (Thomas 1991), located on sheathing scales of buds (Fig. 60) or near the base of stipules covering yound shoots (Lersten and Curtis 1974).

Investigation of the *Alnus* buds showed that the endoplasmic reticulum in young glandular cells is poorly developed. However, at the beginning of the excretory phase it increases, and a tubular agranular structure appears. This change is closely connected with the excretion of flavonoid aglycones and

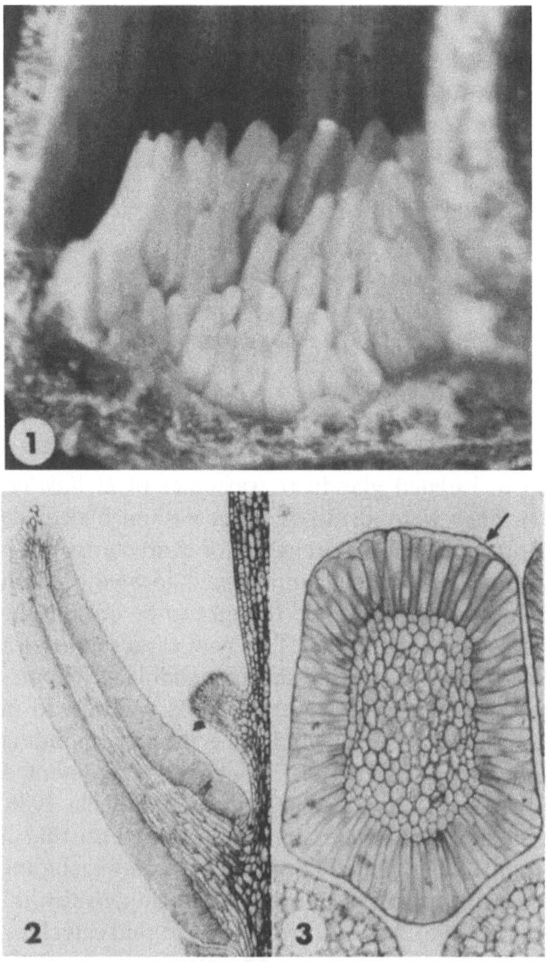

Fig. 60, 1–3. Colleters at the base of bud stipules of *Rhizophora mangle* L. (Rhizophoraceae) **1** Colleters view at the base of stipule dissected from bud (× 20). **2** Median longitudinal section of one colleter (× 66). **3** Cross-section of colleter showing internal mass of cells surrounded by palisade-like secretory epidermis. *Arrow* indicates detached cuticle (× 231). (After Lersten and Curtis 1974)

perhaps of lipophilic terpenoid substances. When the secretion is over, the endoplasmic reticulum has disappeared (Wiermann 1981). Exudates from *Alnus sieboldiana* and *A. pendula* contain a mixture of flavonoid aglycones, phenolic acids, esters and alcohols, and netural phenylpropanoid derivatives (Suga et al. 1972). In buds of poplar, *Populus* (family Salicaceae), lipophilic material including flavonoid aglycones is secreted by the glandular epithelium on the inner side of the bud scales. For the chemotaxonomy of flavonoids, it is shown that in various species of the genus *Populus*, different group of flavonoids prevail. Recently, nearly 30 different flavonoids were isolated from resinous poplar bud exudates. Among them chrysin (5,7-OH flavone), galangin (3,5,7-OH flavonol), and pinocembrin (5,7-OH flavonone) are found in many species of this genus (Wollenweber 1984). Some rare methyl derivatives of apigenin, kaempferol and quercetin, two chalcones, and flavone with a hydroxylic group were also identified. Flavonoids are found in various species of *Populus* in varying amounts. The composition of phenols found in bud exudates in plants of the genus *Populus* varies and can perhaps serve as a taxonomic sign. For instance, flavonoids common to many poplar excreta are lacking in *P. euphratica* Oliv. and are rare in *P. lasiocarpa* and *P. ciliata* Wall. Phenolic compounds in the bulk of their exudates are represented mainly by derivatives of phenylpropenic acids and a number of their esters and ethers. The above-mentioned species differ from each other in the quality of such acids. For *P. euphratica*, this is mainly glyceryl-, monoacetyl glyceryl, and diacetyl glyceryl esters of p-coumaric and ferulic acids, for *P. ciliata* methyl-, and methyl-butyl derivatives of esters of caffeic or acetyloxycaffeic acids, and for *P. nigra* esters of caffeic and isoferulic acids which formed with aromatic and aliphatic alcohols. In contrast, bud exudates of *P. deltoides* Marsh are characterized by their high content of flavonoids and their relative compounds, those of *P. balsamifera* L. by dehydrochalcones. In bud excreta of the hybrid *P.* X *jackii*, major components are flavanones and flavones and their methylated derivatives and dihydrochalcones, but in less quantity than in *P. deltoides*. For bud exudates of plants belonging to the family Betulaceae (Table 8), flavonoids are more characteristic than esters of aromatic acids. They are contained in the geneva *Alnus*, *Betula*, and *Ostrya*. Cinnamic acid derivatives are also present in *Betula nigra*, and stilbene in *Alnus viridis*. The secretions of colleters also contain many phenols. So-called propolis (bee glue), which is a resinous component of waxy mixture collected by bees from plant buds, includes the same flavonoids as the bud exudate (Wollenweber et al. 1987a). Using gas chromatography/mass spectrometry, Greenaway et al. (1987) showed that exudates of poplar buds serve as primary building material for the production of propolis by bees. The major part of the poplar bud exudate enters the propolis without being tranformed. Besides flavonoid aglycones, in propolis, cinnamic and benzoic acids and their derivative cinnamoyl-cinnamate were identified (Wollenweber 1984). The latter is usually characterized as the spring bud aroma of *Populus balsamifera*.

In glandular trichomes also there are taxon-specific substances (Fig. 61). According to Wollenweber (1984), the release of rare flavonoids is observed in

Table 8. Principal phenolic components of bud extracts

Species	Component	Reference
Salicaceae		
Populus nigra	Esters of caffeic acid and aliphatic alcohols	Greenaway et al. (1988b)
P. nigra	Esters of isoferulic acid, aromatic and aliphatic alcohols	Greenaway et al. (1988a)
P. deltoides	Flavonoids: galangin (16-23%, pinocembrin (26–30%), pinocembrin chalcone (up to 21%), pinobanksin-3-acetate (14–25%)	Greenaway et al. (1990a, c)
–	Flavanones (35%) Flavanones methylated or esterified in the 3-d position (30%), flavones (13%).	
P. balsamifera	Dihydrochalcones (70%)	Greenaway et al. (1990c)
P. x *jackii*	Dihydrochalcones (14–15%), flavanones (30–49%), flavones (13–28%), methylated flavanones (4–12%)	Greenaway et al. (1990c)
P. euphratica	Phenylpropenoic acids and their glyceryl, monoacetyl and diacetyl glyceryl esters (~ 70%)	Greenaway et al. (1991)
P. ciliata	Esters of caffeic or acetyloxycaffeic acids (~70%)	Greenaway and Whatley (1991)
Betulaceae		
Alnus koehnei	Flavonoids (apigenin-4′-Me, scutellarein-6,4′-diMe and others)	Wollenweber et al. (1991)
Betula nigra	Flavonoids (apigenin-4′-Me, scutellarein-6,4′-diMe and others)	Wollenweber et al. (1991)
Ostrya carpinofolia	Flavonoids (apigenin-4′-Me, scutellarein-6,4′-diMe and others)	Wollenweber et al. (1991)
Hippocastanaceae		
Aesculus	Flavonoids	Wollenweber et al. (1974)
Rosaceae		
Prunus	Flavonoids	Wollenweber et al. (1972)
Rhamnaceae		
Rhamnus	Flavonoids	Wollenweber et al. (1972)

the genera *Artemisia, Brickellia, Eupatorium, Haplopappus, Beyeria, Flourensia, Ericameria, Encelia,* and *Gerea.* Their names are connected with the genus or species of plant excreting them: alnuston from *Alnus*, coleon from *Coleus*, etc. One example of such phenolic compounds is gossypol, which is contained in pigmented glands of cotton. The toxicity of the palnt is determined due to the presence of this substance. Gossypol is the dimer which is formed from hemigossypol with the participation of peroxidase (Bell et al. 1987). Lysigenous pigmented glands are located on leaves and stems. The toxicity of cotton buds and seeds is also connected with the presence of gossypol in such organs.

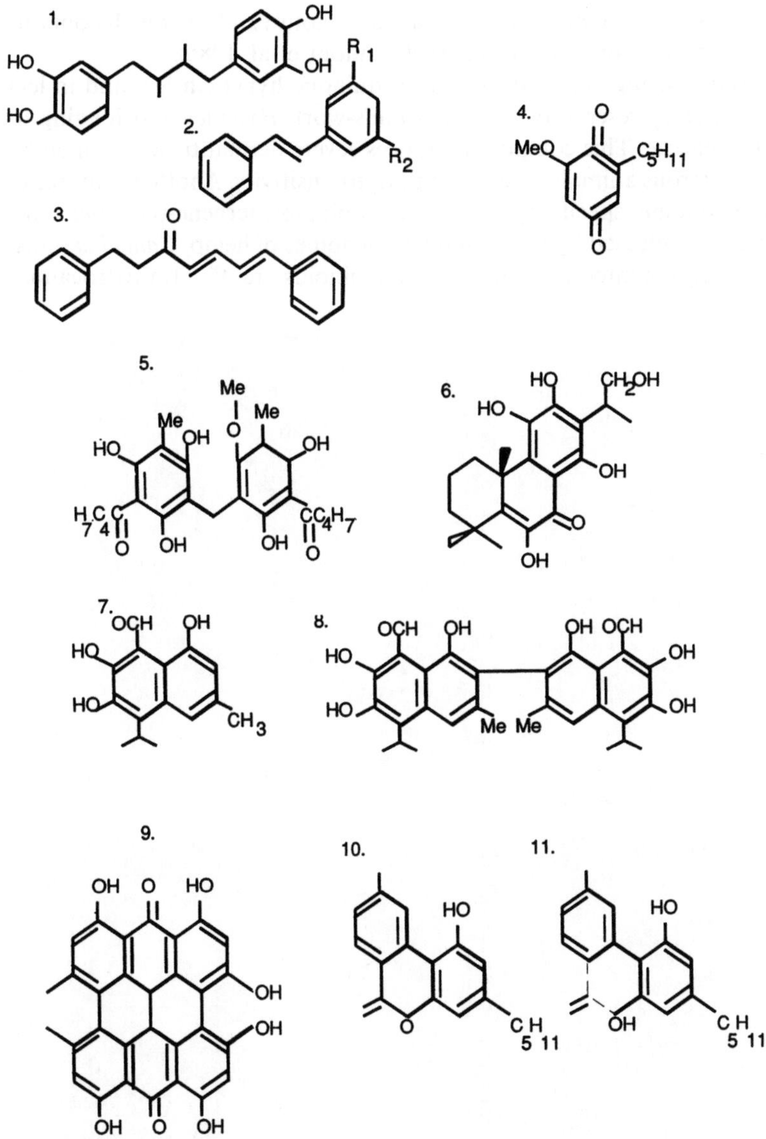

Fig. 61, 1–11. Taxon-specific phenolic substances and their derivatives in glandular hairs. **1** Nordihydroquariaretic acid from *Larrea*. **2** Stilbenes from *Alnus* ($R_1=H$, $R_2=H$; $R_1=OH$, $R_2=OH$; $R_1=OCH_3$, $R_2=OH$; $R_1=OCH_3$, $R_2=OCH_3$). **3** Alnuston from *Alnus*. **4** Primin from *Primula*. **5** Margaspidin from *Dryopteris*. **6** Coleon from *Coleus*. **7** Hemigossypol from *Gossypium*. **8** Gossypol from *Gossypium*. **9** Hypericin from *Hypericum*. **10** Δ^9-Tetrahydrocannabinol from *Cannabis*. **11** Cannabidiol from *Cannabis*

Secretory tissues of the genus *Aloe* eliminate a copious exudate, which contains anthraquinone and anthron-C-glycoside(Beaumont et al. 1985).

A photosensibilizing derivative of anthroquinone hypericin is found in leaf glands and secretory reservoirs of Saint-John's-wort, *Hypericum*, belonging to the family Clusiaceae. This compound induces severe dermatitis when eaten by domestic herbivorous animals, due to their photosensitivity. Another example of high chemotaxonomic specificity are the cannabinoids, terpene-phenolic compounds which are released by the glandular trichomes of hemp, *Cannabis sativa* (Fig. 62). The major components among cannabinoids are Δ^9-tetrahydrocanna-

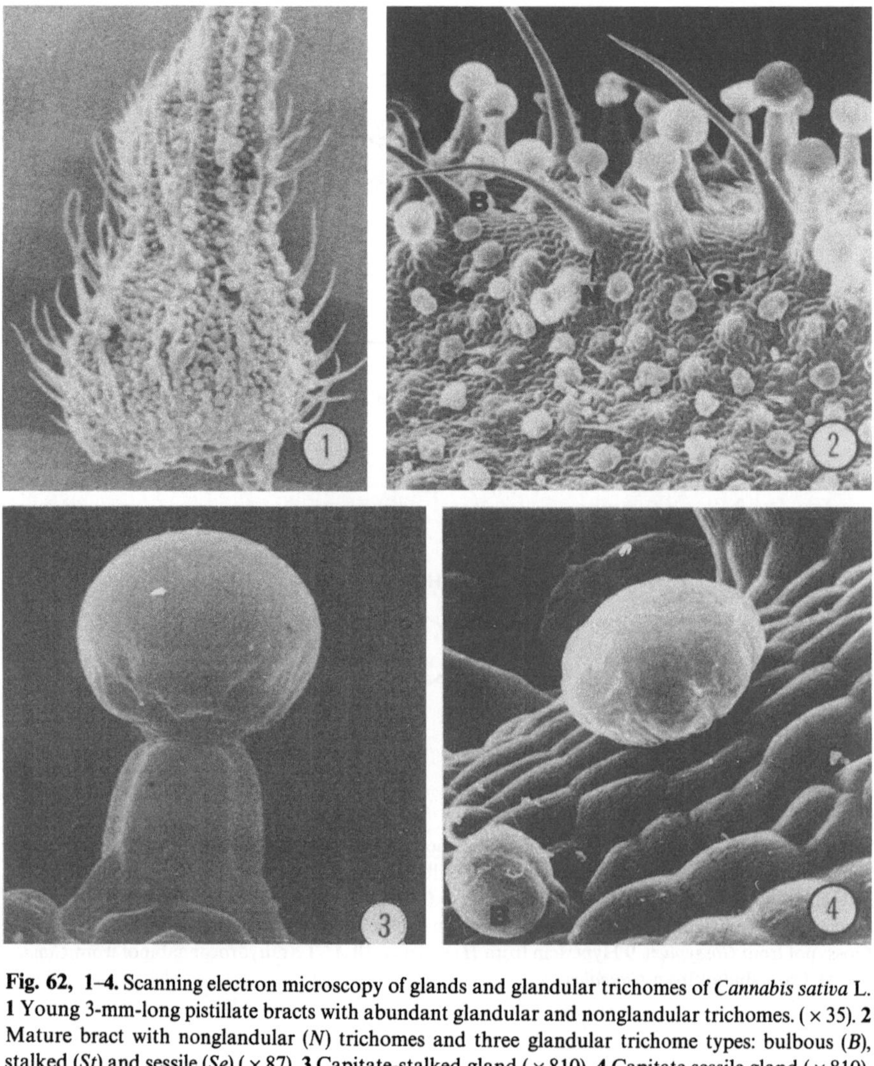

Fig. 62, 1–4. Scanning electron microscopy of glands and glandular trichomes of *Cannabis sativa* L. **1** Young 3-mm-long pistillate bracts with abundant glandular and nonglandular trichomes. (× 35). **2** Mature bract with nonglandular (*N*) trichomes and three glandular trichome types: bulbous (*B*), stalked (*St*) and sessile (*Se*) (× 87). **3** Capitate-stalked gland (× 810). **4** Capitate sessile gland (× 810). (After Mahlberg et al. 1984)

binol and cannabidiol (Fig. 61). the amount of the substances in hemp glands varies with age, dropping from 57 ng/gland for mature glands up to 35 for old and 9 for deciduous organs (Mahlberg et al. 1984). It is possible that the phenomenon also depends on the type of gland. If a stalked gland is located on bracts, the main quantity of the major cannabinoid, Δ^9-tetrahydrocannabinol, is concentrated in it, mainly near the vein. Sessile leaf glands are practically lacking in this component. There is also a difference between stalked and sessile glands. In the former, which are characteristic for bracts, the compound concentration is seven fold higher than in the latter. Plastids are though to participate in the biosynthesis of cannabinoids (Mahlberg et al. 1984). Cannabinoids can act as a drug and Δ^9-tetrahydrocannabinol is more toxic than cannabidiol; the former is also toxic for most of the insects feeding on it.

In glandular hairs, many compounds containing one or more phenolic groups are synthesized. However, all of them, except flavonoids, are formed from common precursors, phenylalanine or shikimic acid. Biosynthesis of flavonoids is unique because their two aromatic rings are formed in different pathways (Fig. 63). Ring A is synthesized from acetate, whereas the precursor for the phenylpropane ring B is phenylalanine, an amino acid which is used for the biosynthesis of protein. The orgin of phenylalanine for the biosynthesis of flavonoids can be by two processes: primary, via the shikimate pathway and secondary, via catabolitic proteolyis, which liberates the necessary amino acid (Margna 1980). When phenols are formed in photosynthesis of chloroplasts,

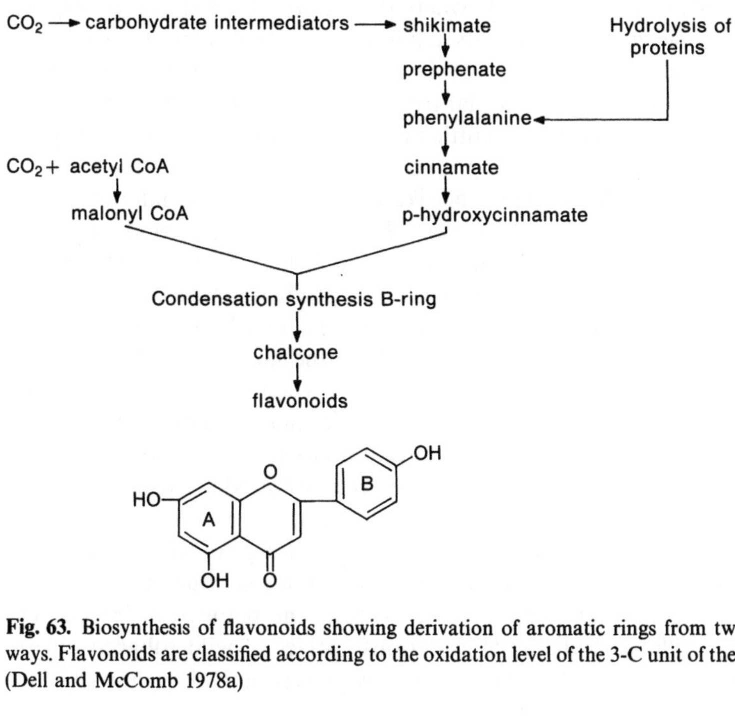

Fig. 63. Biosynthesis of flavonoids showing derivation of aromatic rings from two distinct pathways. Flavonoids are classified according to the oxidation level of the 3-C unit of the flavane nucleus (Dell and McComb 1978a)

their major precursor is obviously phenylalanine, originally synthesizing in the shikimate pathway. However, the main part of the flavonoids and other phenylpropanoids originate from phenylalanine, which is liberated as a result of catabolic proteolysis. The biosynthesis of flavonoids and other phenylpropanoids may be one of the mechanisms of elimination of nonutilized residues of deaminated phenylalanine. This role is thought to be the main one, while another, secondary, one may deal with the production of specific substances possessing physiological activity.

Phenolic compounds being liberated on trichome damage can play a role as deterrents and ensure the immunity or plants to bacterial or fungal infection (Harborne 1988). Excretory compounds are thought to act as allelopathic agents. They can be autotoxic, or can influence neighbor plants (Grodzinskii 1965). Catechols, rutin, and caffeic and chlorogenic acids found in the excreta of glandular trichomes of some plants, such as *Lycopericon esculentum*, are protective compounds against insects (Duffey 1986).

3.9 Secretion of Alkaloids

Electron microscopic studies show that alkaloids are located in both actively growing tissues (meristems, initial cells of phloem and xylem) and in epidermal cells of leaves and stems (Neumann 1985). Alkaloids are usually accumulated in the cell sap of plants (see Chap. 1.4), where they form salts with certain organic acids such as tropic, atropic, and tiglic acids, or are esterificated by alcohols to tropine, nortropine, and scopine (Levitt and Lovett 1985). In some cases, alkaloids are concentrated in glandular trichomes and are released from them to the environment. However, there is little information on this. Only 5% of plant species have been analyzed for the presence of alkaloids.

Data dealing with the secretion of alkaloids by glandular hairs of potato, *Solanum tuberosum*, tobacco, *Nicotiana tabacum*, and wormwood, *Artemisia absinthium*, are summarized in Levitt and Lovett (1985), Neumann (1985), and Zador and Jones (1986). This process was better studied for *Nicotiana tabacum* and *N. stocktonii* leaves, where nicotine, nornicotine, and N-hydroxyacylnornicotine are found (Zador and Jones 1986). Nicotine and nornicotine, but not N-hydroxyacylnornicotine, were present inside the trichomes as well as in other internal parts of the plant. Using 2-^{14}C-labeled nicotine and nornicotine, Zador and Jones (1986) showed that nicotine is produced by the roots and leaves of the plant, then converted in the leaf to nornicotine, followed by trichome transformation of nornicotine to N-hydroxyacylnornicotine, which is rapidly excreted. In some species of tobacco, the enzyme N-acyltransferase, which is located in leaf trichomes and involved to acelate of nornicotine has been found (Huesing et al. 1989). Secreting alkaloids protects the plant from insects. The composition of defensive alkaloids in various species of wild tobacco differes from in the cultured line. N-acylnornicotine is present in leaf trichomes of the wild species of tobacco, *Nicotiana repanda*, *N. stocktonii*, and *N. mesophila*, whereas it lacking or exists only in trace amounts in cultured species (Huesing et al. 1989). The presence of alkaloids is observed in trichomes of other plants. There are gold-yellow glands containing alkaloid lupinine in the sheathing scales, bracts, and perianth of hop, *Humulus lupulus*, (family Cannabiaceae) (Yatsenko-Khmelevskii 1980). The alkaloids hordenine and scopolamine, as well as α-glucosinolate, a derivative of benzylamine (Fig. 64), are found in tomato trichomes (Lovett 1987). Steroid alkaloids of glandular hairs of some species, belonging to the genus *Solanum*, exist in a mixture with essential oils (Fahn 1979). Similar data were given by Nikitin and Pankova (1982) for the alkaloid capsaicin, which is characteristic for the pungent taste of red pepper, *Capsicum annuum* and is located mainly in the epidermal glandular structures of the placenta inside the fruits. The release of the alkaloid berberine by cells of suspension tissue culture of low meadow rue, *Thalictrum minus* (Tabata et al. 1987), is an example of nonglandular secretion. The ability to synthesize berberine is taxon-specific, and is an active process. The secretion of berberine depends on temperature and is blocked by vadanate, an inhibitor of protoplasmic membrane-bound ATPase (Tabata et al. 1987).

Synthesis of many alkaloids occurs both in the chloroplasts (Wink 1985) and the cytoplasm (Deus-Neuman and Zenk 1984). Indole alkaloids are formed in the chloroplasts. For example, some enzymes of vindoline synthesis are thought to be located in chloroplasts of the leaves of *Catharanthus roseus* and to be bound with thylakoids (Luca de and Cutler 1987). A common mechanism of alkaloid formation is the reaction of Mannich, when the interaction of amine

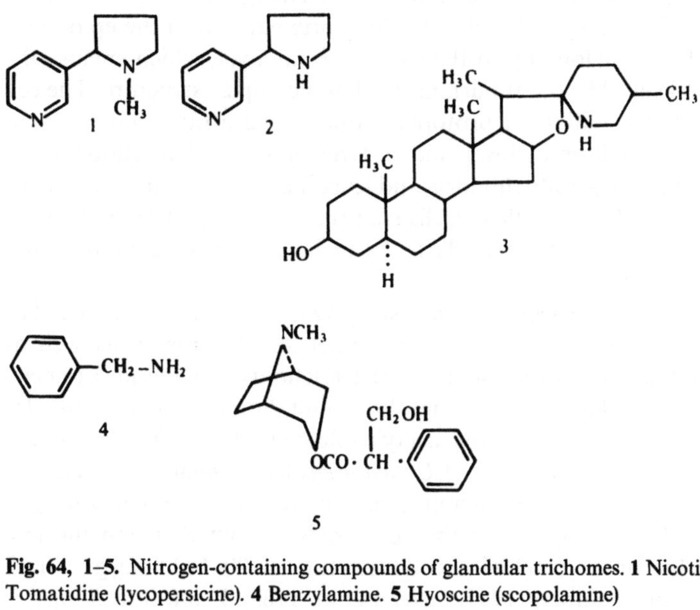

Fig. 64, 1–5. Nitrogen-containing compounds of glandular trichomes. **1** Nicotine. **2** Nornicotine. **3** Tomatidine (lycopersicine). **4** Benzylamine. **5** Hyoscine (scopolamine)

with aldehyde and the carbo-ion takes places, as a result of which the Schiff base forms and is then transformed to an alkaloid (Müntz 1984). The precursors of many amines are amino acids such as lysine in the synthesis of the alkaloid lupinine or asparagine in the formation of nicotine, etc.

The excretion of alkaloids appears to be one form of evacuation of toxic nitrogen-containing metabolites when these are in excess. Alkaloids are also known as repellents and attractants. The excretion of alkaloids plays an important defensive role when an insect attacks the plant. Many plants from the family Solanaceae secrete substances with insecticide activity via the trichomes, such as N-hydroxyacylnornicotine in tobacco (see above) and tomatine, the steroid alkaloid from *Lycopersicon esculentum* (Duffey 1986).

3.10 Secretion of Acetylcholine and Amines by Stinging Trichomes

Amines can be excreted both by special stinging hairs and other glandular structures. Stinging hairs are met in four families of dicotyledons: Urticaceae, Euphorbiaceae, Loasaceae, and Hydrophyllaceae (Levin 1973). In their structure, they are emergences because in their formation not only epidermal cells participate but also more deeply located ones.

Stinging emergences are described in some publications (Esau 1965; Thurnston 1974; Fahn 1979; Akhalkatsi and Vasilyev 1984; Buvat 1989). In stinging nettle, *Urtica dioica*, 2-mm-long stinging hairs cover all the above ground surface of plant, the axis of inflorescence, and the abaxial side of the perianth. Nearly 100 stinging hairs are in 1 mg biomass. The emergences (Fig. 65) consist of the large bottle-like stinging basal cell, which originates from the epidermis, and a number of cells, belonging to the subepidermal layer, which enclose the lower part of basal cell. The basal burning cell has a complex structure. The cell wall is inlaid by silicate bodies, the protoplast contains a central vacuole crossed by a number of cytoplasmic flows and a large nucleus. The endoplasmic reticulum and a number of mitochondria and leucoplasts predominate in such a cell (Akhalkhatsi and Vasilyev 1984). Cells near the base of the stinging hair are rich in cytoplasm and contain chloroplasts. As a whole, their structure is similar to that of other secreting cells.

The composition of the secretion in a stinging cell is rather complex. The biogenic amines serotonin (5-hydroxytryptamine) and histamine, as well as trimethyl-substituted amine or cholinic ester acetylcholine, are found (Emmelin and Feldberg 1947; Collier and Chesher 1956, etc). Some stinging cells also contain formic acid. For the first time, acetylcholine and histamine were found in stinging hairs of leaves and stems of *Urtica urens* by Emmelin and Feldberg (1947). the concentration of acetylcholine measured was 0.018–0.027 $mg\,g^{-1}$ fresh weight, which is as high as in the nerve cells of animals (Emmelin and Feldberg 1947); histamine reached a concentration of 20–143.5 $mg\,g^{-1}$ fresh weight (Emmelin and Feldberg 1947; Saxena et al. 1965). From 3 to 5 $mg\,g^{-1}$

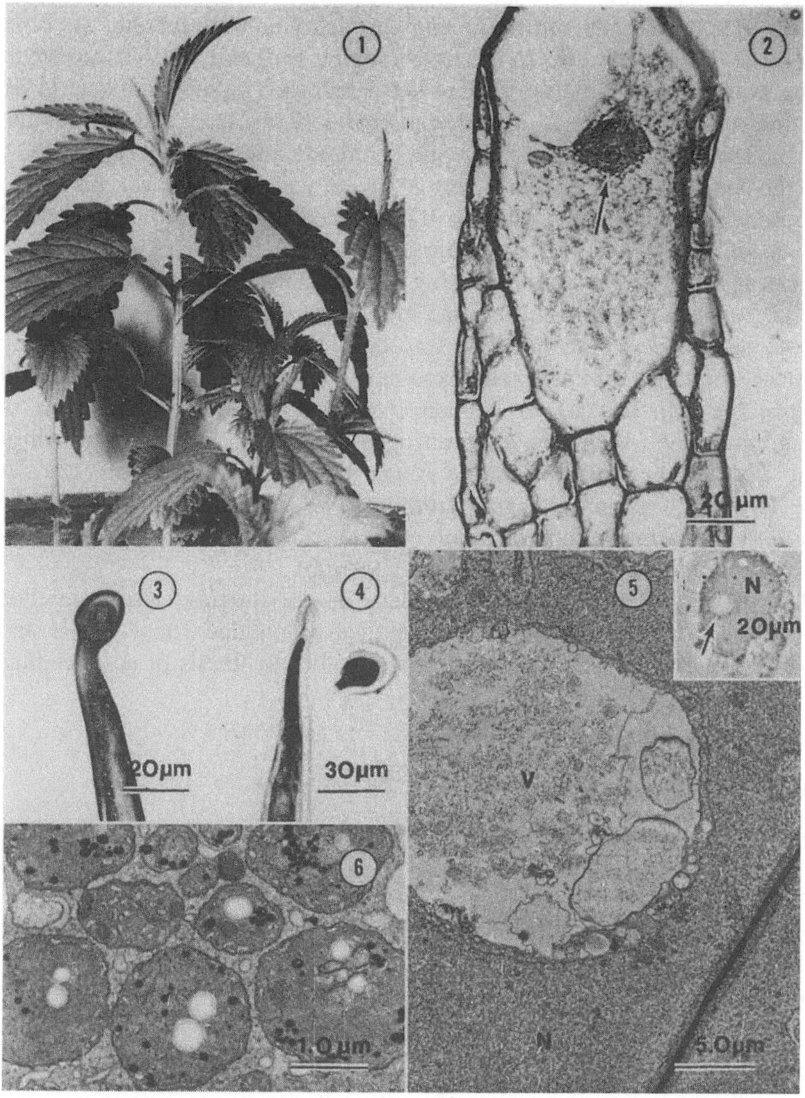

Fig. 65, 1–5. Structure of stinging cells of *Urtica dioica* **1** A mature plant of *Urtica dioica.* **2** The basal region of a stinging cell. Note the large nucleus (*arrow*) and absence of pits in the stinging-cell wall. **3** Whole mount of the intact stinging-cell tip. **4** Whole mount of a dislodged stinging-cell tip. **5** Thin section through a stinging-cell nucleus (*N*) containing a large vacuole (*V*) which is extranuclear in origin. The *dark diagonal* line in the micrograph is a fold in the section. *Inset* light micrograph of a vacuolate (*arrow*) stinging-cell nucleus. **6** Thin section through a stinging-cell chloroplast with a few lamellae, starch grains, and numerous osmophilic bodies. (After Thurston 1974)

fresh weight of serotonin was found in stinging hairs of *Urtica dioica* (Collier and Chesher 1956). Histamine, serotonin, and acetylcholine were peculiar not only to the genus *Urtica*, but also to others belonging to the family Urticaceae: in stinging hairs of the Himalayan plant *Girardinia heterophylla* (Saxena et al. 1966), the Australian species *Laportea moroides* (Robertson and Macfarlane 1957), and the American plant *Cnidoscolus texanus* (Lookado and Pollard 1991), where the concentration reaches 0.2 mg g^{-1} fresh weight. Moreover, histamine has a rather high concentration (up to 1.25 μg/needle) and serotonin (620–1250 μg g^{-1} fresh weight) are present in burning hairs of the leaves of the South American plant *Jatropha urens* belonging to the family Euphorbiaceae (Villalobos et al. 1974). Serotonin is also found in hairs of the fruit of the leguminous plant *Mucuna pruriens*. Acetylcholine, serotonin, and histamine induce burning on animal or human skin when the secretion of a stinging hair touches it, and determine the pain reaction on contact with leaves of these plants. Thus, in burning emergences there are three nitrogen-containing substances which in animal cells play the role of mediators, i.e., participate in the excitation transfer in the form of a chemical signal from cell to cell. More details on the concentration and mechanism of action of these compounds in plants are given in the monograph and review Roshchina (1991a,b).

The synthesis of the secreta takes place in the stinging cells themselves (Fig. 66). the synthesized components are then accumulated in vacuoles and cavities of the endoplasmic reticulum (Fahn 1979). Synthesis of acetylcholine

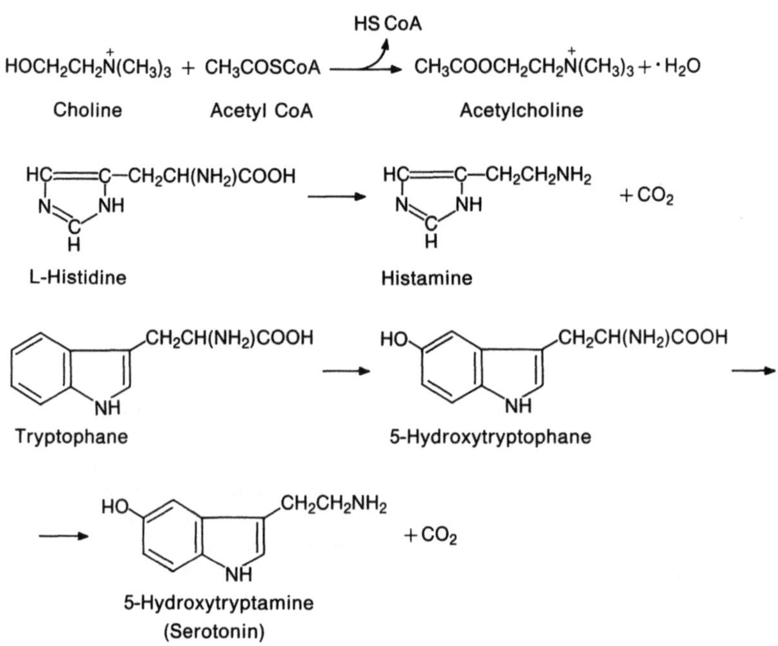

Fig. 66. Scheme of biosynthesis of acetylcholine and biogenic amines found in stinging hairs

(Fig. 66) is carried out with participation of enzyme choline acetyltransferase (EC 2.3.1.6) (Smallman and Maneckjee 1981) from such precursors as acetyl-CoA and choline. Histamine is produced by way of decarboxylation of the amino acid histidine by histidine decarboxylase (EC 4.1.1.22). Synthesis of serotonin includes two steps: tryptophan hydroxylation with the formation of 5-hydroxytryptophan and the following decarboxylation of this product with the participation of such enzymes as tryptophan hydroxylase (EC 1.14.16.4) and tryptophan decarboxylase (EC 4.1.1.28).

The formation of amines throughout the dissimilation of amino acids is a process which is peculiar not only to stinging emergences. It plays an important role in nitrogen exchange of the plant organism. Amines are found in the composition of many plants such as henbane, *Hyoscyamus*, belladonna, *Atropa belladonna*, stramony, *Datura stramonium*, tomato, *Solanum lycopersicum*, and spinach *Spinacia oleracea* (Harborne 1988). the biogenic amine serotonin is identified in embryons of *Juglans nigra*, where it amounts to 180 $mg\,g^{-1}$ fresh weight.

Amines are not the final products of metabolism as is claimed for stinging hairs. Many of them, located in nonsecretory tissues, can be involved in metabolism again and used as building material in the synthesis of various nitrogen-containing compounds. They are used specially easily for alkaloid biosynthesis.

The physiological significance of stinging hairs is though to be in frightening off herbivorous animals by intracutaneous injection of amines. The mechanism of this action depends on the structure of the stinging cell (see Fig. 67) (Fahn 1979). For example, a stinging hair of *Urtica dioica* has a thickened top which is easily removed on contact with an object, and opens the burning end. Sharp brims which stay in place after the decapitation easily puncture the skin, and the content of the hairs penetrates the wound. The apical part of the stinging trichome of *Dalechampia roezliana* has a crystal, which on contact with an animal emerges from the stinging cell and punctures the skin. Toxins of such a hair are injected into the animal tissue and induce a severe burning. In tropical

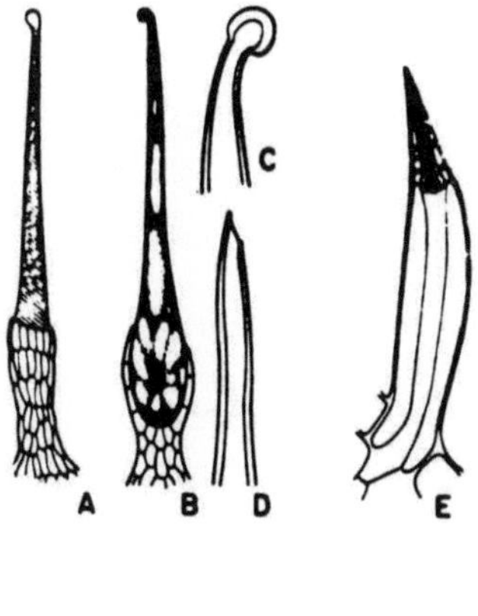

Fig. 67A–E. Stinging trichomes **A–D** *Urtica dioica*. **A** as seen in the light microscope with the focus on the surface of the trichome. **B** as in **A** but with focus on center of the trichome. **C** Intact tip. **D** Upper part of stinging cell after the bulbous tip has broken off. **E** Longitudinal section of stinging trichome of *Dalechampia roezliana*. Crystal marked by *solid black*. (Fahn 1979)

plants belonging to the family Urticaceae, especially such as *Laportea urentissima*, the stinging action of these hairs is very painful and can even lead to syncopes (Robertson and Macfarlane 1957). The protective role of stinging hairs against herbivorous mammals was established in experiments of Pollard and Briggs (1984).

CHAPTER 4

Gas Excretion

4.1 The Pathways of Gas Release

Highly inflamed organic substances released by glandular hairs are eliminated on the leaf surface by evaporation. The volatile compounds are emitted from the inner tissues in two ways, through either the stomata or the cuticle of the epidermis (Fig. 68). Using plasticine as a lubricant Sanadze (1961a) demonstrated that a release of gases occurs mainly via the stomata. Barmore and Briggs (1972) have confirmed this, showing that ethylene diffusion from sweet orange, *Citrus sinensis*, occurred through the lower epidermis of the leaf where stomata are located. Inorganic gases penetrate mainly via stomata, whereas the other pathway, through the cuticle, is less significant. E. Rabinowitch (1953) reported that the rate of carbon dioxide moving through the cuticle is nearly 6×10^{-8} mol cm^{-2} h^{-1}. This gas may penetrate through the stomata more quickly (approximately 5×10^{-6} mol cm^{-2} h^{-1}). Release of volatile substances via the stomata is thought to occur by free diffusion (Ting and Loomis 1965).

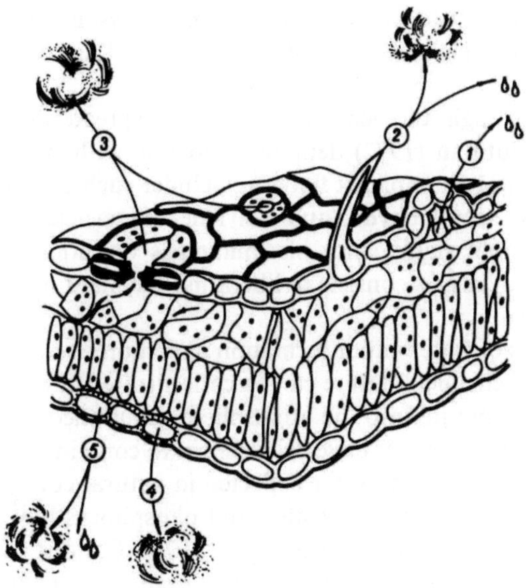

Fig. 68. Main pathways of the substances released from plant *1* Hydathodes (water, salts); *2* glandular hairs and glands (volatile and non-volatile organic substances are released during evaporation and leaching); *3* stomata (gaseous compounds); *4* basic cells of epidermis, cuticle (gaseous compounds); *5* leaching and evaporation from basic cells of epidermis through parts of the cell walls. (Roshchina 1974c)

The processes of adsorption appear to play the most important role in the phenomenon. According to Gorbavtsov (1970), the leaf is a capillary-porous system where the translocation of the substances takes place not only within the gas phase but mainly in the layers adsorbing gases on the surface of the stomata pores. The flow of volatiles through the stomata is described by the author as the sum of the flows in both the gaseous and adsorbing phases. It may be essential that in gas exchange of the leaf with the environment, volatile organic compounds, which are adsorbed on the cuticular surface, can exert an influence at the cuticle hydrophobia level and thus also on their own movement in the adsorbing layers of the cuticle (Gorbavtsov 1970).

The steams of essential oils greatly influence the release of internal gases (Polova and Vicherkova 1986). The effect depends on the vapor concentration and the origin of the oil. For instance, the following sequence, of increasing activity of essential oils has been demonstrated in *Phaseolus vulgaris* as eucalyptus, lavandin, caryophyllus, caraway, and peppermint.

Depending on the concentration, essential oils stimulate or depress transpiration. One of the mechanisms of transpiration depression on higher concentrations of essential oils was thought to be the closure of the stomata (Polova and Vicherkova 1986). Although gas movement through the stomatal pore is basic for high lipophilic substances, penetration via a cuticle is also important. Since the cuticle of the upper side of a leaf is thicker than the lower, the latter is more permeable for gases. Recently, it has been established that, due to the existence of a number of micropores and channels in the leaf cuticle, penetration by metabolites is promoted. The pores and channels of the cuticle in leaves of 27 species belonging to the families Agavaceae, Amaryllidaceae Bromeliaceae, Caryophyllaceae, Ericaceae, Liliaceae, Myrtaceae, Palmae, and Pinaceae, and in stems of *Euphorbia antisyphilitica* were characterized. Absence of pores and channels was observed for leaves of *Dudleya* sp., *Crassula* sp. (Crassulaceae), *Brassica oleracea* (Brassicaceae), *Aloe barbadensis*, *Tulipa gesneriana* (Liliaceae), and *Ficus elastica* (Moraceae).

Gas emission also occurs through the lenticels which are spread over lignified shoots. Chirkova and Gutman (1972) demonstrated this with plant roots under anaerobic conditions (for example, at flooding). Under such conditions, steams vapors of ethanol, acetaldehyde, and ethylene are evacuated through the lenticels. The functional significance of similar gas excretion is considered as an adaptation of anoxia and consists in liberating the plant from the toxic products of metabolism.

Since the cell walls are the sites of water vapor formation and, moreover, on their surface substances of various chemical nature can be concentrated, it may be presumed that the compounds migrate into the atmosphere simultaneously with water vapor. The emitted steams of water have been shown to contain ion-salt particles. In condensates of transpiration water collected in natural conditions from woody plants, traces of ammonium, nitrates, and phosphorus have been found (Gubareva 1962; Novitskaya 1966; Nemeryuk 1970). Cations of Ca^{2+}, Mg^{2+}, NH_4^+, K^+, Na^+, and anions Cl^-, SO_4^{2-}, HCO_3^- were observed in the transpiration water of some agricultural plants (Nemeryuk 1970). The total ion amount in the transpiration condensates collected in the day time varied from 7.93 to 19.72 mg l^{-1} in both plant species and of their habitat. The latter is illustrated by data of Kazarov and Plieva (1989). The stoichiometry between the concentrations of ions emitted with water vapor in a plant of the semiarid zone was: $Na^+ > Mg^{2+} > Ca^{2+}$; K^+ and $HCO_3^- > Cl^- > SO_4^{2-}$, in plants of forest-steppe and subalpine habitats $Na^+ > Ca^{2+} > Mg^{2+} > K^+$. Such trees as *Picea abies* emit volatile sulfur-containing compounds, independently of the presence of sulfates in the soils in both light and dark. H_2S is released mainly with water vapor in transpiration at approximately 1 nmol/per 1 mol of water (Rennenberg et al. 1990). When the stomata of the needle are closed, hydrogen sulfide is excreted for only 1 min and then the process ceases. The main flow of H_2S thus occurs through the stomata. Sometimes, additionally, SO_2 is released and its quantity correlates with the content of sulfate in the soil. Sulfur dioxide is emitted for 2 h at 4.1–10.3 nmol m^{-2} on the leaf area $\times 2.65$. It is thought (Rennenberg et al. 1990) that with a normal (nonenhanced) amount of sulfur in the soil, this process does not suffice for elimination of sulfur excess. For the origin of sulfur volatiles, H_2S can be released in light-dependent processes of sulfate reduction and the desulfohydration of cysteine. In the light, rates of H_2S-release are higher than in the dark. SO_2 may play a role in weak acidation of the cell compartment.

Organic compounds such as sugars and amino acids were found by Novitskaya (1966) in transpiration condensates. Using gas chromatography, more volatile components were observed in the transpiration water by Roshchina (1971). The transpiration condensates collected from leaves of 25-year-old trees of *Acer negundo, Cotinus coggygria, Betula verrucosa, and Populus balsamifera* contained traces of butylene, acetaldehyde, propanal, methanol, ethanol, and propanol (Roshchina 1971). An example of gas chromatographic analysis of transpiration water from woody plants is shown in Fig. 69. The qualitative composition of the samples differed only in minor components, among which methanol was identified. The main differences in composition were observed for butylene, acetaldehyde, and propionic aldehyde, the amount of which was estimated by the height of the chromatographic peaks. The author succeeded in finding only part of the releasing substances for 11 min, followed by a prolonged peak of water, masking many other volatiles (Roshchina 1971). Ethanol was found in the transpiration water of alfalfa, *Medicago sativa*, and crowfoot, *Lotus corniculatus*, grown in the greenhouse and flooded. Twenty to 45 μmol ethanol was accumulated for 24 h in alfalfa, while crowfoot showed only traces of the substance (Barta 1984). In both cases, the transpiration water included products of anaerobic decomposition which appear on poor aeration in experiments with woody plants and on flooding of the root systems of species belonging to the family Fabaceae. Thus, water vapors and volatile organic substances contained in the intercellular spaces are released simultaneously. However, the composition of the steams of various chemicals within plant tissues may differ in quality and quantity from those which are emitted via transpiration. The proportion

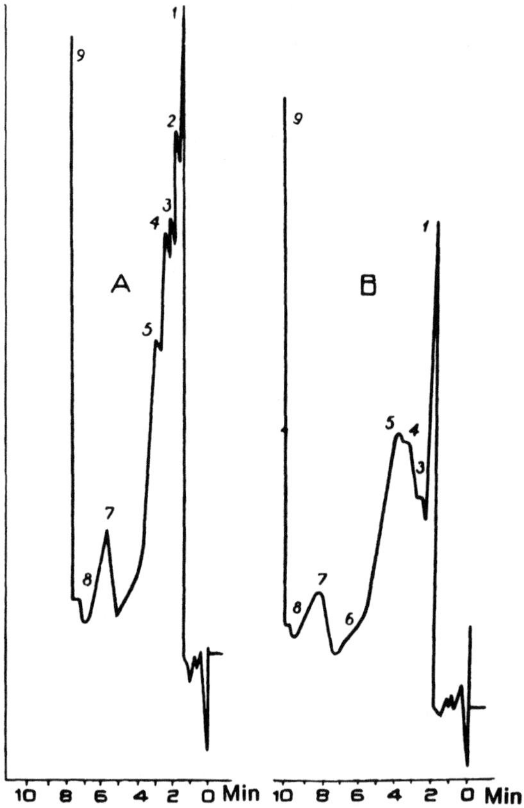

Fig. 69A, B. Gas chromatograms of transpiration water from *Cotinus coggygria* (*A*) and *Betula verrucosa* (*B*): *1* Atmospheric gases; *2* unknown substance; *3* butylene; *4* acetaldehyde; *5* propionic aldehyde; *6* methanol; *7* ethanol; *8* propanol; *9* front of water. (Roshchina 1971)

between them is determined by their different volatility and absorption by cell walls which enter the surface of intercellular spaces. The main pathways of excretion of gaseous and water-soluble compounds are shown in Fig. 68, as well as being described in Chapter 5.

4.2 Volatile Excretions as Complexes of Substances

The composition of volatile plant excretions is very complicated. Each individual component is present in small or trifling quantity, for which reason their observation and identification become possible only due to use of gas- and gas-liquid chromatography and other highly sensitive methods of separation and analysis of organic compounds. By these methods, it was shown that food products, including vegetables and fruits, release up to 200 volatile substances (Mokhnachev and Kuz'min 1966). Among them, 30 volatiles have been identified in the excretion of fresh cabbage, *Brassica oleracea*, 60 in strawberry *Fragaria vesca*, 39 in raspberry, *Rubus idaeus*, etc. Great variability in the terpenes contained in the volatile emissions is characteristic for conifers. Stepanov (1972) found 64 individual compounds in the excretion of conifers in Salair (Sibiria), 60 of which were terpenes (Fig. 70). Isidorov and Zenkevich (1985) analyzed a quantitative composition of volatile organic substances from 22 plant species which are characteristic for forests of the Northern hemisphere. This list included more than 70 compounds of various classes. The total concentration of terpenes in the air of coniferous forests changed from 3.5 to 35 $\mu g\,m^{-3}$. Of great interest are chromatography-mass-spectroscopic analyses of the volatiles released by some plants in the South Crimea (Dmitriev et al. 1988) because the data are used for the selection of plants assortment for sanatorium parks. It has been shown that paraffin, olefin and diene hydrocarbons, terpenes, saturated and unsaturated alcohols, aldehydes and ketones, esters, and chlorine-containing and other compounds are present in volatile excretions of eight species (Table 9).

A release of volatile metabolites was observed in root systems as well. Roots of *Daucus carota* and *Solanum tuberosum* grown by the hydroponic method eliminated organic compounds whose quantity was two fold less than in excreta of the leaves of the same plants. The gas composition of roots did not vary greatly in its components (Dadykin et al. 1967). Only acetaldehyde, propionic aldehyde, and propanol have been identified. The ability of root systems to excrete gas was confirmed in experiments with roots of *Acacia pulchella* which released the main volatile components identified as 2-3-methylbutanol, hexanol, pentanal, 2-3-methylbutanal, and 4-methylacetophenone (Whitefield et al. 1981).

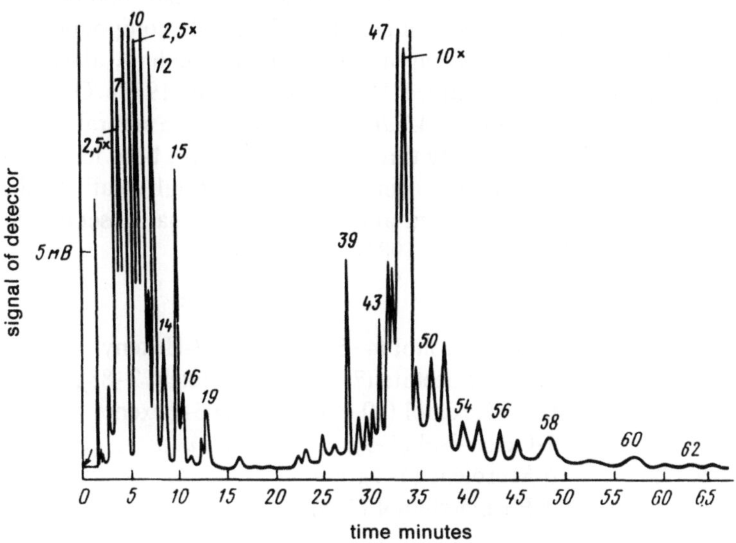

Fig. 70. Gas chromatogram of volatile substances of needle of *Abies sibirica*. *5* Tricyclene; *6* santene; *7* α-pinene; *10* camphene; *11* β-pinene; *12* Δ^3-carene; *14* myrcene; *15* limonene; *19* terpinolene; *39* caryophyllene; *41* γ-murolene; *43* α-murolene; *44–45* humulene + bisabolene; *46* cadinene; *47* bornyl acetate; *50* camphor; *52–53* α-terpineol + borneol. (Stepanov 1972)

Table 9. Major compounds in volatile excretions of some plants. (After Dmitriev et al. 1988)

Plant	Compounds
Laurus nobilis L.	Camphene; Δ^3-carene; 1,8-cineole; limonene; myrcene; α-phellandrene; β-pinene; sabinene
Lavandula angustifolia L.	Camphene; limonene; myrcene; ocimene; β-phellandrene; α-pinene
Mentha piperita L.	Limonene; menthol; 1-menthone; d-menthone; l-menthyl acetate; α-pinene
Monarda fistulosa L.	n-Cymol; myrcene; α-pinene; γ-terpinene; α-tujene
Platycladus orientalis L.	Δ^3-Carene; limonene; myrcene; α-pinene; sabinene; α-tujene
Rosa gallica L.	3-hexenyl acetate; myrcene; α-phenylethanol; β-phenylethyl acetate; α-pinene; sabinene
Rosmarinus officinalis L.	Camphor; camphene; Δ^3-carene; 1,8-cineole; limonene; linalyl acetate; α-pinene; β-pinene
Salvia sclarea L.	n-Cymol; geranyl isobutyrate; limonene; linalool; linalyl acetate; myrcene; ocimene

Fountain and Couchman (1984) showed that the composition of volatile compounds of mature fruits includes a significant amount of various metabolites, part of which form the characteristic fruit smell or aroma. Thus. fruits of *Corynocarpus laevigatus* emitted acetaldehyde, ethanol, methyl acetate, ethyl acetate, methyl-2-methyl butanoate, ethyl-2-methyl butanoate, and traces of ethylene. According to Fountain and Couchman (1984), the smell of fruits depends on the ethyl ester of butyric acid which was the main component. Flowers have great variability in the composition of volatile compounds. In volatile excreta of *Glycine soja*, 50 compounds including alcohols, aldehyde, esters, phenols, etc. have been identified (Nunomura et al. 1976). Gaseous emissions of flowers and pods of alfalfa, *Medicago sativa*, contain several tens of components (Buttery et al. 1982). Thirty three compounds have been identified in flowers, and 31 in pods. Qualitatively (in relative % of total volatiles), the components in blossoms and pods varied and differed from that observed for leaves of *Medicago sativa* (Buttery et al. 1982):

Flowers	Pods	Leaves
Methylbutanol (4–10); (E)-β-Ocimene (Z)-3-Hexenyl acetate (8.1); Decanyl acetate (7)	Benzyl alcohol (4–6); Hexenyl acetate (7–22); (Z)-3-Hexenol (10–40); 1-Octen-3-ol (3–9); δ-Murolene (4–7); 2-Phenylethanol (3–8)	(Z)-3-Hexenyl acetate (72–85); (Z)-3-Hexenol (6–8).

For 24 h, individual flowers of *Odontoglossum constrictum*, *Citrus medica*, *Hoya carnosa*, and *Stephanotis floribunda* released approximately 45–50 μg total volatiles, of which 20 components have been identified (Matile and Altenburger 1988). The major compounds forming the characteristic aroma were methyl benzoate, linalool, and 1-nitro-2-phenylethane for *Stephanotis*, linalool, 1,8-cineole, β-pinene, isopentanol, α-pinene, and methyl salicylate for *Hoya*. Some oscillation in compound release was observed, i.e., a maximum for *Odontoglossum* in the day time, and for *Hoya* in the evening. It is noted that the composition of volatile oils of onion, , *Allium cepa*, was unusual (Kameoka et al. 1984). Its composition included 82 components, among which besides ordinary compounds, such as alcohols, aldehydes and ketones, sulfides, thiolanes, furanones, etc. were also identified. Twenty three–30% of the total amount of sulfur-containing components was dipropyl disulfide. Dobson et al. (1987) studied volatile compounds of flowers and pollen of *Rosa rugosa* and pollen of *Rosa canina*. There are differences between the flower and pollen of *Rosa rugosa* in the composition of the main aliphatic hydrocarbons, terpenoids, and aromatic substances (in relative % of total volatiles):

Flower (> 20%)	Pollen (> 20%)
Citronellol	Geranyl acetone
Nerol	Geranial
Geraniol	2-Undecanone
Geranyl acetate	2-Tridecanone
Benzyl alcohol	Tetradecanol
β-Phenylethanol	Tetradecyl acetate
Methyleugenol	β-phenylethanol
	Methyleugenol

The main part of the volatile components was emitted from the outer oil-bearing envelope of pollen, there was also a difference between pollen *Rosa rugosa* and *Rosa canina*. Only a few common components (6-methyl-5-hepten-2-one, geranyl acetone, pentadecane, and tetradecanol) were found in pollen of both species. Thus, pollen of various species has a characteristic aroma, differing from the odor of whole flowers which is perceived by both humans and insects. The aroma of pollen is an ancient adaptation to attract insects to visit the flowers. Bees recognize such an aroma and include some elements of it in their products.

Gas excretions of celery, *Apium graveolens* (Umbelliferae), are found to contain nine previously unknown substances (MacLeod et al. 1988). Three of them, represent 70% of the total gaseous excreta: 4,7-dimethoxy-5-prop-2-en or apiol 23% (4.27 $\mu g\, g^{-1}$ fresh weight); 3-butylphtalide 22% (3.9 $\mu g\, g^{-1}$ fresh weight), 3-butyltetrahydrophtalide or sedanolide 24% (4.27 $\mu g\, g^{-1}$ fresh weight). The two latter components have a sharp smell, which is characteristic for celery itself (Fig. 71). All three components were found in leaves of parsley, *Petroselinum*; their presence may be a common feature of the family Umbelliferae.

There is little information on the total amount of volatile compounds emitted by a plant. According to Bryantseva (1951), who estimated the values at 100 g dry leaf matter, the walnut, *Juglans regia*, releases 2.16–1.87 $mg\, h^{-1}$

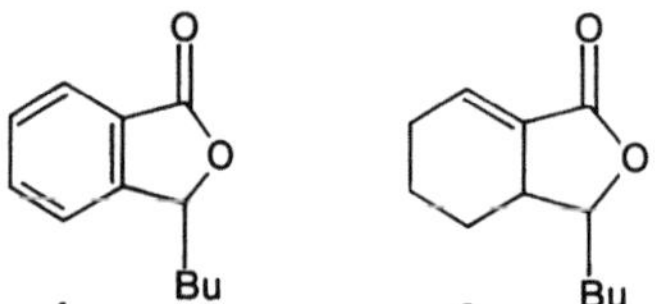

Fig. 71, 1,2. Phtalides in plant excreta. 1 3-Butylphtalide. 2 3-Butyltetrahydrophtalide

organic compounds, Crimean pine, *Pinus pallasiana*, 1.50–1.30, and box, *Buxus* sp, 0.4–0.22. Thus plants emit approximately 0.5–4 g organic compounds per 100 g vegetative dry matter. According to other data (Artem'eva cited by Tokin 1980a), 1 ha of deciduous forest releases 2 kg of volatile organic substances per day (24 h) into the atmosphere, and a coniferous forest 5 kg. Similar information was reported by Protopopov and Chernyaeva (1967). Volatile organic excreta were also found near the crown of woody plants. Kolesnichenko (1976) reported that 1 m^3 of air collected around the crown tops of 40-year-old trees of *Fraxinus* and *Quercus* contains volatile substances equivalent to 13 mg of carbon; air collected under the cover of forest in various regions of Russia has about 10–30 mg. Sage, *Salvia*, from south-western America can reportedly produce volatile organic substances up to 2×10^6 tons per year. On the scale of the biosphere, the values are even greater. Went (1960) reported that the biosphere liberates about 10^8 tons of terpene-like or nonoxidized hydrocarbons into the atmosphere per year.

4.3 Components of Gaseous Excreta

4.3.1 Short-Chain Hydrocarbons

Besides ethylene, which is released by all living plant tissue and is present in the internal air of plants, plant excreta include also other hydrocarbons. Porutskii et al. (1962) found ethylene, propylene, isobutylene, and butylene in excreta of *Linum* and spikes of *Triticum*. Ethylene and propylene were also released by sterilic seeds of cotton, *Gossypium*, pea, *Pisum sativum*, and yellow pine, *Pinus flaveolus* (Vancura and Stotzky 1976). Ethane was identified in the gas phase of tissue culture of carrot, *Daucus carota*, and tobacco, *Nicotiana tabacum* (Thomas and Murashige 1979). Callus of tissue culture of rice, *Oryza*, produced not only CO_2 but ethanol, ethylene, ethane, and also a small amount of acetaldehyde (Adkins et al. 1990). All of these facts indicate the possibility of forming short-chain hydrocarbons by nonspecialized plant cells.

Depending on the plant species, the composition of the released olefins may change significantly. According to Porutskii et al. (1962), in excretions of appletree (*Malus domestica*) leaves, ethylene prevailed, and in pear, *Pyrus communis*, propylene. The species of the family Solanaceae emitted a number of butylene, and plants belonging to the Graminae and Malvaceae mainly ethylene. Luchko et al. (1964) studied the change in composition of unsaturated hydrocarbons produced by fruits of green pea, *Pisum sativum*, at different stages of ripening. The placental period of fruit formation was characterized by propylene release, and seed maturation by ethylene emission.

Intact plants emit low-molecular weight hydrocarbons in so small amounts that it is impossible to study their biological effects. It has been shown, using artificial mixtures, that the physiological activity of individual hydrocarbons belonging to the ethylene range drops with increasing chain length.

4.3.2 Isoprene and Terpenoids

Isoprene (2-methylbutadiene-1,3) has been found in volatile plant excretions by Sanadze (1964, 1969) and Rasmussen (1970). Sanadze et al. (1972), Rasmussen and Jones (1973), Jones and Rasmussen (1975), Tingey et al. (1987) showed that isoprene is a product of normal plant metabolism. Rates of isoprene emission were determined for a wide range of plant organisms including cereals, shrubs, and trees (Evans et al. 1982). Only 37 species of 54 studied released isoprene, the rate varying from 0 to 38.5 $\mu g\ C\ h^{-1} g^{-1}$ fresh weight at 28 °C. The emission of isoprene took place only in light; it began just after 1 min of irradiation of leaf disks and ended after a 1-min period when the light was switched off (Rasmussen and Jones 1973). The rate of gas formation depended on the features of the species, the age of the leaves, and temperature. For woody plants such as *Quercus borealis*, *Eucalyptus gunii montana*, and *Hamamelis jelena*, maximum isoprene emission was observed at 40 °C and minimum at 15 °C. At a temperature optimum of 35–40 °C, young plants emitted more isoprene (~ two fold) than older ones. Rates of isoprene synthesis also depended on the concentration of CO_2 in the environmental air (Sanadze 1964). When the concentration of carbon dioxide increases, leaves of poplar, *Populus balsamifera*, and false acacia, *Robinia pseudoacacia*, released isoprene to a lesser degree, and on saturated concentration of CO_2, emission ceased. The requirements for isoprene synthesis at lower concentrations of CO_2 were later confirmed (Sanadze et al. 1990b) on tissue culture of leaf mesophyll. In contrast, on a significant increase in CO_2, a steady loss ability for biosynthesis and emission of liberated isoprene was noticed, due to the high rate of respiration. A similar dependence of isoprene release on temperature and gas composition of the medium was shown by Monson and Fall (1989). The process of isoprene emission was also not dependent on stomata resistance and can occur through the cuticle (Monson and Fall 1989).

In studies on the pathways of gas biosynthesis isoprene was shown to be one of the products of photosynthesis, for which high light intensity, high temperature, and a constant concentration of CO_2 are needed (Sanadze et al. 1972).

In 1987, Tingey et al. found isoprene to be released not only by angiosperms and gymnosperms, but also by mosses, the process depending on plant species and temperature. In the light , the mosses *Dicksonia antarctica* and *Thelypteris kunthii* form 130 to 500 nmol isoprene m^{-2} s^{-1}. In the dark, the production is 100–1000-fold less than on irradiation. Thus, isoprene is considered as a secondary product of photosynthesis. The amount of isoprene formed varies from 0.02 to 2.6% of total carbon fixed in the CO_2 form on the light. In the dark period, the rate of isoprene emission from leaves of aspen, *Populus tremuloides* Michaux, began to decline immediately, showing that the internal pool of isoprene, or its precursors, is small (Monson et al. 1991). On the contrary, a burst of isoprene was observed within 5 min of post-illumination, and lasted 15–20 min in the four species studied, *Populus tremuloides, Eucalyptus globulus, Quercus alba,* and *Mucuna pruriens* var. *utilis.* Isoprene emission was also not dependent on the rate of CO_2 flux through the reductive pentose phosphate cycle. Based on works of Jones and Rasmussen (1975) and Sanadze et al. (1990a, b), a biosynthetic pathway of isoprene has been proposed (Fig. 72). According to this concept, the original precursor of isoprene is acetyl ~ CoA whose two molecules are initially transformed into acetoacetyl ~ CoA, following which, as a result of the addition of one more acyl unit, 3-oxy-3-methylglutaryl ~ CoA, a compound with a branched chain consisting of six

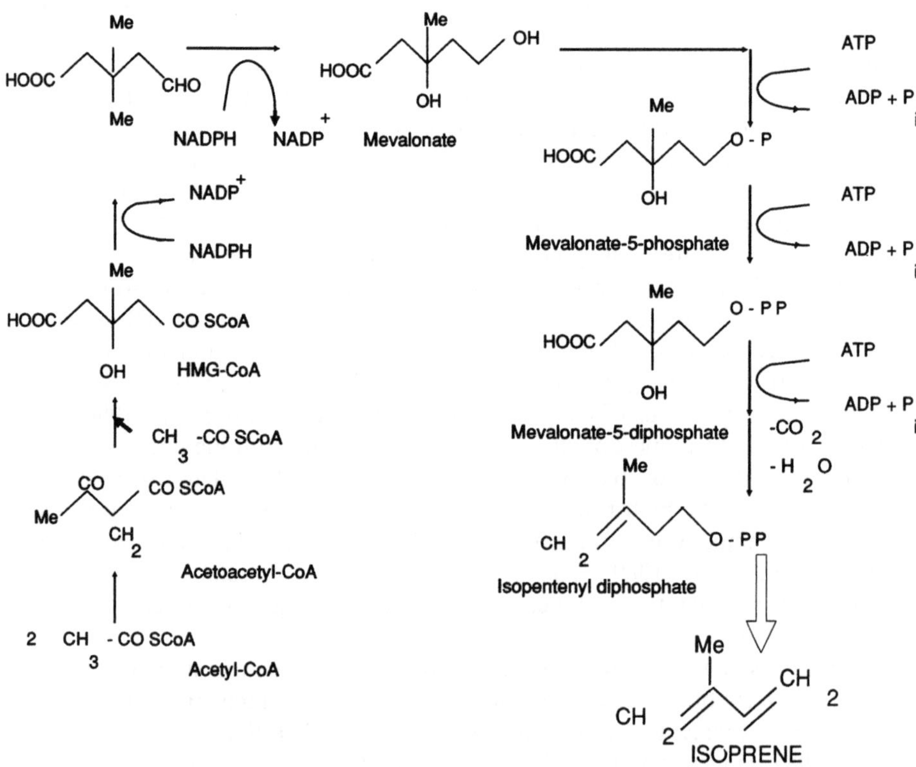

Fig. 72. Proposed biosynthetic pathway of isoprene. (After Jones and Rasmussen 1975; Sanadze 1990)

carbon atoms, appears. The compound serves as an intermediate step in isoprene biosynthesis. The next step consists of the transformation of 3-oxy-3-methylglutaryl to mevaldic acid, a redox process which is dependent on NADPH. Then mevaldic is reduced to mevalonic acid, which is a primary product in the chain of reactions inherent exclusively to isoprenoid metabolism. Then phosphorylation of the mevalonic acid takes place, forming phosphomevalonic acid, which is again phosphorylated by transformation into pyrophosphomevalonic acid. The carboxylic group of pyrophosphomevalonic acid disappears as CO_2 and water are also liberated; a double bond arises forming pyrophosphate, which is a direct precursor of isoprene.

According to Sanadze et al. (1972), the excretion of isoprene by other plant species is connected with photosynthesis itself. Since isoprene synthesis is a light-dependent reaction, one consider to need for that reductant NADPH and ATP as energy source to produced in photosynthesis (Sanadze 1990; Sanadze et al. 1990a, b). The original precursor for isoprene formation, as well as for many terpenoid compounds, is acetyl ~ CoA. A close connection between photosynthetic carbon metabolism and synthesis of the same compounds forming from acetyl ~ CoA, for instance, fatty acids, polyprenols, carotenoids, etc. is thought to occur in the chloroplasts. When synthesis of these substances and photosynthetic assimilation of CO_2 are inhibited, a temporary excess of acetyl ~ CoA appears, that, according to the hypothesis of Sanadze (1990), can increase the rate of formation and emission of isoprene. However, earlier Jones and Rasmussen (1975) has shown that in some genera (*Hamamelis*, *Quercus*) the mediators of phosphorespiration, glycolate and glycine, are included in isoprene to an even greater degree than in acetate or mevalonate. Perhaps the same correlation between the release of isoprene and photorespiration exists in leaves of the European aspen, *Populus tremula*, (Monson and Fall 1989) and some other species. In 1990, Hewitt et al. found that leaves of the herbaceous plant, the giant reed *Arundo donax* L. emit isoprene, and that this process does not depend on photorespiration. Thus, synthesis of isoprene may occur from acetyl ~ CoA via mevalonate or glycolate to isopentenyl diphosphate.

Since isoprene is a normal metabolism product of photosynthesized tissues, it is thought to be in the internal air of the plant which assimilated CO_2, although up to now this gas was found only in the air surrounding the plant. Attempts to indicate it in the inner air of the plant are as yet unknown. Recent data have shown that isoprene is present in the air (respired) breathed by humans and some animals (Gelmont et al. 1981). Among the gases of breath, ethane and pentane, which are products of peroxidation of fatty acids, have been found as well. The main gas is isoprene. Within 24 h an individual releases 2–4 mg of isoprene. Thus, isoprene is released by both plant and animal organisms. However, the main origin of this gas in the biosphere are plants.

Interesting data on the concentration of isoprene in the forest atmosphere of the Amazon are given by Rasmussen and Khalil (1988). A significant amount of isoprene is emitted by the tropical forest throughout the whole year. The highest volume is found under the forest cover, where it reaches 8 parts per billion parts of air. At a height of 300 m above ground, the concentration became only 1–2 ppb. In lower layers of atmosphere concentration of isoprene varies strongly as well. It is highest in the day, the maximum being at midday, and very low at night. The global emission of isoprene calculated on the basis of a release of the gas by plant leaves is approximately 3.5×10^{14} carbon per year.

The interest in the problem of the release of isoprene by plants lies in the leading role of this hydrocarbon in the photochemical formation of ozone in the troposphere of urbanized regions (Trainer et al. 1987; Chameides et al. 1988). Isoprene also participates in production of CO, oxygenated hydrocarbons, and organic acids (Jacob and Wofsy 1988). The estimated global emission of isoprene is very high, ca. $350–400\times10^{12}$ g carbon per year (Zimmerman et al. 1978; Cicerone and Oremland 1988; Rasmussen and Khalil 1988).

Terpenoids, as widespread natural derivatives of isopentenyl diphosphate, are the main part of the volatile excretions of conifers (Stepanov 1972; Yokouchi and Ambe 1984; Stepen' and Sukhinin 1985), of essential oils of sage, *Salvia leucophylla*, and sagebrush, *Artemisia californica* (Muller 1965). Terpenoids are thought to be normal components of plant cells and their synthesis occurs in common nonspecialized cells, which has been confirmed by experiments on isolated tissue cultures. For example, Berlin et al. (1984) reported that cells of tissue culture of *Thuya occidentalis* emit a great quantity of monoterpenes, up to 3 mg g^{-1} dry weight per 24 h. The excretions included α-pinene, β-pinene, myrcene, limonene, and terpinolene. The production of monoterpenes by tissues of rosemary and lavender is shown by Webb et al. (1984), who demonstrated taxonomic specificity in monoterpene composition in *Lavandula angustifolia* and *Rosmarinus officinalis.* Unlike lavender, rosemary contains mainly α-pinene and 1,8-cineole and lacks β-ocimene. In lavender, only β-pinene is found, and β-ocimene predominates. The number of terpenes depends on the stage of tissue differentiation. The most intensive production of monoterpenes was found, in particular for rosemary, in mature leaves up to 140 μg g^{-1} fresh mass, compared with tissue culture 0.18–0.3, and seedlings 4.9–24.

Synthesis of terpenes takes place in plastids. In the glands of the exocarpus of fruits of *Poncirus trifoliata*, it occurs in the chloroplasts (Heinrich et al. 1980). Leucoplasts are capable of synthesizing the monoterpene hydrocarbons as well, because they have a complete enzymic system for synthesis of isopentenyl diphosphate, a precursor of terpenes (Gleizes et al. 1983). Leucoplasts isolated from the exocarpus of *Citrofortunella mitis* were able to form precursors of terpenes. The process begins with acetyl ~ CoA participation, and until the formation of mevalonic acid, passes through the same stages as for isoprene synthesis (see Fig. 72), then continues by the usual pathway for terpenoids (see Fig. 50). In flowers, synthesis of monoterpenes takes place in the chromoplasts, which has been confirmed in experiments with these organelles isolated from flowers of the common daffodil, *Narcissus pseudonarcissus* L., which synthesized α-pinene (2.8%), myrcene (56.4%), linalool (22.9%), and ocimene (17.1%) from [1-^{14}C]-isopentenyl diphosphate and [1-$^{2}H_2$]-isopentenyl diphosphate as substrates (Gleizes et al. 1983). Terpenoids comprise the main part of the volatile excretions of conifers, and are easily observed in the air around them. The saturation of the air by terpenes depends on various factors, and under particular conditions can achieve significant values (Stepanov 1972; Yokouchi and Ambe 1984; Stepen' and Sukhinin 1985). The dependence of terpene production on the taxon of the plants is also shown.

Studies done in Siberia (Stepen' and Sukhinin 1985) demonstrated that the amount of terpene derivatives is decreased in the order: cedar, *Cedrus* > pine *Pinus* > fir *Abies* > spruce *Picea* of the trees studied, with α-pinene dominating. Some components were specific only for certain genera and species. Terpenes like thujone, α-phenchene, and α-phellandrene were inherent only in spruce, *Picea* and tricyclene in pine, *Pinus*. Thus, there is the possibility of a significant saturation of the air environment around conifer plants by volatile terpene compounds, among which species-specific substances have been identified.

Since monoterpene emission depends on temperature and irradiation (Yokouchi et al. 1983; Yokouchi and Ambe 1984), seasonal changes in concentrations of monoterpenes are observed in a pine forest, monoterpene excretion being lower in winter and higher in summer and autumn. These oscillations correlate with temperature variability. Maxima in α-pinene were observed in August, October, and November; the average concentration in May was 0.12 ppm (Yokouchi et al. 1983). The emission of monoterpenes also changes depending on light intensity (Yokouchi and Ambe 1984). Monoterpenes released by 2-year-old plants of *Pinus* consisted mainly of α- and β-pinenes, myrcene, and β-phellandrene. Maximum excretion was at a midday, i.e., it was irradiation-dependent. Thus, daily and seasonal oscillations are observed in the quantities of volatile emissions. Japanese researchers (Tanaka et al.1989) reported that red light irradiation stimulated terpenene accumulation in common thyme, *Thymus vulgaris*. The amount of carvacrol and cymol was doubled terpinene and thymol increased to three- to four-fold under 48 h exposure of the plant to light. This observation shows the photoreceptory nature of photoregulation in the formation of terpenoids as a participation of the phytochrome in the phenomenon (Tanaka et al. 1989). Mihaliak and Lincoln (1989) have shown that terpenoid synthesis is dependent on the nitrogen nutrition of the plants. The concentration of mono- and sesquiterpenes in plants of *Heterotheca subaxillaris* grown under nitrogen-deficient conditions was higher than under nitrogen-rich. The carbon for such terpenoid synthesis originated from CO_2, fixed in photosynthesis, and or from photosynthate which was transported from older to younger leaves. Based on these data, the authors (Mihaliak and Lincoln 1989) advanced the hypothesis that defensive terpenes are formed under limitation of nitrogen nutrition.

The rate of monoterpene emission was determined for some conifers (Evans et al. 1982) as 0.01 to 3.53 μg C/n per g fresh weight. According to Zimmerman et al. (1978), terpene production by plants calculated for data received for US vegetation is 4.8×10^{14} C per year.

4.3.3 Aldehydes and Ketones

Aldehydes are often found in plant excretions, but they are still poorly studied. Since, like many other volatile substances, aldehydes are synthesized and emitted in very small amounts, for their separation and identification highly sensitive methods of gas chromatography-mass spectroscopy and functional analysis are used. By these methods, the presence of aldehydes in the air around plants grown in closed chambers has been found (Sanadze 1961a; Dadykin et al. 1967). Aldehydes, like other volatile products of intracellular metabolism, can be distilled (with water vapor). Ethanol and propanal were found in the composition of the transpiration water collected in nature from adult trees (Roshchina 1971), whereas in their internal air butanal could sometimes be identified (Roshchina 1974b,c). Pentanal and butanal were found in leaf condensates of cowberry, *Vaccinium vitis idaea*, (Skvortsov and Smirnova 1972). Acetaldehyde, acetone, propanal, hexanal, and benzaldehyde were identified in the composition of volatile excretions of soybean, *Glycine soja* (Nunomura et al. 1976). Germinating under sterile conditions, seeds of pea, *Pisum sativum*, kidney bean, *Phaseolus vulgaris*, maize, *Zea mays*, and longleaf pine, *Pinus palustris* emit ethanal and formaldehyde (Vancura and Stotzky 1976). The release of aldehydes by seeds begins with the appearance of the first root, and the maximum emission is observed within between 24 and 48 h after germination.

Aldehydes may be formed by nonspecialized plant cells, since they are met in the gas phase of tissue cultures. Acetaldehyde has been found in tissue cultures of 25 plant species (Thomas and Murashige 1979). When volatile excretions of 3-week-old callus in tobacco, *Nicotiana tabacum*, were analyzed, among other compounds formaldehyde and acetaldehyde were found (Gaal 1984).

The pathways of aldehyde synthesis in plants have been poorly studied. Figure 73 shows that substrates for the formation of low-molecular weight aldehydes can be compounds with amino groups – amines and amino acids. The reaction occurs according to the equation:

$$R - CH_2NH_2 + H_2O + O_2 = R - CHO + NH_3 + H_2O,$$

and is catalyzed by aminooxidases or aminooxidoreductases (EC 1.4.3.4) and (EC 1.4.3.6) contained in the active center, correspondingly flavin and pyridoxal phosphate. For example, acetal, isobutanal, isopentanal, 2-methylbutanal, methional, and phenylacetic aldehyde are shown (Motoda 1979) to form from alanine, valine, leucine, isoleucine, and methionine in extracts of green leaves of tea, *Thea sinensis*, beans of coffee, *Coffea arabica*, and cacao, *Theobroma cacao*, on the addition of polyphenoloxidase (EC 1.14.18.1). Since some amino acids can be used for the formation of aldehydes, the latter are also considered as products of protein degradation. There are data which show that aldehydes of seeds are specifically bound with protein and, on heating, are released from them (Chiba et al. 1979).

As substrates for the synthesis of aldehydes, alcohols can also serve

$$R - CH_2OH + NAD\ (P)^+ = R - CHO + NAD\ (P)\ H + H^+.$$

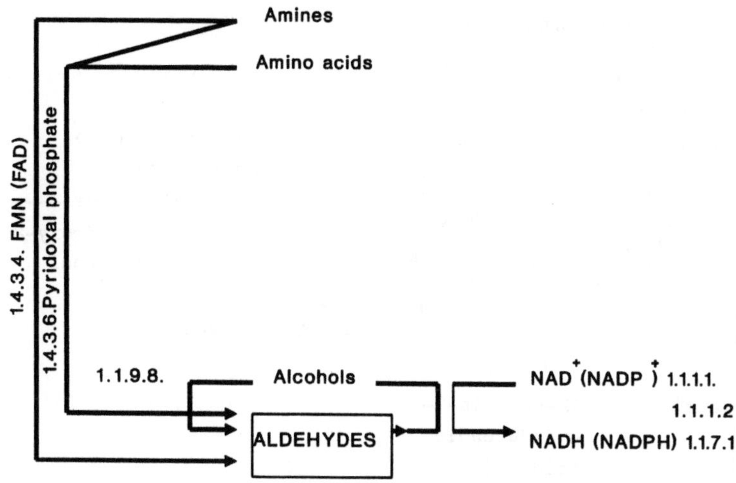

Fig. 73. Pathways of aldehyde synthesis. (Roshchina and Roshchina 1983)

This reaction is catalyzed by the enzymes alcohol dehydrogenases or alcohol NAD (NADP) oxidoreductases (EC 1.1.1.1; 1.1.1.2; 1.1.7.1). The first two enzymes ae universal, but the third is more active in catalysis of the formation of aldehydes with > 4 carbon atoms. The production of aldehydes from alcohols is also catalyzed by alcohol dehydrogenase (EC 1.1.9.9.8), which, instead of NAD^+ ($NADP^+$), uses other electron acceptors. Some authors (Hatanaka et al. 1978 a, b) suggested that, besides the above-mentioned pathways, aldehydes may be synthesized from organic acids, and the process is catalyzed by special enzymatic system. Thus, in shoots of English rye-grass, *Lolium perenne*, aldehydes of p-hydroxybenzoic, vanillic, and syringic acids have been found (Hartley and Buchan 1979). Linoleic and linolenic acids are substrates for the formation of aldehydes and alcohols on membrane damage (Hatanaka et al. 1978a). This process is catalyzed by acylhydrolases (Fig. 74). Nonenzymatic formation of the aldehydes n-hexanol and 12-hydroxy-2E-dodecenal by isomerization of unsaturated C_{18}-alcohols has also been demonstrated in the fraction of isolated *Thea* chloroplasts (Hatanaka et al. 1989). In green tissues, the production of C_6-unsaturated aldehydes such as hexenal reaches 208–1524 $\mu g\ g^{-1}$ fresh weight or even higher (Hatanaka et al. 1989).

The pathways for the formation of acetaldehyde, propanal, and butanal from pyruvate in acetic, propionic, and butyric fermentations were first established for microorganisms and could be proposed for plants as well (Luckner 1977). Recently, a new pathway for butanal synthesis in strawberry cell suspension cultures has been shown (Yeong-Ching Hong et al. 1990). This aldehyde is formed from α-keto-valeriate with the participation of decarboxylase.

The presence of aldehydes in plants and their release is determined by their role as universal intermediates for the common metabolism of any living

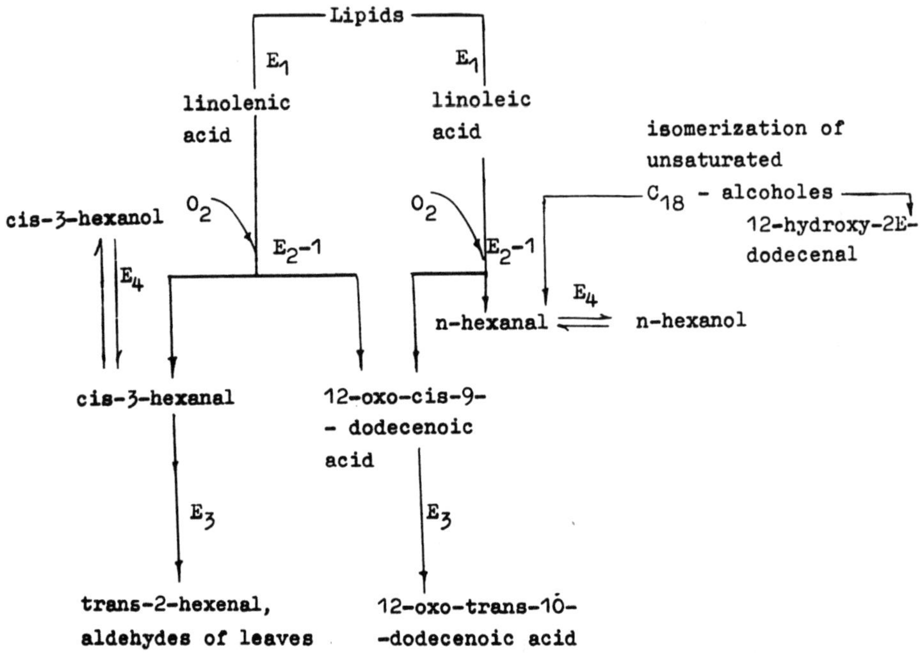

Fig. 74. Biosynthesis of C_6-aldehydes and alcohols. E_1 lipolitic acyl hydrolase; E_2 enzymic and/or nonenzymic isomerization factor; E_4 alcohol dehydrogenase. (After Hatanaka et al. 1978a, b; 1989)

organism. As substrates for the formation of organic acids and alcohols, they can be included in the subsequent metabolic processes (Durmishidze 1977).

Besides aldehydes, in plant excreta acetone has also been found. Murphy (1985) showed that germinating seeds of lima bean, *Phaseolus limensis*, soybean, *Glycine max*, water melon, *Citrullus lanatus*, sunflower, *Helianthus annuus*, and three pine species, *Pinus pinea*, *P. lambertiana*, and *P. edulis* release acetone. The quantity of acetone emitted was especially high in species with a high lipid content, the maximum, 4×10^{-6} M, of acetone formation is similar to that of animals. It is synthesized from acetoacetate at the ketosis of fatty acids (Metzler 1977) with participation of acetoacetate carboxylase (EC 4.1.1.4). Whether aldehydes and ketones have any biological activity and can take part in biochemical interactions between organisms will be considered below. A number of facts indicate this possibility. Propionic aldehyde is the most physiologically active and toxic for humans, so that its admissible concentration in the closed atmosphere of buildings should not exceed 1 mg m^{-3}. Unlike propanal, the admissible acetaldehyde concentration limit is higher, 5 mg m^{-3}. Saturated aldehydes inhibit some physiological processes in plants and microbes. One mM formaldehyde or acetaldehyde irreversibly inhibits cell division of *Escherichia coli*, and retards or stops germination of seeds and seedlings growth of various plants (Kolesnichenko 1976; Schauenstein et al. 1977). Acetaldehyde at a concentration of > 1 mM depresses photosynthesis (Kolesnichenko 1976).

Alkanals can act on cell respiration. For instance, formaldehyde inhibits the oxidation of succinate, glutamate, α-oxyglutarate, and pyruvate, which is coupled with phosphorylation (Schauenstein et al. 1977). Alkanals also depress ATPase and the reversibility of electron transport by ATP in mitochondrial particles. Other aldehydes (acetaldehyde, propanal, isobutanal, isopentanal) change the oxidative processes in mitochondria, acting on the appropriate dehydrogenases, and the metabolism of pyruvate. These effects are partly reversible. Acetaldehyde irreversibly inhibits nitrogenase, hydrogenase, CO_2-fixation, and growth of cyanobacteria in *Anabaena cylindrica*, but has no effect on the light-dependent uptake of O_2 (Slatyer et al. 1983).

Aldehydes induce sharp shifts in the water-holding capacity of cells and the permeability of cytoplasmic membranes (Roshchina and Roshchina 1970; Roshchina 1974a). The toxic effect of aldehydes is determined by their ability react with the amino- and sulfhydryl groups of proteins, inducing conformational changes (Schauenstein et al. 1977). However, the strength and character of the action depend on their structure. lipophility, and the polar features of the molecules. Most active among the saturated alkanals is formaldehyde. As the number of carbon atoms increases, the inhibitory capacity of aldehydes is decreased (Schauenstein et al. 1977).

Based on many biological effects induced by aldehydes, it could be supposed that the substances play a certain role in allelopathic interactions between plants. The plant donor of aldehydes acting on the proteins of another plant, called an acceptor, appears to be more competitive. On the other hand, the plant acceptor has defensive mechanisms against the action of the plant donor. For example, such a plant acceptor may take up the excreted aldehydes and include them in its own metabolism by way of redox conversions, as a result forming organic acids and alcohols (Durmishidze 1977).

4.3.4 Low-Molecular Alcohols

Low-molecular mass alcohols were found in many plant excretions. Main component of these excreta is ethanol which is released in higher amounts than other alcohols: butanol, isobutanol, allyl, and hexyl alcohols (Glinka and Reinhold 1962; Rakitina 1970). It should also be noted that in intact organs and whole plant of cowpea, *Vigna savi*, cabbage, *Brassica oleracea*, and lettuce *Lactuca sativa* ethanol, butanol, and methanol were also found (Masquelier and Vitte 1967). Aldehyde ethanol and ketone acetone were also identified in them.

Dadykin et al. (1967) studied the volatile excretions of vegetables, from the perspective of space flights, using gas chromatography. Some ten different compounds were identified in the gas emissions of leaves of red beet, *Beta vulgaris*, carrot, *Daucus carota*, tomato, *Lycopersicon esculentum*, and potato *Solanum tuberosum*., among them methanol and ethanol. Unlike leaf excreta of the same plants, in root release of carrot and potato these substances were two fold less. Moreover, propanol was present in them (Dadykin et al. 1967).

By gas chromatography, steams of ethanol were found in the internal air collected from shoots of six woody species and in the transpiration water of the same plants (Roshchina 1971, 1973a, 1974a). Low-molecular mass alcohols were contained in air emitted through the lenticels (Chirkova and Gutman 1972). Additionally ethanol, methanol, propanol, and butanol were eliminated. Alcohols with more carbon atoms in the molecule, methylbutanol and hexanol, were identified in gas excreta of *Acacia pulchella* (Whitefield et al. 1981). Palma-Fleming and Kepner (1983) found hexanol, hexenol, terpineol, etc. among volatiles released by California live oak *Quercus agrifolia*.

The appearance of alcohols known as products of anaerobic fermentation in gas excretions of woody plants is induced by hypoxia which arises in tissues either on limitation of O_2 or due to the structure of the tissues where the gas is transported. An illustration of the latter may be the poor accessibility of the woody plant's cambium, being separated from the atmosphere by layers of external bark and secondary phloem. It is not surprising that ethanol was found in the cambium of many woody plants (Kimmerer and Stringer 1988). Since it has been identified in xylem sap as well, this alcohol may be liberated from the cambium with the transpiration flow. The production of ethanol was correlated with a high activity of cambial alcohol dehydrogenase and pyruvate carboxylase (Kimmerer and Stringer 1988). Thus, the presence of ethanol and enzymes of alcohol fermentation in the cambium shows that aerobic respiration is limited in O_2 under these conditions.

Under anaerobiosis, ethanol and acetaldehyde can also appear in usually aerated organs of plants. Fifty species of woody and herbaceous plants grown in an N_2-atmosphere are shown to release ethanol and acetaldehyde (Kimmerer and MacDonald 1987). Leaves of poplar, *Populus*, produced more ethanol than the roots. Unlike leaves of woody plants, leaves of herbaceous plants such as maize, *Zea mays* and soybean, *Glycine soja* L., grown in the greenhouse in N_2 atmosphere, produced a lesser amount of these products of anaerobiosis, including ethanol. Hampered aeration in closed chambers during tissue cultivation of cherry tree, *Prunus avium* L. also led to the release of ethylene and CO_2 (whose concentration can achieve 30% within 30 days). As a result, after a few days, the developing shoots emitted ethanol and acetaldehyde (Righetti et al. 1990).

Methanol and ethanol have been also found in fruits of *Citrus* plants (Lund et al. 1981) as in excreta of many other plant tissues. However, the main part of the alcohol elimination probably occurs through the root systems. Excretion of alcohols via leaves is approximately four fold lower (Lund et al. 1981). The accumulation of ethanol is associated with the apical parts of the roots, to which access of O_2 is difficult due to the absence of intercellular spaces. According to data of Bolton and Erikson (1970); the excretion of ethanol via the roots is 1/5 of the total alcohol concentration, but via the leaves only 1/20. Alcohol can be transported from roots to leaves by transpiration flow.

A significant amount of volatile substances, including low molecular alcohols, can be released by seeds and seedlings. The elimination of volatile compounds begins simultaneously with the appearance of the first root, and in most seeds remains at a high level form 24 to 48 h. Organic volatile substances determined by gas chromatography composed 5% of the atmosphere of germinating seeds. Ethanol and methanol, as well as acetaldehyde, propionic aldehyde, formaldehyde, and formic acid, have been found in the atmosphere of closed chambers with seeds and seedlings of kidney bean, *Phaseolus vulgaris*; pea, *Pisum sativum*; cabbage, *Brassica oleracea*; maize, *Zea mays*; cotton, *Gossypium* sp.; cucumber, *Cucumis sativus*; radish, *Raphanus sativus*; pumpkin, *Cucurbita pepo*; tomato, *Lycopersicon esculentum*; pine, *Pinus sylvestris*; and alder, *Alnus* sp. (Vancura and Stotzky 1976). Excretion of alcohols and aldehydes during the germination of seeds was also shown in experiments by Berestetskii et al. (1978). The release of ethanol, methanol, and a small amount of acetone have been demonstrated in germinating seeds of *Phaseolus vulgaris* (Doireau 1972). There is a quantitative variability in the composition of gaseous metabolites of germinating seeds between various plant species on one hand and between angiosperms and gymnosperms on the other. The release of alcohol is possible for any nonspecialized plant cell under certain conditions. For instance, ethanol is constantly produced by tissue cultures of 25 species, as well as CO_2, ethylene, ethane, and acetaldehyde (Thomas and Murashige 1979).

Glycolysis, the tricarboxylic acid cycle, and the pentose phosphate cycle are thought to take place during seed development. These processes are not of equal significance at all stages of seed germination. The formation of acetaldehyde and ethanol are explained by glycolysis under partial anaerobiosis (Gaal 1984). The accumulation of nonoxidized products of metabolic exchange depends on the usually poor supply of oxygen to the tissue, which induces glycolysis. In the glycolytic process, the glucose molecule is transformed via ten subsequent reactions to two pyruvate molecules. Pyruvate can be utilized in three ways: (1) be included in the Krebs cycle and then oxidized to CO_2 and H_2O, (2) be reduced to lactate under anaerobiosis, (3) be transformed to ethanol and CO_2. Under a combination of conditions favorable for glycolysis, such as limitation of air access and heightened temperature, alcohols, mainly ethanol, as well as lactic acid, are accumulated in the tissue. The production of a great amount of lactate is rarely found in plants. It is accumulated through lactic fermentation in only some few species, for example, pea, *Pisum sativum*; potato, *Solanum tuberosum*; rice, *Oryza sativa*; wheat, *Triticum vulgare*; and cabbage, *Brassica oleracea*. The process of alcohol fermentation begins as pyruvate splitting, then CO_2 and acetaldehyde are formed. The latter is reduced, transforming to ethanol with participation of NADPH and alcohol dehydrogenase. The addition of the water oxygen to acetaldehyde via dismutation oxidizes the substance to acetic acid. The hydrogen of NADH, reduced in the process, may carry the rest of the acetaldehyde, forming ethanol again. Ethanol is not a final product of metabolism. Being quickly included in metabolic exchange, it can enter the reactions of the Krebs cycle. Exogenous ethanol added to the sample is transformed to organic acids, amino acids, sugars, volatile hydrocarbons, lipids, etc. Ethanol infused in the fruits of tomato, *Lycopersicon esculentum*, undergoes deep conversions, accompanied by the formation of carbon dioxide and ethylene (Grineva 1975).

The biological effects of alcohols have been studied by some authors. Alcohols are thought to influence the physiological processes only at relatively high concentration. Petruzzelli (1984) found stimulation of seed germination of Aleppo pine, *Pinus halepensis*, in the dark by 0.5–1% ethanol. A stimulatory effect on *Rumex crispus* seeds has been observed on treatment with 0.5–2.0 μM methanol and 0.03–0.3 M propanol (Taylorson 1984). However, acetaldehyde and ethanol in 0.1 M concentration inhibited seed germination in the same species. The formation of citric acid in the fungus *Aspergillus niger* was stimulated by 1–4% methanol, but the culture growth of the species decreased (Georgiev and Aleksieva 1990). Perata et al. (1986) noted that ethanol at a concentration of 8.5 mM depressed growth in Jerusalem artichoke *Helianthus tuberosus* by 50%. This alcohol decreased the viability of protoplasts isolated from tissue culture of carrot, *Daucus carota*, and at a concentration of 10 mM inhibited embryogenesis ten fold. High concentrations ($10^{-2} - 10^{-1}$M) of ethanol, propanol, butanol, and acetone influence the energetic process stimulating ATPase activity of the coupling factor (Toxobaeva et al. 1987).

4.3.5 Volatile Nitrogen-Containing Substances

Plants grown in a normal atmosphere can release ammonia (Farquhar et al. 1979). Experiments were carried out with cereal cultures, which emitted this gas on aging, even while still photosynthetically active. According to Farquhar et al. (1979), the evolution of NH_3 by leaves of *Zea mays* depended on the nitrogen supply. The rate of the process was 0.00–0.005 nmol $m^{-2} s^{-1}$ under high N-nutrition, increasing to 0.57 nmol $m^{-2} s^{-1}$ under low nitrogen concentration, especially when senescing leaves were present on the same plant. Some ammonia usually is emitted by the blossoms of plants belonging to the family Araceae (Smith and Meeuse 1966; Robacker et al. 1988). Hogan et al. (1983) found active urease (EC 3.5.1.5) in leaf tissues of soybean, *Glycine soja*, and wheat, *Triticum vulgare*, that catalyzed the reaction of ammonia formation form urea. The activity of urease was increased in the presence of ethanol, propanol, and isopropanol. Maximal rates of ammonia release were observed under n-propanol, and equalled 34 μmol g^{-1} fresh weight for soybean, *Glycine soja* (control rate without alcohol 7.4 μmol g^{-1} fresh weight and 12 μmol g^{-1} fresh weight for wheat, *Triticum vulgare* (control rate without alcohol 2.57 μmol g^{-1} fresh weight).

The biological effects of ammonia have been studied in some few works. According to Prutenskaya et al. (1967), the increase in NH_3 concentration from 0.096 to 24% reduces respiration in seedlings of garden cress, *Lepidium sativum*, to practically O. In the opinion of these authors, visible changes occurred in the permeability of cell membranes, having a nonspecific character. However, under natural conditions, the concentration of NH_3 in the air does not achieve toxic values.

Plant nitrogen can be eliminated in the form of volatile oxides such as NO_3, NO_2, and NO, under both normal and stress conditions (Dean and Harper

1989). The oxides of plants are reduced to ammonia on the participation of nitrate- and nitrite reductases. NADH, NADPH, and reduced ferredoxin may serve as electron donors in these reactions:

$$NO_3^- \rightarrow NO_2^- \rightarrow NH_3.$$

In 1979, Klepper showed that a significant amount of nitrogen oxides is released from the leaves of soybean, *Glycine soja* L., treated by herbicides which inhibit photosynthesis. These herbicides block the light-dependent reduction of NO_2^-, but do not depress the reduction of NO_3^-. As a result, nitrogen dioxide is accumulated in the plant and is then released. The formation of NO, as Klepper (1979) proposed, occurs due to the chemical reaction of the accumulated NO_2 with plant metabolites. The release of nitrogen monoxide by the leaves of soybean was also shown later by other authors (Wei et al. 1987). Harper (1981) proposed that the emission of NO is a result of enzymatic reduction of nitrogen dioxide, because on infiltration of NO_2^-, as a substrate, into the leaf and then boiling, NO is not released, whereas the unboiled infiltrates eliminate it. Other evidence of an enzymatic character for NO emission is connected with the mutants which cannot excrete nitrogen oxides due to the absence of nitrate reductase (Nelson and Ryan 1983). Quantitative determination of nitrogen oxides by gas chromatography and mass spectroscopy showed that the amount of NO emitted is nearly 30–40 μg of the total nitrogen and during the season reaches 45 kg of nitrogen per 1 ha under field conditions (Mulvaney 1984). The significance of the excretion of nitrogen oxides for plant life is not yet clear. This is surprising, because this process is connected with the loss of the nutritional element that the plant needs.

Besides inorganic nitrogen-containing substances, aminoid compounds are released by both nonspecialized and specialized secretory cells (see also Chaps. 2.3.2 and 3.10). Emission of easily volatilized substances is thought to be connected with the flower aroma, which plays the role of an attractant for insect pollinators. The fragrance helps them to recognize a flower of a certain species whose nectar or pollen is needed for the their food. Volatile amines, for example, are found in condensates of blossoms of some species belonging to the family Araceae (Smith and Meeuse 1966; Robacker et al. 1988):

Arum dioscorides	*Arum italicum*	*Dracunculus vulgaris*	*Hydrosme rivieri*	*Sauromatum guttatum*
2-Aminoethanol	Agmatine	Diethylamine	Agmatine	Agmatine
Methylamine	Cadaverine	Dimethylamine	2-Aminoethanol	2-Aminoethanol
1,2-Propane-diamine	Diethylamine	Ethylamine	Ethylamine	Cadaverine
Scatole	Dimethylamine	Indole	Histamine	Diethylamine
	Ethylamine	Isobutylamine	Isobutylamine	Dimethylamine
	1,6-Hexadiamine	Methylamine	Methylamine	Ethylamine
	Methylamine	Trimethylamine	Scatole	Indole
	Putrescine			Isobutylamine
	Scatole			Methylamine
				1,2-Propaneamine
				Putrescine
				Trimethylamine

Among these, 16 compounds have been identified. The analyzed species differ in their amine composition. The presence of histamine in *Hydrosme rivieri* is of special interest to researchers, because it is known as a neuromediator in animals. The amount of this substance in thc plant is 1 μl ml^{-1}.

The aroma of the blossoms in the Araceae is created by several amines and free ammonia. The odors of lizard arum, *Sauromatum guttatum*, are unusual because in the same flower the aromas are both sweet and unpleasant obnoxious rotting components, originating from different parts of the flower. The sweet flavors, like other flower scents contain terpenoids, including mainly limonene and α-pinene. The smell of decomposition is created by amines and free ammonia. It is interesting that the sweet aroma of *Sauromatum guttatum* is spread by the clavate organ of the spathe tube of the inflorescences, but the aminoid fecal smell by the inflorescence itself (Robacker et al. 1988). Volatile amines are easily observed with the help of filter paper soaked in HCl. If amines are present, a white smoke or fog formed by the salts of ammonia and HCl appears around the inflorescence. Among the plants with an obnoxious smell created by volatile amines are: hogweed cow parsnip, *Heracleum sphondylium*, bear's-foot, *Helleborus foetidus*, cuckoopint, *Arum maculatum*, and hawthorn, *Crataegus oxyacantha* etc. Analysis of the odorous substances in these plants showed monoterpenes and monoamines ranging from methylamine to hexylamine (Fig. 75). The components of the mixture with an unpleasant smell are scatole, indole, and aliphatic organic acids.

The volatile amines methylamine, ethylamine, n-propylamine, and isopropylamine are released during the growth of plants belonging to the lower steps of evolutionary growth. Amines are eliminated by the cells of various species of green algae and cyanobacteria (Klochenko et al. 1990). The main pathway for the formation of amines is the decarboxylation of appropriate amino acids, for instance isoamylamine from leucine, isobutylamine from valine, etc. Volatile amines can be included in metabolism again and be used as building material for

Monoamines

CH_2NH_2 methylamine

$CH_3CH_2NH_2$ ethylamine

$CH_3(CH_2)_2NH_2$ propylamine

$CH_3(CH_2)_3NH_2$ butylamine

$CH_3(CH_2)_4NH_2$ amylamine

$CH_3(CH_2)_5NH_2$ hexylamine

Diamines

$NH_2(CH_2)_4NH_2$ putrescine

$NH_2(CH_2)_5NH_2$ cadaverine

Indoles

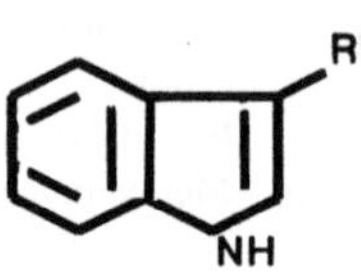

R = H indole

R = CH_3 Scatole

Fig. 75. Components of the unpleasant aminoid smell in plants the synthesis of various nitrogen-containing compounds, especially for alkaloid production (see Chap. 3.9). Many aminoid compounds with an unpleasant smell of decomposing protein play an essential role in attracting insect pollinators, mainly flies and dung beetles.

4.3.6 Carbon Monoxide and Hydrogen

Among the gaseous compounds released by plants are CO, NH_3, H_2, and nitrogen oxides.

The elimination of CO (see also Chap. 2.2.3) was first found on analysis of the gas composition of pneumocysts of brown algae (Langdon 1917). In the first 50 years of our century, the ability to release CO has been established for higher plants as well. According to Wilks (1959), the largest amount of eliminated CO observed for such a plant as *Medicago sativa* was 2.1 mg/100 g tissue and the smallest, as for *Lactuca sativa* and *Carya* sp., only 0.001–0.003 mg/100 g tissue. Leaves usually emit more monoxide than stems. For instance, leaves of *Axyris* emitted 0.212 mg CO per 100 g tissue, but stems only 0.022. In *Medicago sativa*, gas elimination in the first case was five fold higher than in the second.

The release of CO is not necessarily observed only under stress. It often takes place under normal conditions in the light and in the presence of oxygen. However, in 1987, Siegel and Siegel showed in the common mushroom, *Agaricus*, that seeds and seedlings of some angiosperms can release carbon monoxide in the dark on exposure to an atmosphere with a low concentration of oxygen (5%) and high carbon dioxide (Siegel and Siegel 1987). Rates of CO emission by higher plant seeds varied from 0.23–0.28 nmol $g^{-1} h^{-1}$ for *Phaseolus aureus* and *Brassica oleracea* var. *capitata* to 0.41–0.51 nmol $g^{-1} h^{-1}$ for *Raphanus sativus* and *Cucumis sativus*. Basidiomycetes released carbon monoxide at a similar rate per hour. Autoclaving seeds sharply reduced CO emission, approximately ten fold. These data show that saprophytes and autotrophs have a light-dependent pathway of CO formation, which is especially noticeable at night or in a forest where light is limited.

The emission of carbon monoxide in woody plants was studied by Bauer et al. (1979). Maximum CO amount, 16×10^{-9} g $m^{-2} h^{-1}$, was demonstrated for pine, *Pinus sylvestris*, plane-tree maple *Acer pseudoplatanus*, oak, *Quercus robur*, and European beech, *Fagus sylvatica*, under a light intensity of 800 W m^{-2}. The formation of CO by plants calculated by Bauer et al. (1979) is 0.7×10^{14} g $year^{-1}$. Possible pathways of carbon monoxide formation have been poorly studied as yet. CO can form non-enzymatically because heating of the tissue does not stop its emission (Loewis and Delwiche 1963). The origins of carbon monoxide in plants may be a phenolic substance (gallic acid, pyrogallol) in tissues of alfalfa, *Medicago sativa*, and common eelgrass, *Zostera marina* (Loewis and Delwiche 1963), serine in *Physalia physalis* (Wittenberg 1960), and Δ-aminolevulinic acid in cyanobacteria (Troxler and Dokos 1973). A small amount of CO was observed in fractions of chlorophyll (Wilks 1959). This phenomenon is similar to that which occurs in mammals, where the emission of carbon monoxide is explained by the oxidation of a methylic group in the hemoglobin molecule on the addition of ascorbic acid (Sjöstrand 1952). In model experiments, emission of CO was shown to occur on enzymatic degradation of rutin (Simpson et al. 1960). Plants themselves release $0.2–2.0 \times 10^{14}$ g of CO per year, but more of this gas, $4.2–13.3 \times 10^{14}$ g per year, is formed in the atmosphere as a result of the oxidation of isoprene and terpenoids emitted by plants (Zimmerman et al. 1978). An obligate stage of isoprene conversion in the air is the formation of formaldehyde, which is decomposed under the action of light to CO. Terpene oxidation is thought to be under the action of ozone or free oxygen radicals on their double bonds. Terpene rings may be broken into segments, which leads to the formation of short-chain aldehydes, from which carbon monoxide and hydrogen are then also formed. It cannot be excluded that in some cases CO appears as the result of the oxidation of aerobic methanotrophs, which, oxidizing CH_4 to CO, make a certain addition to the atmospheric production of carbon monoxide (Zimmerman et al. 1978). Carbon monoxide may be assimilated in light (Krall and Tolbert 1957; Chappelle and Krall 1961). The ^{14}C of the molecule is included in serine and glycine (Delwiche 1970). Plants are thought to play an important role in the purification of the atmosphere from carbon monoxide.

Information on the emission of hydrogen by plant cells appeared during the first 40 years of our century in works with green algae (Gafron and Rubin 1942) and isolated chloroplasts (Boichenko 1946). This effect was observed both in light and in the dark in a nitrogen atmosphere in the absence of CO_2. The emission of H_2 by green leaves in the light has been shown using gas chromatography (Sanadze 1961b). The correctness of the interpretation of the results of these experiments was doubted. Hydrogen was proposed to be a product of metabolism of the bacteria accompanying plant life, but not of the plants themselves. However, plants grown under sterile conditions also emitted hydrogen in light. The mechanism of the phenomenon was unclear. Boichenko (1946) showed on isolated chloroplasts that the formation of H_2 takes place with the participation of hydrogenase via the fermentation of glucose: however, the emission of hydrogen by chloroplasts in the light can also occur without hydrogenase in the medium (Mal'tsev et al. 1986). Boichenko (1946) proposed that if CO_2 is absent from the medium, the hydrogen of water is not included in the carbon dioxide reduction but is liberated as free H_2. Stress conditions are also being considered as factors inducing a by-pass of the normal electron and proton transport paths at photosynthesis. When photosystem 1 or its components are damaged and there is no NADP photoreduction, the main electron transfer occurs via the alternative pathway (Mal'tsev et al. 1986). The ratio between the alternative (stress) and main (basal) pathways may be changed because hydrogen is not only released, but is also taken up in the light.

The list of plants studied whose chloroplasts are able to emit hydrogen is still small: pea, *Pisum sativum*: goosefoot, *Chenopodium album*, dead nettle, *Lamium album*, maize, *Zea mays*, wheat, *Triticum vulgare*, and spinach, *Spinacia oleracea* (Mal'tsev et al. 1986). The similarity in mechanism of H_2-formation between chloroplasts and green algae confirms the possibility of an evolutionary symbiogenesis of unicellular algae and eukaryotic heterotrophic cells. It should be noted that the increase in both atmospheric hydrogen and CO (see above) occur due to oxidation of the isoprene and volatile terpenoids formed by plants (Zimmerman et al. 1978). The phenomenon can also partly take place within the plant; however, this supposition needs special study.

4.4 The Significance of Gas Excretion

The physiological role of gas excretions by plants is diverse. Obviously, the common significance of the release of volatile substances, as well as of nonvolatile, is in the elimination of excessive (aldehydes, alcohols, etc.) or toxic (components of membrane decomposition such as saturated hydrocarbons, ammonia) products of metabolic exchange.

More specific is the defensive function of gas excretion. It is known that gas excretion is closely connected with transpiration, which protects the plant from overheating. Volatile low-molecular terpenes and essential oils in the composition of water vapors can influence the rate of transpiration. It has been shown that under their action, the rate of water evaporation by leaves is decreased. Low concentrations of volatile essential oil (eucalyptus, peppermint, clove or caryophyllus, caraway) stimulated transpiration in kidney bean, *Phaseolus vulgaris*, and sunflower, *Helianthus annuus*, whereas at higher concentrations it retarded the process (Polova and Vicherkova 1986). The effect of the vapors of essential oils was stronger if the air humidity was lower. The proposed mechanism of the phenomenon is in the regulation of closing or opening of stomata (Polova and Vicherkova 1986). Thus gas excretion is one manifestation of adaptation to soil-climatic conditions.

The defensive role of volatile products is clearly seen in cases of phylogenetic accommodation in order to attract or scare off insects. Low-molecular mass terpenes released by flowers are signals attractive for insect pollinators (Kullenberg and Bergtröm 1975), and some toxic terpenes can frigthen off insect attack. Thus excretory terpenoids serve as olfactory or taste repellents (Vasilyev 1977; Harborne 1988). Terpenoids which frighten insects off can also inhibit growth and development of the larvae. Elliger et al. (1976) found that two volatile diterpenes 16-kauren-19-oic and trachyloban-19-oic acids (Fig. 76), depress the growth and development of some Lepidoptera. The physiological effects of other terpenoids secreted by glandular hairs are shown below after Kelsey et al. 1984).

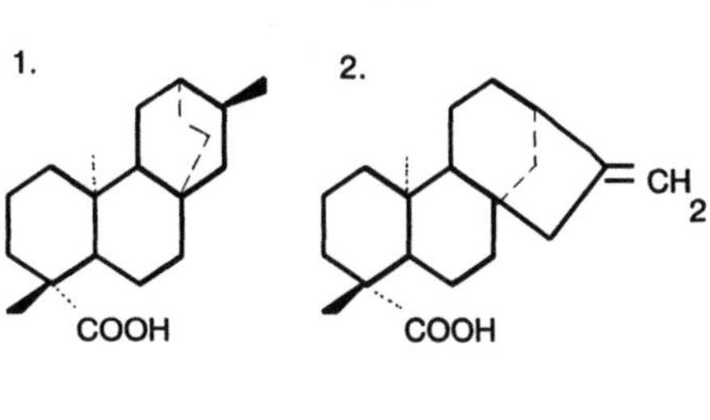

Fig. 76, 1,2. Trachyloban-19-oic acid (**1**) and 16-kauren-19-oic acid (**2**) in plant excreta

Substance	Effect
Monoterpenes	
Camphene	Insecticide
Menthol	Repellent
Menthone	Inhibitor of growth
Myrcene	Insecticide
Nepetalactone	Repellent
α-Pinene	Repellent
Sesquiiterpenes	
β-Caryophyllene	Growth inhibitor
β-Selinene	Insecticide

It should be noticed that some insects, e.g., termites, like plants, also excrete a mixture of mono- and diterpenes as a defense against predators and competitors. Their main component is as high as 90% α-pinene and the remaining 10% includes sabinene, limonene, terpinolene, and some nonidentified components (Everaerts et al. 1990). It seems that there is some convergence between products of gas excretion in both plants and animals. It may well play some role in adaptation.

Volatile terpenes can participate in biochemical interactions 'plant–plant', which is called allelopathy. Muller et al. (1969) explained the inhibitory action of sage (*Salvia*) shrubs on the grasses, growing nearby, by the excretion of toxic terpenes such as 1,8-cineole and camphor. The manifestation of the effects depends on their concentration, for which reason a depression of growth is observed under increased amounts of gaseous excreta.

Interaction of gaseous plant excreta with components of atmosphere, including of volatile pollutants is a question of special interest for both plant life itself and human habitat. Organic excreta such as alchohols, aldehydes, and kctones undergo a cleavage by ozone in the following reaction (Criegee 1959):

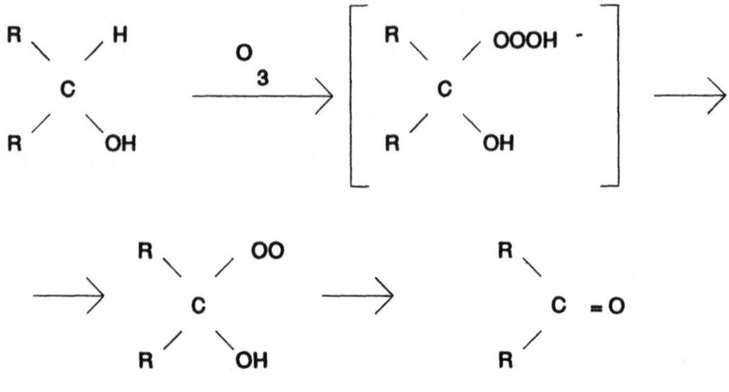

(1)

As a result of this event, oxidized products arise. A high capacity to react with ozone is known for unsaturated compounds having double bonds, in particular ethylene, isoprene, terpenoids, etc. (Razumovskii 1979). Ozone reacts rapidly with similar substances, forming ozonides which are no less toxic than ozone itself (maximal permissible dose for human 10^{-5} %). In this reaction ozone contact gaseous product released and cleave (or check off) C=C bond forming peroxide or other links.

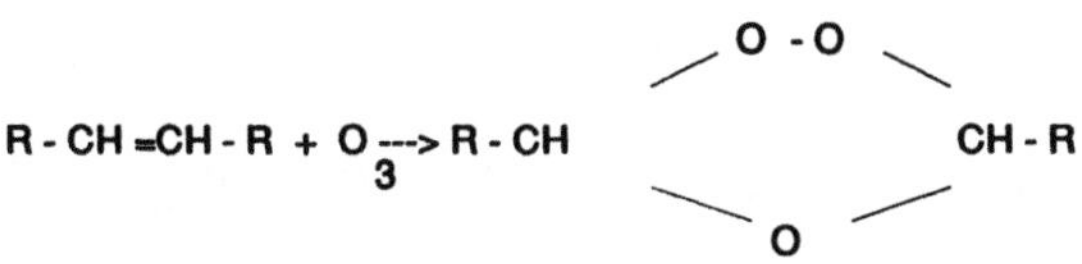

(2)

Ozonides can be easily broken spontaneously, transforming to strongly oxidized compounds with shorter carbon chains, usually aldehydes or ketones and free radicals. For aromatic volatiles, they react with ozone converting to various oxigenated derivatives, also including free radicals, as seen below:

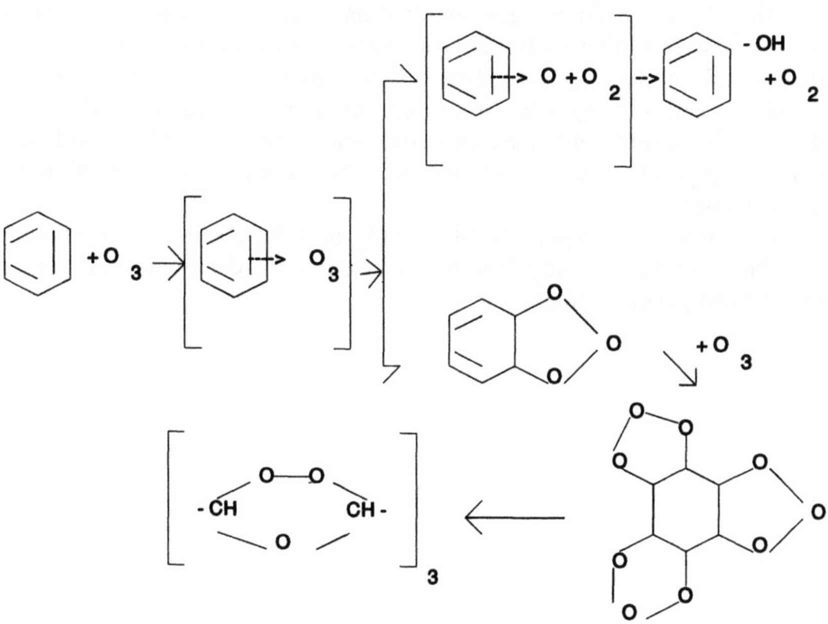

(3)

It should be noted that polycylic aromatic hydrocarbons are produced by plants more sensitive to ozone than monocyclic ones. Moreover. the ozone linkage with the aromatic nucleus preserves the ozonolytic cleavage of other aromatic rings. Phenols are lacking the hydrogen of hydroxylic groups and their aromatic nucleus is broken under ozone:

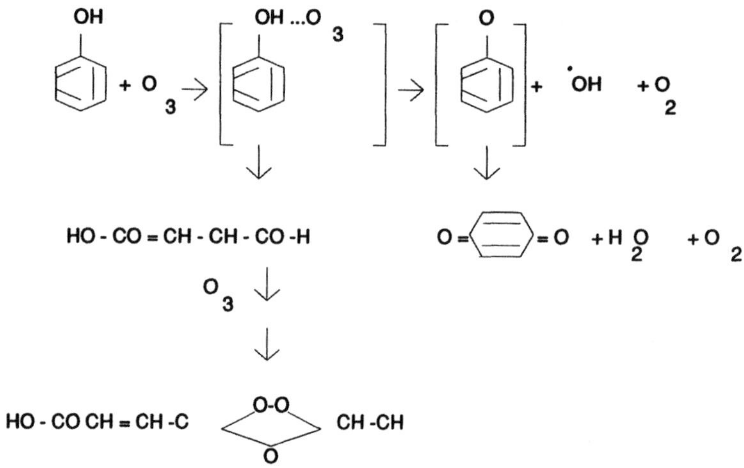

(4)

Dmitriev (1972) considered gas excretion as a protective reaction of plants against ozone excess (after storm, electric charges in the atmosphere, excessive ultraviolet and γ-radiations, etc.). This is especially seen in coniferous forests, which release many terpenoids. Moreover, this is concerned with either the negative or the positive influence of ozone and ozonides, forming with the participation of plant excreta and often being more toxic than ozone itself on the human organism.

Gaseous pollutants, especially SO_2, NO_2 and NO, can also react with volatile plant excreta, forming sulfur- and nitrogen derivatives, but this question has as yet been poorly studied.

CHAPTER 5

Leaching

The release of water-soluble substances depends to a great degree on the structural organization of the plant cell separated from the environment by two barriers: the cellulose-pectic wall and the plasmatic membrane. In this chapter we shall consider characteristics of cell wall construction, the composition of leachates, and the mechanism of leaching.

5.1 The Cell Wall as a Phase of Leaching

The structure, chemical composition, and physiological role of the plant cell covers are analyzed in detail in the works of Frey-Wyssling and Mühlethaler (1965), Miroslavov (1974), Holloway (1982a,b), and Jeffree (1986). The walls of the epidermal cells, through which plant substances are leached, have a complicated organization (Fig. 77). As is known, they consist mainly of cellulose, hemicelluloses, pectins, and glycoproteins. Molecules of cellulose are organized and combined in microfibrillas which form the framework of the cell wall. The sizes of the intermicellar and interfibrillar spaces available for diffusion are from 1 to 10 nm, which far exceeds the dimensions of the substances transported from plants (Lüttge and Higinbotham 1979). Intermicellar spaces can be decreased due to deposition of matrix substances such as pectic ones. Nevertheless, water and soluble compounds can move within these lacuna or intervals. Salt transport depends on the chemical structure of pectin, which contains chains of polygalacturonic acid. Carboxylic groups of it are either esterified (methoxy endings) or free. Free carboxylic groups form negative fixed charges in the cell walls. The protein (glycoproteids) fraction of the covers also has negative fixed charges. Therefore, anions hardly pass through the free space of the cell but, on the contrary, cations are attracted by negatively charged cell walls. Thus, transport of substances through the cell wall is determined by the pores in the cellulose framework and the fixed electrical charges of the ion groups in the cell wall matrix.

The state of the cell wall and primarily of the cuticle, which consists of a sequence of layers of different morphological structure, is of great significance for the process of leaching (Fig. 77). Usually, the cuticle has three layers. The outer surface layer is a wax named the epicuticular. Below lies a cuticular

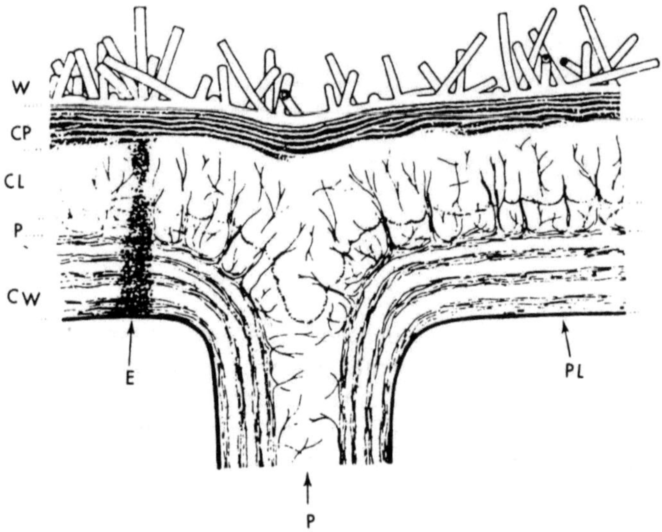

Fig. 77. The generalized structure of a plant cuticle. *E* Ectodesma; *P* pertinacious layer and middle lamella; *CW* cell wall; *CP* cuticle proper, showing lamellate structure; *Cl* cuticular layer or reticulate region, traversed by cellulose microfibrils; *W* epicuticular wax; *PL* plasmalemma. (Jeffree 1986)

laminar layer, which can be of differing thickness. For instance, in clivia, *Clivia miniata*, it is 0.5–1 μm and in dodder, *Cuscuta campestris*, 30–40 μm (Jeffree 1986). The laminar cuticular layer is often submerged in wax (Goodwin and Mercer 1983). The last internal layer of the cuticle is represented as a mixture of cutin, wax, polysaccharides of the cell wall and, perhaps, traces of protein. This layer borders on the pectin streak, which is equivalent to the middle lamina and serves as its continuation.

Cutin is the main polymer of the cuticle. It is a complicated mixture of fatty hydroxyacids bound together by esteric links forming a three-dimensional structure. Acids such as 10,16-dihydroxyhexadecanoic acid and its isomers are found more often than C_{18}-trihydroxy acids, known for example for agave, *Agave americana*, and spinach, *Spinacia oleracea*. Similar acids are absent in different species of gymnosperms (Holloway 1982a, b). It should be noted that in various species, the structure and composition of cutin can differ significantly. However, the physical and chemical characteristics of cutin are identical. They are resistant to chemical destroyers, acid hydrolysis, and oxidants. In alkali media, cutins are hydrolyzed.

Epicuticular waxes form a smooth or amorphous pellicle (film) including various crystallic structures such as tubes, fibrillas, laminas, ribbons, sticks, or granules of imperfect contour (Jeffree 1986). Tubular types of waxes are present in most plants and consist of assymmetric secondary alcohols, mainly n-nonacosan-10-o1, and their homologs, and β-diketones such as hentriakontan-14,16-dion and tritriacontan-16. Waxes are soluble even in very non polar dissolvents: petrol, hexane, or chloroform. The role of epicuticular waxes is as defense against abundant evaporation. Their removal increases their permeability to water 500-fold (Lüttge and Higinbotham 1979). In many gymnosperms and some angiosperms, such as *Andromeda polifolia*, wax filaments partly cover the stomata in order to prevent water transport to the internal air space (Jeffree 1986).

Since transpiration also occurs through the cuticle with epicuticular wax, the cuticular layers cannot be completely nonpermeable for hydrophylic particles. Permeability to water in plant cuticles was studied by Becker et al. (1986). On the isolated cuticular lamina of ten species of woody and shrubby plants, it was demonstrated that if the cuticle becomes thicker, the permeability to water and diffusion coefficients are increased. Therefore, thin cuticles are found to be more effective barriers for water than thick ones. This may depend not only on the loose position of the structural elements but also on the water content in the cell wall. Water is an important component of the cell wall and forms part of the gel-like structure of pectins. The changes in water content of the cell wall induce reversible shifts in the matrix structure and in the interaction between microfibrillas and matrix. Water also plays the role of a dissolvent and chemical reagent (Goodwin and Mercer 1983). As dissolvent, it influences the permeability of the cell wall for other molecules and ions. The permeability is higher when more water is in the cell wall. As a chemical reagent, it participates in the hydrolysis of glycosidic bonds. Thus, the hardest part of the cell wall, the main barrier for water and soluble substances, is the cuticle, with wax on its surface. However, according to many investigators (Schnepf 1964; Miroslavov 1974; Vasilyev 1977), the cuticle is easily permeable for various compounds because the movement of water and water-soluble metabolites appears to occur through dendrites consisting of the hydrophylic substance, pectin. Some authors (Mueller and Carr 1954) came to the conclusion that there are submicroscopic pores in the cuticle. This has been confirmed in experiments where the structure of the leaf surface was studied by using collodion replicas (Hall 1967). It is also possible that the pores are on junctions between cells. Kolpikov (1969) reported that the evacuation of liquid and bubbles of gases through such pores has been observed under the microscope under natural conditions. It has been suggested (Golik1965) that exodesms take part in nonstomata gas and water exchange; although they do not pass through the cuticle but within these threads, the substances are transported to the cuticle before excretion. The elimination of substances of course, depends on the presence of wax film on the cover tissues of leaves, because it is connected with the moistening of the leaf surface. Besides numerous hairs, of different degrees of complicity, which makes the diffusion of water-extracted metabolites more difficult, are localized on the leaf epidermis. All these features of the structure of the epidermal surface are attributed to the outer covers of the main epidermal cells, but they are the same for secretory organs such as trichomes which originate from the epiderma. Unlike leaching from aboveground organs of the plant, the elimination of the water solution from roots occurs more easily because the cell wall is lacking in cuticle.

Usually, compounds in the free space of the cell and not directly controlled by cell membranes are leached first. Their evacuation occurs according to the laws of diffusion, and most researchers consider it to be a passive process, dependent on the difference of concentrations. However, Grineva (1975) supposed for root systems that the release of substances may take place both via exactly diffusion way and by an active mechanism supported by metabolic energy. Tukey and Morgan (1964) showed that leaching from cells is faster during the first hours than observed later. Moreover, when leaching lasts longer, more cations (three to eight fold for Ca^{2+} in 24 h) are eliminated than are primarily contained in the plant leaves. This shows that at the beginning, the substances are evacuated from the free space and then the periplasmatic part of the cell is replenished by products from the protoplast (Kursanov 1976). Then a redistribution of metabolites in the whole organism occurs. Leaves and roots which lost a significant amount of water-soluble compounds during the leaching are supplied again with the same products flowing from other parts of the plant (Kursanov 1976). The leaching and the redistribution of the substances are connected first of all with the change in permeability of the plasmalemma. This can be observed, for example, when bivalent cations are removed. The same result can be seen in rain after a drought period, because under these conditions leaves undergo osmotic and temperature shock (Shapovalov 1973).

Perhaps the permeability of an internal cytoplasm membrane such as the tonoplast is also changed when it rains. Some substances, for instance phenols, found in leachates from leaf surfaces, may appear to be due to the increased permeability of the tonoplast. There is indirect evidence for the possible efflux of phenolic substances from the vacuole throughout the intact tonoplast (Stom 1979).

5.2 Leaching of Salts

Leaves, young shoots, roots, fruits, and flowers easily lose mineral compounds on becoming moistened. This phenomenon is named exoosmosis or leaching salts.

Tukey and Morgan (1963) studied leaching in 180 species of deciduous and coniferous trees. Plants were capable of releasing all macro- and microelements into water. The authors came to the conclusion that practically any plant can leach many substances. By using radioisotopes, it has been found (Tukey et al. 1958) that ions are leached in the following sequence:

$$Na > Mn > Ca > Mg > S > K > St > Fe > Zn > P > Cl.$$

Leaf leachates often have an alkalic reaction that is explained by the content of magnesium and potassium carbonate and bicarbonates in the exudations (Harr et al. 1984).Among 20 studied species of the family *Malvaceae*, at least 19 have pH 9.5 for the solutions collected from the upper or lower leaf surfaces. Due to leaching, the leaves of many plants are able to neutralize acid rains. This has been demonstrated for the first time (Hutchinson and Adams 1987) in experiments with artificial overhead irrigation of leaves of arctic grass, *Artemisia tilesii*, radish, *Raphanus sativus*, sunflower, *Helianthus annuus*, and sugar beet, *Beta vulgaris*. The highest neutralizing activity was marked for leaves of first three species. The same authors (Adams and Hutchinson 1987) showed that the neutralizing ability of the plants studied does not depend on the age of the leaves. However, it is more evident for cotyledons than for mature leaves. Moreover, the duration of acid rains is of much greater significance for the phenomenon.

The degree of leaching from various plants is different. For example, Ca^{2+} and K^+ are leached from leaves of kidney bean, *Phaseolus vulgaris*, maize, *Zea mays*, pumpkin, *Cucurbita pepo*, four fold faster than from leaves of potato, *Solanum tuberosum*, and tomato *Lycopersicon esculentum*, and five- to six-fold faster than from leaves of *Beta vulgaris* (Tukey et al. 1958). Phosphorus, magnesium, and nitrogen as nitrates are practically not leached from *Triticum vulgare* and maize, *Zea mays*. In water infusions from maize, nitrogen as well as ammonium, potassium and calcium have been found in smaller amounts than in those from wheat (Olifer 1972). The leaching depends not only on the plant taxon, but also on the physiological state of the plant. In young leaves this process is minimal, but increases when the leaf matures, achieving its maximum in leaves beginning to age (Tukey et al. 1958).

In nature, inorganic salts are leached by rain, dew, and fog. The amount of leachate can reach dozens of kilograms of substances per year per hectare. About 8–9 kg ha^{-1} of Ca and K, 2–3 kg ha^{-1} of Mg and N are leached within 4 months of planting in linden, *Tilia cordata*, pine, *Pinus sylvestris*, and mountain ash, *Sorbus aucuparia*, grown in suburbs of Moscow and the Moscow region (Mina 1965). According to Dalbro (1955), 1 ha of apple orchard loses through the leaves under various atmospheric precipitations about 20–30 kg of K, 10.5 kg of Ca and 9 kg of Na per year. Significant losses of substances by plants are noted by other authors as well (Sviridova 1960; Olifer 1972; Sokolov 1972). The highest leaching of mineral elements, 45% dry weight, is observed under rain in moist tropical forests of the Himalayas. Evergreen forests, and forests where foliage usually falls, in moderately wet regions of the USA, lose two fold less mineral elements, 20–25% dry weight (Amthor 1986). Carbon, nitrogen, and to a lesser degree, potassium are mainly leached by rain. Phosphorus is released in the smallest amounts in all types of forest. The leaching salts are thought to be a nutrient medium for algae and lichens of the forest canopy. Thus, the leaching of nutrient substances by rain is an important component of nutrient cycles in forest ecosystems (Parker 1983).

Ion leaching has perhaps a perfect physiological meaning. It is proposed that it facilitates the uptake of mineral elements by the roots and their movement to stems and leaves. This is experimentally shown for Ca^{2+}, whose transfer to leaves of kidney bean, *Phaseolus vulgaris*, is enhanced (Mecklenburg et al. 1966).

5.3 Leaching of Organic Compounds

Atmospheric precipitations percolating through the crowns of trees and above ground parts of herbaceous plants, besides mineral compounds, are reached also by organic matter. The loss of assimilates (mainly in the form of sugars) through the leaf surface of aboveground parts of the plant is nearly 400–800 kg ha^{-1} $year^{-1}$ (Tukey et al. 1958; Tukey 1970). In 24 h of continuous rain or fog, leaves of kidney bean, *Phaseolus vulgaris*, which easily lack water-soluble substances, can lose, in the form of saccharides, upto 5–6% dry weight (Tukey 1970). The liquids leached from leaves of quaking aspen, *Populus tremuloides*, contain sugars, mainly fructose and galactose (Wildman and Parkinson 1981). From exudates of the leaf surface and extracts of the cuticle of tobacco, *Nicotiana tabacum*, some components containing esters of sucrose (6-o-acetyl-2,3,4-tri-o-acyl)-α-D-glycopyranosyl-D-fructofuranosides), esters of wax, fatty alcohols, and α-4,8,13-duvatriene-1-ols (DTT-monols): *cis*-abienol or (Z-λ-12,1-diene-8α-o1) and α- and β-4,8,13-duvatriene-1,3-diol (DVT-diols) were found (Menetrez et al. 1990). They influence the germination of sporangium of *Perenospora tabacina* and conidium of *Alternaria alternata*. These processes are inhibited by *cis*-abienol (10 μg cm^{-2}) but not by α- and β-duvantrienemonols, esters of sucrose and carbohydrates. It should be noted that abienol had no effect on the sporangium in the presence of sucrose esters or carbohydrates. The germination of the sporangium was blocked completely by α- and β-duvatrienediols (Menetrez et al. 1990).

Many amino acids, whose chemical nature is determined by the plant species, are also easily leached from leaves (Tukey 1970). Experiments under sterile conditions showed that the number of amino acids in excretions of germinating seeds of wheat and pea reaches respectively 21 and 23. After 72 h of germination, the total amount of amino acids in the samples of pea was 11-fold higher than in those of wheat. Other organic acids can be leached as well. Wasicky (1959) reported that within 2-h of rain in Brazil, leaves of *Rhododendron* lose 34% and in some days even up to 70% ascorbic acid.

Among substances missing in leaves are growth inhibitors of a polyphenolic nature. The possibility of flavonoid elimination by water flux is determined by the form and site of their localization in the cell. Charest et al. (1986), using electron microscopy and chromatography, showed that in leaves of *Chrysosplenium americanum* methylated flavonoids are bound with the cell wall of the mesophylic epidermis. However, glycosidated flavonoids are localized mainly in protoplasts of the upper and lower epidermis, as has been demonstrated on isolated leaf protoplasts of *Pisum sativum* (Weisenböck et al. 1986). Protoplasts of the upper epidermis contain 50% and of the lower surface 35% flavonoglycosides of their total amount in cells. The lowest amount (only 14%) of the compounds was in protoplasts of the cell mesophyll. Flavonoglycosides were represented as quercetin-3-triglycoside and its ester with p-coumaric acid.

Woods (1960) reported information on the leaching of hydrojuglone by rain from leaves of black walnut, *Juglans nigra*, and walnut, *J. regia*, and its toxic action on the surrounding plants. Experiments carried out in Voronezh dendropark (Roshchina 1974a,c) showed that some samples of water, flowing down in the rainy period from the surface of the shoots of 20–25-year-old trees, contained substances of polyphenolic nature. They included the pyrogallolic group, as in the liquid probe of the smoke tree, *Cotinus coggygria*, and staghorn sumac, *Rhus typhina*; or protocatechuic groups, as in the leachates from pear, *Pyrus communis*, and walnut, *Juglans regia*. More polyphenolic compounds were found if the rainfall solutions are collected after a continuous period of dry weather. The leachates had no biological activity because they contained only a small amount of the substances. The chemical nature of a few of them has been identified. According to Komissarenko et al. (1968), leaves of smoke tree, *Cotinus coggygria*, besides the main component gallotannin, contain more than seven substances of phenolic nature. In leaves of *Rhus aromatica*, gallocatechin and free gallic acid are found, and in foliage of pear, *Pyrus communis*, the quinone glycoside arbutin whose aglycone is hydrojuglone (Kretovich 1971). Kaempferol-3-arabinoside (juglanine) and quercetin-3-α-α-arabofuranozide were isolated from leaves of *Juglans regia* (Bandyukova et al. 1968). It is possible that some kind of transformation products of the compounds are leached also by water. Leachates from the surface of whole plants which possess allelopathic activity showed the presence of a high number of phenols and terpenoids, well-known growth inhibitors – abscisic, caffeic, and p-coumaric acids (Buta and Spaulding 1989) whose quantity differed two fold in various cultivars of fescue, *Festuca arundinacea* var. Reber and *F. arundinacea* var. Kentucky (Table 10). In leachate of thorn apple, *Datura stramonium*, the alkaloids scopolamine and hyosciamine, inhibitors of growth, were found.

Leachates received from the surface of seedlings of four subtropical Taiwan grasses, *Brachia mutica*, *Digitaria decumbens*, *Imperata cylindrica*, and *Panicum repens*, were studied for seed germination and growth of seedlings in comparison to others of the same species and test plants such as lettuce, *Latuca sativa*, and Italian ryegrass, *Lolium multiflorum* (Chou 1989) Retardation of growth has been observed in *Brachia mutica* and *Panicum repens* treated by leachates of *Digitaria decumbens*. Autoinhibition by their own leachates was noted for *Digitaria decumbens* and *Brachia mutica*. In *Imperata cylindrica*, the leachate has no any inhibitory effect on its own seedlings whereas it was active in the test species. Using HELC, six phytotoxic phenols were found in the analyzed leachates, among them being ferulic, p-coumaric, 2,4-dihydrobenzoic, vanillic, p-hydroxybenzoic, and p-hydroxyphenylacetic acids in concentrations from 0.3 to 2.5×10^{-7} mol g^{-1} per sample. The total amount of phenolic acids was ca. $0.8 \times 10^{-7} - 7.75 \times 10^{-7}$ mol g^{-1} per sample.

Leaching of aldehydes by leaves is also possible. Indeed, formaldehyde (0.1–0.1 mg/l) has been found in rainwater (Junge 1963). Acetaldehyde and propionic aldehydes, as well as ethanol and propanol (Table 10), were observed in leaf infusions of woody plants (Roshchina 1973a).

Table 10. Examples of individual substances found in artificial washings from surfaces of above-ground parts of plants and their infusions

Compounds	Plant family and species	Organ	Amount of substance (μM)	Reference
Phenols	Graminae	Washings	0.01–0.150	Buta and
Abscisic acid	*Festuca arundinacea*	of leaves		Spaulding (1989)
Caffeic acid	var. Reber		9–11	–
p-Coumaric acid	var. Kentucky		14–23	–
–	*Adenostoma fasciculatum*	Infusion of leaves	–	Muller and Chou (1972)
Hydroxybenzoic	–	–	–	–
	–	–	–	–
	–	–	–	
Ferulic acid	–	–	–	
Syrenic acid	–	–	–	
Phloridzin	–	–	–	
Hydroquinone	–	–	–	
Alkaloids				
Scopolamine	Solanaceae			
	Datura stramonium	–	1600–33 000 (~ 1%)	Lovett et al. (1981)
Hyoscyamine				
(L-atropine)	–		900(~ 0.3%)	–
Aldehydes	Betulaceae			
Acetaldehyde	*Betula verrucosa*	Infusions of leaves		Roshchina (1973a)
Propionaldehyde	–	–		–
Alcohols	Fabacea			
Ethanol	*Robinia*		–	–
Propanol	*pseudoacacia*	–		–

Growth stimulators such as gibberellins are also found in leaf excretions of *Chrysanthemum morifolium* at low temperature. The leaching of cholinic ester acetylcholine from leaves and leaf stinging hairs of the common nettle, *Urtica urens*, by rain, fog, and dew were proposed as well (Tukey 1966).

A number of works are devoted to the study of leachates from leaves or their infusions collected under laboratory conditions. In most cases, the chemical nature of the leaching substances has not yet been found, although their biological activity has been shown in biological tests. Grodzinskii (1965), using biological test, has demonstrated the presence of growth retardants in water infusions from 250 plant species. The highest content of the inhibitors was in leaves, the lowest in roots, with stems in the intermediate position. Strong inhibitors of seed germination of growth of seedlings of wheat, *Triticum vulgare*, were leached by artificial rain from leaves of the weed *Parthenium hysterophorum*. Inhibitors leaching by rain are also contained in hairs covering above ground parts of the plant. (Kanchan and Javachandra 1980). According to Bovey and Diaz-Colon (1969), inhibitors are in water infusions of many tropical and subtropical plants. An attempt to identify the active matter of the samples (Grodzinskii 1965; Roshchina 1974a,c) showed that in most cases inhibitors are of phenolic nature. Water infusions from leaves of *Adenostoma fasciculatum* contain a mixture of phenols and phenolic acids, among which hydroquinone, phloridzin, hydroxybenzoic, sirenic, ferulic, and coumaric acids were identified (Muller and Chou 1972). Using gas chromatography, acetaldehyde, propionic aldehyde, ethanol, and propanol have been found in water infusions of leaves of birch, *Betula verrucosa*, false acacia, *Robinia pseudoacacia*, poplar, *Populus balsamifera*, mountain-ash, *Sorbus aucuparia*, and other woody plants (Roshchina 1973a). The microtraces may not be specific to plant species, and appear after exposure of the leaves to water. It is possible that they are formed in plant tissues logged in water due to insufficient aeration or belonging to functional products of the microorganisms developing in the infusions.

Leaching of various compounds can also take place in root systems (Rovira 1969a,b). The excreting surface of roots is very large due to the root hairs, whose ultrastructure can maintain a high level in the exchange of metabolites, and whose thick cell walls are close to the soil. All compounds which are taken up from the external media or synthesized in the plant can be released through the root systems. According to Vancura (1964), at the beginning of root growth, root eliminations amount to 7–10% of all aboveground plant mass. It should be noted, however, that studies of plant root excretions under natural conditions are difficult, due to the surrounding rhizospheric microflora, which may influence the content and quantity of the leachates. This is one reason why the main investigations have been done in water cultures, sometimes under sterile conditions. Rao (1990) reviewed information dealing with root excreta. According to his review, the quantity of the root exudate is 9% dry weight of celery plant roots, 15% ^{14}C entering seedlings, 7–10% shoot weight, 1% increase in dry weight of the whole plant, 1.6–2.9% root carbon, 50% dry weight of root (for example, of apple tree) for first vegetation, and 7% products of photosynthesis in 4-week-old maize plants. The most widespread compounds released by root systems are amino acids whose composition depends on the plant species. In root leachates of *Cucurbita pepo*, from 9 to 11 amino acids have been found, determined by paper chromatography (Sabinin 1955). In other test objects their amount was higher. Roots of *Pisum sativum* released 20, *Avena sativa* 15, and *Fagopirum sagittatum* 14 free amino acids (Rakhteenko et al. 1977). As a whole, the average number of the substances leached by the individual plant during vegetation, in μg, was: 152.6 for pea, *Pisum sativum*, 69.1 for oat, *Avena*, and 69.5 for buckweat, *Fagopyrum* (Rakhteenko et al. 1977).

The differences in quantity and quality of the amino acids excreted by various cultivars of leguminous plants may influence the specific invasion of the nodule bacteria into root hairs Kravchenko et al. (1987) came to this conclusion on studying the content of amino acids in root exudates of four types of pea, which differ in the effectiveness of symbiosis with nodule bacteria. They found 14 different amino acids, among which asparagine and glutamine dominated, while the amount of the others varied significantly, depending on the type.

Sugars in root leachates were observed by many authors. Most often these were glucose, aldose, arabinose, xylose, and raffinose (Rovira 1971). Some authors (McDougall and Rovira 1970; Johnen and Sauerbeck 1977) showed that up to 30–50% of the carbon produced in photosynthesis is excreted through the roots. According to Umarov (1986), low molecular mass carbohydrates can play a defensive role in repressing synthesis of microbial exoenzymes at the initial stage of plant pathogenesis. Guzev et al. (1990) showed that the release of low-molecular mass substances such as sugars, formed mainly in photosynthesis, is an effective defensive mechanism against microbial destruction of plant tissues. This is proposed to have a certain ecological significance. The experiments demonstrated the influence of plant root excreta on the cellulolytic activity of soil microorganisms. A possible mechanism of the defensive phenomenon is in a repression of the synthesis of hydrolytic enzymes.

Organic acids, such as oxalic, malic, fumaric, pyruvic, and acetic acid, are also often found in root exudates (Rovira 1971; Grineva 1975). Release of organic acids at anaerobiosis was studied by Grineva (1975). A quantitative determination of the organic acids released was performed by Kloss et al. (1984). The exudates of roots and root tissues from sterilized seedlings of *Diplachne fusca* determined on agar and nutrient solution included the following acids, mg/40 ml of solution: mainly malic, 550, and to a lesser degree, citric 350; fumaric 100, and succinic 90.

An important study of root exudates of normal and chlorotic plants of maize, *Zea mays*, was carried out by Elgala and Amberger (1986). For the collection of samples they used the original technique of sand-water culture (lower part of the roots was immersed in the solution and the upper was in the sand). Under these conditions, chlorotic plants excreted more citric, pyruvic, glutaric, α-ketoglutaric, tartaric, and glyoxylic acids. The difference in the content of the exuded amino acids was also found: such amino acids as cysteine, isoleucine, leucine, and lysine were identified only in root excretions of plants suffering from iron deficiency. The release of a noticeable amount of organic acids may help to recover from this stress, because the exuded acids dissolve salts of Fe absorbed and precipitated on the root surface.

The connection of secretory process in roots with mineral nutrition is confirmed by experiments where the influence of complete nutrient media and phosphorus deficit were analyzed (Lipton et al. 1987). Under phosphorus deficit, the total amount of organic acids in the root exudates of alfalfa, *Medicago sativa*, was increased to 48.92 $\mu mol\,g^{-1}$ dry mass. In particular, that quantity of acids was 1.5–2-fold higher in stressed samples than with normal mineral nutrition, the amount of malic acid even decreasing by 20–25%. The increase of organic acids in root excretion is thought to be a certain defensive mechanism to overcome unfavorable mineral nutrition conditions, as observed above for Fe deficiency. The acid exudation by roots can be an adaptive feature. Indeed, citric acid is excreted by roots of white lupin, *Lupinus albus* L., and is precipitated on the root surface in the form of calcium citrate if the plant grows on calcium-rich soils. The pH of the exudates reached 4.8, which induces the solution of $CaCO_3$ from the calcified soils (Dinkelaker et al. 1989).

Among various organic compounds which are found in root exudates, polyphenolic substances attract special attention. Djordjevic et al. (1987) showed that roots of white clover, *Trifolium repens*, released hydroxylated phenols including 7,4-dihydroxyflavone, coumarin, umbelliferone, and isoflavone formononetin. As also for marigold, *Tagetes patula*, their roots exuded two benzofurans and four thiophenes, and five-chain sulfur-containing heterocyclic compounds (Fig. 78) (Tang et al. 1987). These secondary metabolites were shown to be toxic for various organisms, especially for microorganisms. Due to phenols such as scopoletin, coumarin, and phenolic acids, root exudates of wild oat, *Avena fatua*, inhibited the growth of roots and coleoptile of spring wheat, *Triticum aestivum*, seedlings (Perez and Ormeno-Nunez 1991). Moreover, bacteria able to metabolize phenols are present in the rhizosphere.

Flavonoids are an important class of compounds found in root exudates (Rao 1990). Some of them are taxon-specific, for example:

Arachis	3,4-Dihydroxyflavone
Malus domestica	Phloridzin
Lens culinaris	4′,7-Dihydroxy-3-methoxyflavone
	3′,4′,7′-Trihydroxy-3-methoxyflavone
	3′,4′,7′-Dihydroxy-3-methoxyflavone
Glycine max	Coumestrol
	Daidzein (4′,7-dihydroxyisoflavone)
	4′,7-Dihydroxyflavone

About 3% of the total amount of isoflavonoids in *Cicer arietinum* L. and *Phaseolus aureus* can be excreted to the environment (Rao 1990). The quantity of

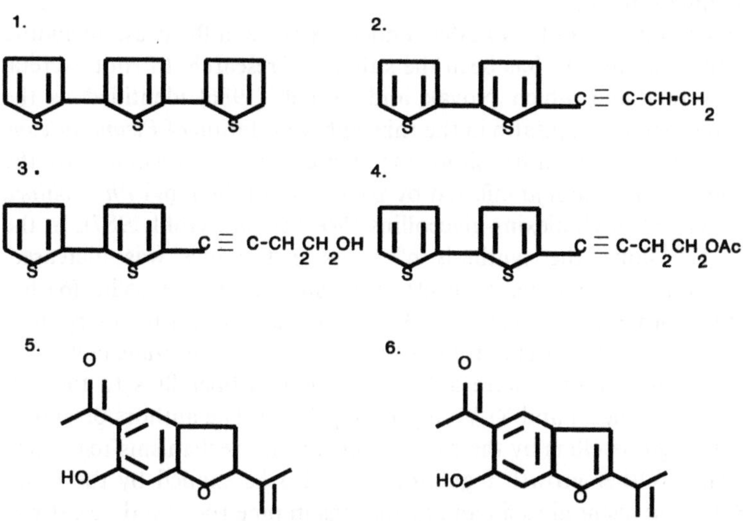

Fig. 78, 1–6. Thiophenes (**1–4**) and benzofuranes (**5,6**) in root excreta of *Tagetes patula*. **1** α-Tertienyl, **2** 5-(3-Buten-1-yl)-2,2′-bitienyl, **3** 5-(4-hydroxy-1-butinyl)-2,2′-bitienyl, **4** 5-(4-Acyl-1-butienyl)-2,2′-bitienyl, **5** 6-Hydroxy-2-isoprop-enyl-acetyl (coumaronon or dihydroeuparin), **6** Euparin. *Ac* acyl flavonoids exuded by the roots is rather high, although it often varies. For this reason, they can influence plants and microorganisms inducing allelopathic effects, acting on the biological fixation of nitrogen, recycling of nitrogen, the activity of nitrate-fixing and nitrogen-fixing bacteria, the symbiotic fixation of nitrogen serving as defensive compounds against parasites (Rao 1990).

For the first time (Lundegardh and Stenlid 1944) found 3′,4′-dihydroxyflavone in root exudates of peanut, *Arachis*, which was later confirmed by Petril and Chrastil (1955). Of the total phloridzin, 3.5% is excreted in nutrient solution by the roots of the apple tree (Borner 1959, 1960) and this tree also exudes coumestrol and daidzein, which are about 20% of the total root isoflavonoids (D′Arcy-Lameta 1986). In root exudations of water-grown plant of Japanese sweet flag, *Acorus gramineus*, phenylpropanoids have been found (Greca et al. 1989), among them 1,2-dimethoxy-4-(1′Z-propenyl) benzene (1) and their derivatives, which can have allelochemical effects on green algae and cyanobacteria.

Hairy root cultures of *Lythospermum erythrorhizon* Sieb. et Zucc (Boraginaceae) secrete the naphthoquinone pigment, shikonin, which possesses antibacterial activity and stimulates the formation of the granulated tissue that is used in medicine and cosmetics (Shimomura et al. 1991). The production and excretion of this red pigment depend on the type of solid medium.

The exudation of phenols is very important for plant-soil-microbial systems because of their effects as allelopathic compounds and plant growth regulators on one hand and defensive substances in plant pathogen interactions on the other (Siqueira et al. 1991). In the latter case, phenolic compounds play the role of signaling molecules, acting either as inducers for virulent genes in pathogenic conditions, in particular *Agrobacterium*, or as inducers or repressors, as in *Rhizobium*-legume symbiosis.

The composition of root exudates under extreme influences, including infection by microorganisms, has been the subject of research. In root excreta, phytoalexins were found which Novak and Stanek (1986) identified as the already known phytoalexin pisatin in the rhizosphere solution of *Pisum sativum*, whose appearance and quantity show the degree of pea resistance to the pathogen. Roots of *Glycine soja* infected by zoospores of the fungi *Phytophtora megasperma* secreted phytoalexins-glyceollins (Werner and Hohl 1987). In the presence of Mg^{2+}-ions, the synthesis of stress metabolites was increased, whereas Ca^{2+}-ions in the some concentrations suppressed glyceollin (6a-hydroxyphaseolin) formation. In some works, the release of hormones by root systems has been studied. Roots of tomato, *Lycopersicon esculentum*, cultivated on Hogland nutrient solution were able to eliminate gibberellins to the surrounding media (Couillerot and Yi Young-Byung 1985). The authors presumed that exudation of gibberellins by the roots is one of the mechanisms regulating its level in plants. Other growth regulators can also be exuded by the roots. Gogala (1988) found jasmonic acid and cytokinins in pine root exudates. It was shown that this first substance is a strong inhibitor (even at a low concentration, 100 μg/1) of mycorrhiza growth, while cytokinins, on the contrary, stimulated fungal growth. In root exudations of cotton, *Gossypium*, a complex of the cytokinins zeatin, zeatin riboside, and isopentenyladenineadenine was found (Ding Jing and Shen Zhen-de 1985), with zeatin dominating.

In exudations of the high-molecular mass fraction of roots, polysaccharides, uronides, and proteins have been identified (Collet 1975). Enzyme invertase and amylase were found in root excretions of wheat *Triticum vulgare*, pea, *Pisum sativum*, and maize, *Zea mays* grown under sterile conditions (Krasil'nikov 1958). In root exudates of seedlings of *Lactuca sativa*, oxidases which can oxidize catechin were identified (Tan and Kubo 1990a). The solution of catechin around the roots was colored, forming a yellow-red product with an absorbance maximum at 420 nm, and the thermal treatment blocked the effect. Homogeneous (by 80%) protein with a molecular mass of 20 kdaltons was isolated from the exudates. The K_m of this oxidase of catechin was 1.82 mM at t° = < | > 30 °C and pH 8.2. In root excretions of other plants such as alfalfa, *Medicago sativa*, tomato *Lycopersion esculentum*, and maize, *Zea mays*, catechinoxidases were also found (Tan and Kubo 1990a, b). In exudates from maize roots, protein of 6.3 kdaltons was isolated, in whose presence oxidation of catechin by seedlings was blocked (Tan and Kubo 1990b). It is suggested that oxidases decrease the growth-inhibiting effect of the natural phenols, catechol and hydroquinone, by oxidizing them (Tan and Kubo 1990a). Vitamins were also observed in root exudates of wheat, pea, and maize. (Krasil'nikov 1958; Rovira 1971).

In root exudates, terpenoid compounds are present, among which some significantly influence the growth of surrounding plants (Table 11). They are mainly sesquiterpenes and sesquiterpene lactones (Fig. 79). All sesquiterpene lactones (10^{-9}–10^{-5} M) stimulated the germination of seeds of obligate root parasites (Fischer et al. 1989a, b), and they are able to inhibit or stimulate this process for seeds of other nonparasitic plants. Some parasites cannot germinate at all if the root excreta of the host plant are absent from the soil. For example, strigol, which is contained in the excreta, may stimulate germination of obligate root parasites of cereals such as *Striga asiatica* up to ten fold in comparison with control. (+)-Strigol found in cotton root exudate induces the germination of

Table 11. Terpenoids of root exudates stimulating or inhibiting plant growth.

Substance	Plant	Reference
Burrodine	*Ambrosia dumosa*	Fischer et al. (1989a, b)
Guayaulin A	*Parthenium argentatum*	–
7-α-Hydroxy-3-desoxysaluzazine	*Podochaenium eminens*	–
Confertiflorin	*Ambrosia confertiflora*	–
Desacetylconfertiflorin	–	–
Dihydroparthenolide	*A.artemisifolia*	–
Parthenin	*Parthenium hysterophorus*	–
Strigol	*Gossipium hirsutum*	Cook et al. (1966)

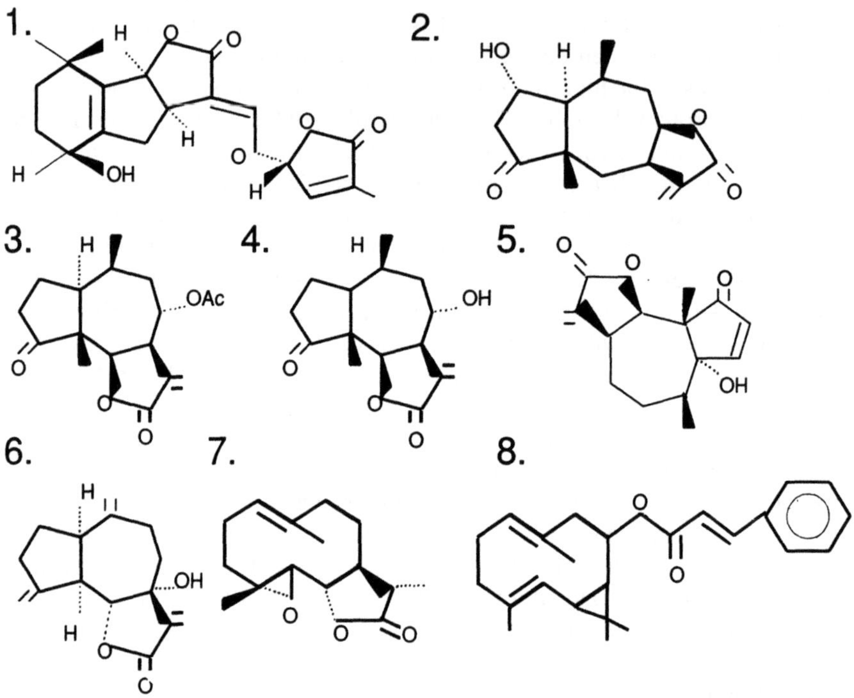

Fig. 79, 1–8. Sesquiterpenes stimulating or inhibiting plant growth. **1** Strigol, **2** Burrodine, **3** Confertiflorin, **4** Desacetylcofertiflorin, **5** Partenin, **6** 7-α-Hydroxy-3-desoxyzaluzanin C, **7** Dihydroparthenolide, **8** Guayulin A

parasites such as *Striga asiatica* (L) Kuntze and *Orobanche aegyptiaca* Pers, acting at concentrations lower than 10^{-11} M. Parasites from the genera *Striga*, *Alectra* (family Scrophulariaceae, and *Orobanche* (family Orobanchaceae) significantly decrease the cereal yield in tropical and subtropical countries. Their seeds live in the soil for many years until exudates of the host plant stimulate their germination former Cook et al. (1966, 1972) first isolated (+)strigol from cotton root exudates and identified it. These authors also showed that natural (+)-strigol stimulates seed germination in *Striga asiatica* by 50%. The activity of artificial (−)-strigol was even higher, an effect that can be achieved at a concentration of 10^{-16} M (Hauck and Schildknecht 1990).

The leaching of various substances also takes place at seed germination. More often in seed exudates there are sugars, including both mono- and disaccharides, amino acids, and organic acids. The substances are found in leachates from germinating seeds of pea, *Pisum sativum*, soybean *Glycine hispida*, common vetch, *Vicia sativa*, and kidney bean, *Phaseolus vulgaris* (Amoros and Durand 1964). By paper chromatography, the release of 14 amino acids (lysine, arginine, glutamine, leucine, valine, asparagine, etc.), four sugars (glucose, fructose, saccharose, arabinose) and three organic acids (malonic, citric,fumaric)

Table 12. Phenols in leachates from seeds and seedlings

Substance	Plant	Part of plant	Reference
Acids			
Protocatechuic	*Pterocarpus santalinus*	Seeds	Venkataramaiah et al. (1984)
Homoprotocatechuic	–	–	–
Floretic	–	–	–
Melilotic	–	–	–
Vanillic	–	–	–
–	*Brachia repens*	Seedlings	Chou (1989)
–	*Digitaria decubens*	–	–
–	*Imperata cylindrica*	–	–
–	*Panicum repens*	–	–
Ferulic	–	–	–
–	Brachia repens	–	–
–	*Digitaria decubes*	–	–
–	*Imperata cylindrica*	–	–
–	*Panicum repens*	–	–
o-Coumaric	*Brachia repens*	–	–
–	*Digitaria decubens*	–	–
–	Imperata cylindrica	–	–
–	*Panicum repens*	–	–
–	*Pterocarpus santalinus*	Seeds	Venkataramaiah et al. (1984)
Isoferulic	–	–	–
Salicylic	–	–	–
2,4-Dihydroxybenzoic,	*Brachia repens*	Seedlings	Chou (1989)
p-Hydroxybenzoic,	*Digitaria decubens*	–	–
p-Hydroxyphenyl acetic	*Imperata cylindrica*	–	–
	Panicum repens	–	–

by seeds of *Pinus rubra* have been shown (Agnihotri and Vaartaja 1969). In nutrient media where pea seeds germinated, 20 amino acids and some peptides were determined (Boulter et al. 1966). Amino acids and sugars were found in exudates of seven sorts and hybrids of rice, *Oryza sativa* (Leelavathy 1970). Twenty one amino acids are present in the excretions of maize seeds and 23 of lupin. Among them glutamine and α-alanine predominated (Berestetskii and Kravchenko 1980a, b). When seeds germinate scutellum, ordinary cotyledon in cereal grains eliminates glucuronic, phosphoric, and citric acids (Shchiparev et al. 1976). It is presumed that acidation of the endosperm in the first hours of swelling and germination is one of the important triggering mechanisms of caryopsis germination.

In leachates of seeds and seedlings, phenolic substances were also identified (Table 12). In infusions from seeds of sandalwood padauk, *Pterocarpus santalinus*, about 17 phenols are found. Melilotic acid and esculetin leach more slowly than other compounds. In leachates of four subtropical grasses from Taiwan, six phytotoxic phenols were identified by Chou (1989). Their amount varies from 0.3 to 2.5×10^{-7} mol g^{-1} per sample while the total quantity of phenolic acids was 0.8–7.75×10^{-7} mol g^{-1} per sample. It is thought that phenolic compounds of seeds promote plant embryo resistance to microflora and maintain their dormancy.

Alkaloids are also released by germinating seeds. They have higher physiological activity than other excreted substances. When seeds of thorn apple, *Datura stramonium*, swell, scopolamine and hyoscyamine are eliminated. On the first day, leaching was 0.12% of the total amount of alkaloids. The growth of some plants was depressed by 0.1–1% scopolamine (Lovett and Potts 1987). Mature seeds of tea, *Thea*, released caffeine (1,3,7-trimethylxanthine) and theobromine (3,7-dimethylxanthine), localized in the seed coat (Suzuki and Waller 1987). On the swelling of the seed and peel destruction, caffeine is transformed into a water-soluble form. The extracts retarded the growth of roots and stems in a test plant. Alkaloids found in the seed coat appear not to be used for embryo nutrition, but to serve as a defensive barrier between embryo and environment. During seed germination, caffeine is liberated gradually and does not inhibit the growth of seedlings.

Other biologically active substances are also found in seed exudates. Sterile seeds of *Brassica oleifera* eliminate thiamine, biotin, pantothenic, nicotinic, and folic acids (Bloudeau 1966). Toxic nonprotein amino acids are also released. In fact, the elimination of these amino acids begins just after imbibition in seeds of wisteria, *Glycine wightii*, incubated in distilled water. The nonprotein amino acid canavanine, which is an analog of arginine and inhibitor of seedling growth, appears in the exudations after 60 h of imbibition (Bell 1972). Canavanine, in the amount 4–6% fresh mass, is the canavalia bean *Canavalia ensiformis* (Bell 1972). The ecological function of such toxins in seeds appears to be as a defense against being eaten by herbivorous animals.

Thus, under rain, fog, artificial overhead irrigation, soaking, growing on water solutions or water-logging (flooding), a certain quantity of ions and organic substances can be released from plant tissues. Practically all water-soluble substances contained in plants can be leached. The total amount of the components extracted by water reached 0.7-3% total plant mass, as is observed for tobacco, *Nicotiana tabacum*, and maize, *Zea mays* (Rovira 1959).

5.4 Dependence of Leaching on External Factors, Phase of Development, and Anatomy of Plants

The release of various substances is considered as a normal function of the plant organism (Kirillova 1964; Grodzinskii 1965, and many others). However, under optimal conditions, the amount of the leached substances is rather small. Damage and treatment by various irritators such as sublethal temperature, changes in pH of medium up to 3.5, nonstable solutions, etc. strengthen the release of compounds from plants (Belikov and Kirillova 1958; Kirillova 1964; Tukey 1970). The above-mentioned changes are a cause of the increase in membrane permeability, the reaction of a living cell to injury.

As has been shown in some experiments, the leaching of mineral and organic compounds from leaves depends on many external factors. Thus, in leaves of *Coffea arabica*, the loss of K^+-ions is connected with variability in temperature (Arens and Arens 1970). The lowest excretion was observed at 8 °C, and with an increase of temperature up to 23–30 °C, the process was enhanced several times. Light strengthened the leakage of organic compounds, especially sugars (glucose), from leaves as well as of phosphorus and sulfur (Tukey et al. 1958). Leaching also depended on the pH of the solution by which the plant was moistened. Intact leaves of kidney bean treated by solution with varying pH released different amounts of mineral substances (Evans et al. 1981). At low pH, leaves become more permeable for Ca^{2+}, nitrate, sulfate, and at pH 5.7 lose more K^+ and Cl^-.

The quantitative and qualitative composition of root excretions varies for plants grown on soils of different fertility. According to Kaverzneva (1986), the excetory action of pine roots in loamy soils is markedly higher than in sandy loam. If the soils are richer, the roots release more amino acids and carbohydrates. The total average amount of carbon leached to the soil from roots was 250–300 kg ha^{-1}. Organic acids, terpenoids, and the enzymes of nitrogen exchange, urease and glutaminase, have also been found in the root exudates.

Leaching is dependent on the phase of plant development because the direction of common metabolic exchange is changed during vegetation. Leakage of sugars from leaves of spurge, *Euphorbia esula*, and mother chrysanthemum, *Chrysanthemum indicum*, becomes weaker in the period of flower bud formation. Then it increases extremely in the period of flowering, and drops again at the end of vegetation. Young, actively growing tissues are more resistant to the action of rain than older ones. Since the leakage of substances is enhanced in the flowering period, rains depress the development of some flower buds of apple trees (Shapovalov 1973). Direct determinations have shown a dependence of sugar leaching on the stage of fruit ripening as well. The leakage of sugars increases from 4 mg per day for nonmatured fruit to 84 mg per day for ripe ones. When rains are continuous and frequent, this loss influences the yield (Tukey 1970).

The content and amount of leached compounds depend not only on the chemical composition of the interior of the plant, the intensity and direction of metabolic exchange, but also on anatomo-morphological features of the plant cover tissues. Certainly, we cannot expect direct correlations between the amount of a substance in the leaves and its efflux into environment.

The study of the composition of plant excretions is also necessary to take into account their important role for epiphytic microflora (Krasil'nikov 1958). Some part of the substances found in leakages, leachates, nutrient solutions, etc. are perhaps products of the living functions of microorganisms themselves.

Indeed, the increase in ethanol concentration and change in the composition of other leached components observed under a rise in temperature and/or long duration of experiments should be attributed to the more active physiological function of living organisms inhabiting on plant surface.

5.5 Physiological Meaning of Leaching

Continuous cloud bursts can remove a significant amount of substances, leading to the retardation of plant growth and development. Cholodnyy (1932) found that rain decreases the harvest of wheat, *Triticum* sp., and barley, *Hordeum vulgare* at the stage of milk ripeness by more than 30%, due to leakage of carbohydrates. The yield of winter wheat and tomato, *Lycopersicon esculentum*, can be affected by artificial overhead irrigation which causes the loss of metabolites (Panenko and Pilipenko 1981). The yield of fruits is better in a dry season than in a rainy one, when the loss of sugars occurs (Tukey 1970). These and some other facts led to conceptions that retardation in plant development and decrease in harvest are a result of the loss of necessary metabolites due to rain. Moreover, the release of the necessary substances inhibits plant growth, while addition of bivalent cations can restore the process.

Leaching of metabolites can also be of positive significance because in this manner plant tissues can be freed of an excess of substances. Moreover, leakage of mineral compounds from aboveground part of the plant facilitates their uptake and movement to the leaves from the roots. Indeed, loss of Ca^{2+} by leaves accelerates the ion uptake by roots and its subsequent transport along the stem (Mecklenburg and Tukey 1964). Simultaneously with the release of mineral elements and sugars, other compounds, which possess a bactericidal activity and play a certain role in plant immunity, can be eliminated from the free space of the cell. Leaching of phytohormones, inhibitors, alkaloids, and phenols enriches the environment by biologically active substances whose absorption and accumulation by soil may cause allelopathic effects (Grümmer 1955; Grodzinskii 1965; Ivanov 1973). Root exudates are essential for the interaction with rhizospheric microflora. Ethanol, acetaldehyde, and other metabolites released by the root serve as the origin of carbon for microbes inhabiting the root surface, whose growth depends on the intensity of the excretory process (Azarova 1983). In other cases, root exudates can also play the role of attractants for microorganisms in the rhizosphere. For example, they have been shown to serve as a signal attracting free-living nitrogen-fixing microorganisms which inhabit the roots of maize, *Zea mays*, and wheat, *Triticum vulgare*, and as a result, an increase in bacterial motility by 25–100% was observed. Amino acids such as alanine, asparagine, glutamine, lysine, serine, and valine contained in root leakage can also play the role of attractants, acting at concentrations of $10^{-6} - 10^{-5}$ M (Ariel and Lemos-Pastrana 1980b).

CHAPTER 6

The Elimination of Substances in Response to Extreme Factors

Until recently, the excretory function of a plant under extreme factors has not been discussed in the literature. Stress situations arise at various injuries of the organism when either an extremal power is higher than the threshold for normal adaptation or variations of the acting factor are too sharp for the plant to maintain homeostasis. The term "stress" was introduced for the first time by Selye (1956) to designate animal reactions to various sharp changes in the environment. The reciprocal reaction is considered to be a defensive one which can be directed to the preservation of the individual. Now the term stress has become widespread and is used for all living organisms including plants. The stress factors can be both abiotic (mechanical, chemical etc) and biotic (invasion of pathogens and others). In such cases, defensive mechanisms of plants where the release of substances plays an important role are triggered. The excretory function as a chain of physiological and biochemical reactions of the cell in response to stress will be described in this chapter. Moreover, special attention will be given to the composition of plant excreta as a possible indication of a stressory state of the organism.

6.1 Injuries to Membranes Under Stresses

Various factors by external media can induce injuries to the plant organism resulting in the elimination of secondary metabolites, which can be considered as the liberation of physiologically active substances.

Such examples, taken from the review of Matile (1987), are illustrated in Fig. 80. In the plant cell, potentially toxic secondary compounds and enzymes, responsible for their transformation to poisons, are normally, separated from each other. In *Sorghum bicolor*, the cyanogenic glycoside dhurrin is localized in the vacuolar sap of the epidermal cells, while two enzymes, β-glycosidase and hydroxynitril lyase, which can decompose the glycoside to hydrocyanic acid (HCN), are in the cytozole and chloroplasts of the mesophyll. The interaction between dhurrin and these enzymes occurs only after destruction of both compartments by mechanical injury, for instance, on feeding by herbivorous animals.

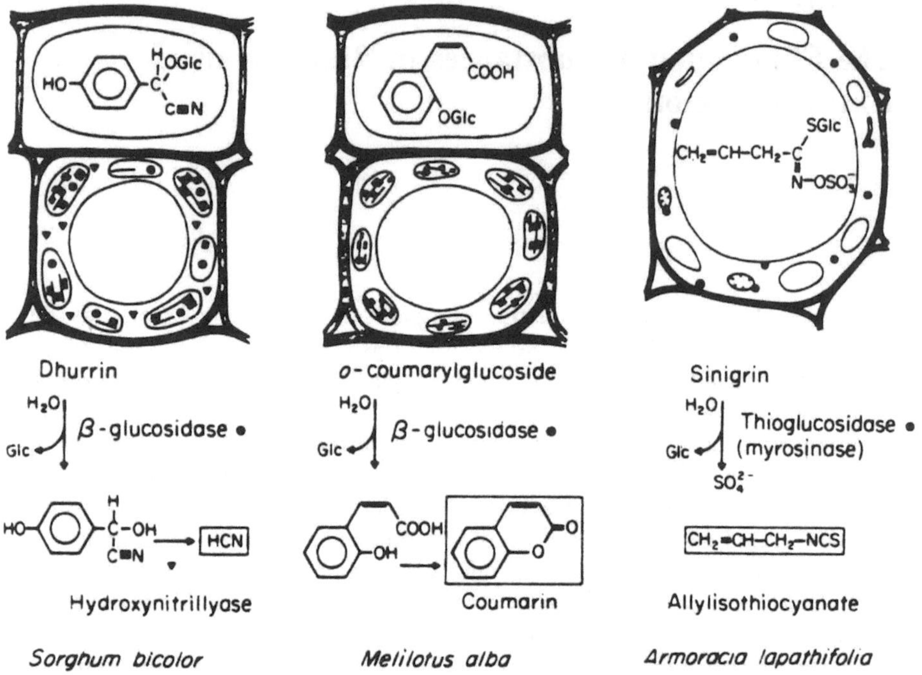

Fig. 80. Differential localization of potentially toxic secondary metabolites and of the enzymes responsible for the release of toxic compounds. *Glc* glucose. (Matile 1984, 1987)

Unlike in *Sorghum*, vacuoles of epidermal and mesophyllous leaf cells of sweet clover, *Melilotus alba*, contain the glycoside o-coumaric acid but the enzyme *β*-glycosidase, which cleaves off this substance to toxic coumarin, is localized in the outer layers of the mesophyll cell walls. In this case, two membranes, the tonoplast and the plasmalemma, separate such a glycoside from the appropriate hydroxylase. The breaking of these barriers in injured tissue leads to the interaction of the cyanogenous glycoside with the enzyme, which results in the formation of coumarin, responsible for the characteristic odor of newmown hay. Finally, another example of the different compartmentation of the precursors of toxins and enzymes, transforming them into toxins, is in horse radish, *Armoracia lapathifolia*, the glycosinolate sinigrine; located in the vacuole, it is separated from thioglucosidase, which can convert the substance to isothiocynate only by the tonoplast. This enzyme is localized in the cytozole. The decomposition of sinigrine occurs after injury to cell integrity, when thioglucosidase comes into contact with the glucosinolate. As a result, the characteristic stinging taste and smell of horse radish, determined by the products of sinigrine decomposition, appears. Thus, since normally secondary compounds are stored in the compartment, their elimination to the external media is possible only after disturbance of the barrier function of the tonoplast, plasmalemma, or other membranous structures (see also Chap. 1.2). In this connection, possible mechanisms of membrane injuries under stress should be considered. The lipids of membranes are the most sensitive to damage, and are degraded first (Gailliard 1978). For instance, when potato tubers are cut, 20% of the endogenous membrane lipids are lost during the first few seconds and 40% within 1 min. This may be due to deacylation of the membrane phospholipids and galactolipids. As a result, fatty acids are liberated. In intact cells they are found in small quantities because free fatty acids are toxic. Free fatty acids can be attacked by oxidative enzymes. Among the products of oxidation are hydroperoxides of fatty acids, carbonic compounds, and other oxigenated derivatives. For instance, in tomato, *Lycopersicon esculentum*, the formation of the aldehydes nonal and hexenal from linolenic acid occurs according to the hydroperoxide type of reaction with the participation of the enzyme lipooxigenase (Galliard 1978).

Other sites of membrane damage are the protein components forming complexes (receptory, catalytic, etc.). Under the action of non-favorable factors, changes in ion permeability, which is modulated by receptors, and in the activity of enzymes participating in the energetic and metabolic reactions, take place. One of the most sensitive to damage is considered to be the process of photophosphorylation. Indeed, the decrease in ATP level in a cell destroys the barrier function of the membranes. Finally, damage to the plant organism leads to shifts in the whole metabolism where all reactions interconnect with each other. For this reason, reactions which do not result from an injury are thought to be the exception rather than the rule.

Damage to membranes is induced by a deviation (an increase or decrease) from the optimal temperature that stimulates the efflux of organic substances from living cells. In experiments by Belikov and Kirillova (1958), exoosmosis of organic matter from scales of onion and leaves of *Elodea* increased more at 35–45 °C than at 17–20 °C. Analogous effect has been observed for roots of yellow pond-lily, *Nuphar luteum*, and cucumber, *Cucumis sativus*. The increase in membrane permeability occurred in hypertonic solutions due mainly to leaching of the cations Ca^{2+} and Mg^{2+}, which stabilize membranous systems. The addition of magnesium and calcium salts (5×10^{-4} M) to distilled water decreases the release of both organic and inorganic substances (Belikov and Kirillova 1958). A disturbance in membranous permeability is usually observed when the plant is attacked by pathogenic fungi and microorganisms which also produce toxins. The mechanism of the action of similar substances is unclear in most cases. Membrane permeability is changed also under other factors. Finally, the metabolism of the whole plant is altered due to the action of low or high temperatures, pesticides, etc.

6.2 Metabolites Released Under Stress

The term stress or stress-induced metabolites is used for substances which, as a rule, are absent under normal conditions and formed from inactive precursors under stress. Similar compounds are usually synthesized de novo. However, under stress, plant exctreta may include also substances liberated from compartments on damage, or products of their hydrolysis and oxidation. In spite of much progress in the study of stress compounds as a whole, there are few works devoted to the chemical nature and mechanism of the production of substances released when the plant is wounded or damaged. Recent data in this field is presented here.

6.2.1 Ethylene

Release of ethylene by plants as a reaction to various abiogenous and biogenous stresses is a widespread phenomenon. Ethylene is known as a phytohormone produced by normally functioning plant tissue (see Chaps. 2.2.2 and 4.3.1). However, when extreme factors influence the plant, the production of this gas is increased. More ethylene is also produced on aging of flowers of species such as *Petunia hybrida.* This process is also activated if the stigma of the pistil is wounded (Whitehead et al. 1984). Mechanical injury and inoculation of *Bipolaris sorokiniana* led to an intensive release of ethylene, an increase in the amount of 1-aminocyclopropane-1-carboxylic acid (ACC), and the activity of ACC-synthase of the leaf laminae of meadow-grass, *Poa pratensis*, L. (Coleman and Hodges 1987). When the plant suffers damage mechanically, a rise in ACC activity is observed after 1 h, in the intensity of ethylene production after 2 h, and in the quantity of ACC after 3 h. Similar rates of ethylene emission were observed under γ-radiation (Laurigaudierre et al. 1990). The gas release under such condition increases from 0.3 (nontreated tomato fruits) to 2 nmol g^{-1} fresh mass h^{-1} (treated ones) as a result of stimulation of ACC-synthetase. Like mechanical damage. γ-radiation enhanced ethylene production, probably, due to stimulation of PNA translation. However, a dose of > 1 kR (1000 J kg^{-1}) blocks the activity of the ACC-synthetase itself. Infection by *B. sorokiniana* first greatly increases gas emission (after 36 h), then the activity of ACC synthase (after 72 h). The Maximum accumulation of ACC in this case is observed later – after 96 h. The amount of ACC and the activity of ACC synthase was essentially higher in plants infected by *B. sorokiniana* than in wounded ones.

The production of ethylene is enhanced when there is a disturbance in the total gas regime of the tissue. On flooding, the deficit of O_2 stimulates synthesis of the phytohormone. According to Tang and Kozlowski (1984), the amount of ethylene in seedlings of American elm, *Ulmus americana*, red ash, *Fraxinus pennsylvanica*, longbeak eucalyptus, *Eucalyptus camaldnisis*, and other species amounts to 0.4–6.0 to 20 nmol g^{-1} dry mass in the flooded plants and only 0.2–0.3 nmol g^{-1} dry mass in unflooded ones. Analogous phenomena are observed in the stem bark of sunflower, *Helianthus annuus*, tomato, *Lycopersicon esculentum*, and kidney bean *Phaseolus vulgaris*, where flooding stimulated ethylene synthesis which, in turn, increased cellulase activity and the formation of aerenchyma (Kawase 1981). A deficit of O_2 induced synthesis of ethylene and also the appearance of aerenchyma in roots of *Zea mays* (Atwell et al. 1988).

The production of ethylene is stimulated by an increase in CO_2 concentration in the air surrounding the plants. This process is facilitated, and the activity of the enzymes participating in ethylene formation is also enhanced, accelerating the transformation of 1-aminocyclopropane-1-carboxylic acid (ACC) to ethylene in leaf disks of tobacco, *Nicotiana tabacum* (Philosoph-Hadas et al. 1986). This has been confirmed by experiments on axial organs of cockle, *Xanthium pennsylvanicum* (Esashi et al. 1986). A maximum increase in ACC and its conversion to ethylene was observed at a CO_2 concentration from 1 to 10% (depending on the O_2 concentration in atmosphere). When seeds of the same plant swell, carbon dioxide increases the ethylene formation, acting on two stages of its biosynthesis: the stage of ACC production and that of ACC transformation to ethylene. Interaction between the main pathways of CO_2 metabolism, C_2H_4 activity, and plant growth is described by Grodzinski and Woodrow (1989). CO_2 (50–1500 $\mu l\, l^{-1}$) stimulates ethylene emission by the photosynthesizing leaves of plants of both C_3 and C_4 types. It acts at the stage of transformation of ACC to ethylene by intensifying the synthesis of the ethylene-forming enzyme. Ethylene can change photosynthesis and respiration, thus modifying plant development and productivity both under natural conditions and in the closed growth chamber. However, the action of ethylene on CO_2 metabolism is indirect and differs from the effects of carbon dioxide itself on C_2H_4 formation. Ethylene acts mainly as trigger to some processes, later leading to changes in CO_2 metabolism. The differing resistance of stomata, photosynthesis, respiration, and carbon distribution are due to the indirect effects of ethylene which show themselves as changes in epinasty, absorption of light, abscission of flowers, state of leaf surface, and development of roots. Ozone is believed to be one of the gases inducing the appearance of stress-induced ethylene.

Ozone induces the emission of stress-induced ethylene when the plant is exposed for 3 h in air containing only 25×10^{-6}% O_3 (Craker 1971). Treatment by ozone (588 $\mu g\, m^{-3}$) for 3 h in the greenhouse also stimulated C_2H_4 formation (Pell and Puente 1986). Ethylene synthesis also depends on the extent of plant damage by carbon monoxide. The main path of ethylene emission from the leaf under stress is considered to be gas diffusion through the stomata. This is confirmed by experiments of Rodecar and Tingey (1986), who exposed soybean, *Glycine soja*, tomato, *Lycopersicon esculentum*, *Eucalyptus*, and ivy, *Hedera helix*, in chambers with 0.75–0.95 $\mu l\ O_3/1$. The enhancement of ethylene release, fivefold for soybean and 6.3-fold for tomato, occurred from both surfaces of the analyzed leaf. Unlike these species, for ivy and eucalyptus it was increased 3.6–3.8-fold, and emission was observed only from lower side of the leaf, where there are stomata in these species.

Ethylene elimination was also stimulated in response to the increase of SO_2 concentration in air to 3 $\mu l\, l^{-1}$ (Kimmerer and Kozlowski 1982). The same effect was observed either on treatment of the plant by insecticide powder, cadmium, chlorine, or under the influence of automobile exhaust gases (Kimmerer and Kozlowski 1982) and artificial acid rain with pH 2.8–5.5 for 1 h under greenhouse conditions (Pell and Puente 1986). Hence, ethylene release is presumed to be a symptom of stress induced by pollution and other unfavorable factors. In connection with this hypothesis, the possibility of ethylene determination as an indication of the damaging action of acid rain has been studied (Arny and Pell 1986). The experiments were done on 3–4-week-old plants of potato, *Solanum tuberosum*, radish, *Raphanus sativus*, and soybean *Glycine soja* in the greenhouse under artificial rain made of solutions with pH from 2.8 to 5.6. The increase in ethylene production was observed only after repeatd treatment by the solution with pH 2.8. Based on these observations, the conclusion has been reached that ethylene is not a sensitive indicator of plant reaction to acid rain (Arny and Pell 1986). However, according to Wolfenden et al. (1988), whose experiments fixed the increase in the rate of ethylene emission under air pollution and acid rain, gas release could be a diagnostic test.

Among other factors influencing ethylene production is water stress. Water deficit, as well as excess of moisture, induces an increased synthesis of this hydrocarbon during the preclimacteric period (George et al. 1982). Moreover, a similar effect is observed at high salt concentrations in the soil (Chrominski et al. 1986). It should be verified in future whether the amount of ethylene can serve as a test for stress induced by salting or the disturbance of the water balance.

Deficiency in mineral nutrition is also one of the causes of stress-induced ethylene emission. Moreover ethylene inhibits the formation of nodules in leguminous plants inoculated by *Rhizobium*. There is a clear correlation between the concentration of nitrates and the amount of ethylene released (Ligero et al. 1987). The increase of nitrates in the nutrient solution to 9.2–11.5 $mEq\, l^{-1}$ stimulates the elimination of ethylene by the roots.

Ethylene produced under stress can play a role in the regulation of growth and resistance to diseases. On fungal infection, the amount of C_2H_4 in leaves of barley, *Hordeum sativum*, and peanut, *Arachis* sp., is increased (Walther et al. 1981), as well as in seedlings of wheat, *Triticum vulgare*, inoculated by the fungus *Fusarium* (Wurzer-Fassnacht and Hoffman 1986). The treatment of infected plants by preparations reducing the infection leads to a decrease in ethylene release to the level of the noninfected individuals (Wurzer-Fassnacht and Hoffman 1986). The formation of ethylene was observed by Ben-David et al. (1986) in leaves of red pepper, *Capsicum annuum*, infected by *Xanthomonas campestris*. The intensity of ethylene elimination is enhanced 35-fold in comparison with the control (105 $pmol\, g^{-1}\, h^{-1}$). On infection, young leaves of pepper produced more ethylene than mature, less sensitive ones. The main sites of gas formation are the peripheral zones of leaf lamina and necrotic spots (Ben-David et al. 1986). According to Abeles (1972), the primary function of stress-induced ethylene is to facilitate the abscission of organs damaged by disease, drought, and other extremal factors. Ethylene can play the role of a factor of resistance to the pathogen due to its hormonal activity. This gaseous stress-induced phytohormone appears to be an signal to the plant in order for the organism to activate the defensive mechanisms in response to pathogen invasion (Ecker and Davis 1987). Ethylene regulates the expression of genes coding enzymes of the phenylpropanoid biosynthetic pathway, such as L-phenylalanine ammonium lyase (EC 4.3.1.5) and 4-coumarate CoA-ligase (EC 6.2.1.12); the enzyme of biosynthesis of flavonoid glycosides, chalcone synthase (EC 2.3.1.74); glycoproteins, the protein components of cell walls, and enzymes participating in biosynthesis of hydroxyproline. The effects of C_2H_4 on the enzymes differ from those observed after wounding of the cell wall. This phytohormone is able to regulate the gene expression of the litic enzyme chitinase, which is synthesized in response to pathogenic invasion. Chitinase participates in lysis and degradation of the fungal cell wall. The activity of the enzyme is increased in the presence of endogenous ethylene. Ethylene obviously participates in the control of gene expression of chitinase at the transcription level (Broglie et al. 1986). The emission of stress-induced ethylene and the synthesis of phytoalexins are two of the defensive mechanisms against pathogens. Treatment of the plant with ethylene promotes the appearance of female features. Ethylene possibly belongs to the hormones controlling the sex of the plant (Sriram and Ram 1984). The gas is eliminated by female flowers of hemp, *Cannabis*, in marked amounts (46.67 ml g^{-1} fresh mass h^{-1}) and in lesser ones by males (11.73 ml g^{-1} fresh mass h^{-1}). Ethylene is thought to control sex expression because the use of the pesticide ethephon can induce the appearance of hermaphrodites and female flowers on male plants. The main mechanism of the phenomena is proposed to consist in stimulation of ethylene emission (Sriram and Ram 1984).

The mechanism of the formation of stress-induced ethylene is still unclear. Roberts and Osborne (1981) suggest that cofactors, activating enzymes, which are included in this process, are released after the stress-induced disturbance in membranous permeability. However, at stresses such as the disintegration of water exchange, production of ethylene is enhanced earlier than any changes in membrane permeability, in particular leaching of electrolytes could be observed. Moreover, C_2H_4 emission is decreased when the efflux of electrolytes is facilitated. All these facts show that membrane integrity is necessary for the biosynthesis of ethylene. Until recently, it was unclear in exactly which cell compartments this gas is synthesized. The participation of the tonoplast in the process is proposed. To verify this hypothesis, experiments were carried out with vacuoles isolated from horse bean, *Vicia faba* (Mayne and Kende 1986). When the tonoplast is broken by osmotic shock, detergents, ethanol, or a freezing-thawing procedure, inhibition of ethylene synthesis took place. A similar effect was observed if either the uncouplers of oxidative phosphorylation, carbonyl cyanide m-chlorophenylhydrazone and 2,4-dinitrophenole, or the ionophores, valinomycin and nigericin, were used in the treatment of vacuoles. Thus, the tonoplast is confirmed as being one of the sites of ethylene biosynthesis.

According to Kasperska and Kubacka-Zebalska (1989), the emission of stress-induced ethylene by leaves of soybean, *Glycine max*, is an indicator of tissue injury, and correlates with lipooxigenase activity. The latter is stimulated by minor damage, especially induced by leakage of electrolytes $< 40\%$. It should be noted that if leakage of electrolytes is more than 40%, the lipooxigenase is depressed. Inhibition of the enzyme by n-propylgallate also blocks ethylene emission. Therefore, the synthesis of stress-induced ethylene needs activation of the systems participating in the generation of free radicals.

6.2.2 Ethane and Other Simple Hydrocarbons

Ethane is most often formed as the main gaseous hydrocarbon after homogenization of the plant tissues, which in an undamaged state, release predominantly ethylene. Elstner and Konze (1976) showed that the highest amount of ethylene (379 $pmol\,h^{-1}\,g^{-1}$ fresh weight) is eliminated by intact leaf disks of sugar beet. After the crushing of the disks, this emission is strongly decreased to 33 $pmol\,h^{-1}\,g^{-1}$ fresh weight. In contrast, ethane production is low (1 $pmol\,h^{-1}\,g^{-1}$ fresh weight) in untreated disks, and stimulated to 86 $pmol\,h^{-1}\,g^{-1}$ fresh weight in crushed ones. Callus tissue of carrot, *Daucus carota* L., possesses the ability to eliminate both ethylene and ethane (Maitra and Sen 1989). However, ethane formation is mainly connected with cell disorganization, as it appears to be a product of the oxidation of unsaturated lipids. Regenerating tissue produces markedly less ethylene whereas the formation of ethane is blocked completely. According to Biedinger and Schnabl (1991), ethane production can be an indicator of the regeneration potential of electrically fused sunflower protoplasts. If the concentration of ethane increases, the regeneration potential becomes smaller.

Unlike plants which release ethylene and ethane under one or other kind of stress, leaves of tobacco, *Nicotiana tabacum*, plunged into a solution of Na NSO_3 (0.2–50 mM), eliminate a high amount of propylene and propane, but less ethylene and ethane (Yung-jen-Chou et al. 1985). During homogenization of the leaves, the production of ethane is noticeably increased. The meaning of the release of propylene and propane is still unclear.

The following experiments showed that reversed dependence between ethylene and ethane formation in plants is observed on freezing cereals in liquid nitrogen, on water deficit and hypoxia, or mechanical injury and anaerobiosis of etiolated seedlings of wheat, *Triticum* (Kimmerer and Kozlowski 1982). Ethane is produced as a result of the chemical influence on cells, as well, for example, in the presence of salicylhydroxamic acid, as is characteristic for *Vicia faba* in light and dark. The amount of gas released in the light was higher than in the dark – about 300 $pmol\,mg^{-1}$ chlorophyll (Schnabl et al. 1983). The formation of ethane by seedlings of red pine, *Pinus resinosa*, and paper birch, *Betula papyrifera*, in an SO_2 atmosphere was fixed by Kimmerer and Kozlowski (1982). In such cases, this emission was continued lineally in relation to the area of necrosis on the surface. Thus the appearance of the gas does not depend on the intactness of tissue and perhaps has no compartment in leaf cells. According to Kimmerer and Kozlowski (1982), independent formation of ethane and ethylene can indicate the existence of two different ways of formation for both gases. Other authors (Peiser et al. 1982; Schnabl et al. 1983) consider ethylene and ethane to be formed from common precursors. They may be products of the peroxidation of fatty acids.

The biological effects of ethane have been poorly studied. It is known that the gas can decrease the stimulatory action of ethylene on bud formation and growth of leaves (Lefebvre 1978). There are data showing that ethane is an active repellent; it is contained in resinous excretions of pine and is toxic for some beetles. This simple compound is highly effective. Only 0.6 μl of this gas is formed by 1 g of pine needles, but it is enough to prevent their being eaten (Harborne 1982).

Under stress, other short-chain hydrocarbons are released. Under treatment with paraquat, microalgae emit not only ethane and ethylene but propane as well (Bochler-Kohler et al. 1982). The hydrocarbons ethane, ethylene, propane, pentane, and pentene are formed by photosynthetic structures of cyanobacteria from unsaturated fatty acids of the plant membranes. Thus the appearance of volatile hydrocarbons is an effect of degradation of the photosynthetic membranes (Sandmann and Boger 1982). This supposition has been confirmed in experiments with exogenous fatty acids, such as oleic, linoleic, linolenic, and arachidonic, where it was demonstrated that the quality and ratio of forming low-molecular hydrocarbons depended on added fatty acid which was oxidized according to the peroxide mechanism. The highest amounts of ethane (6.31 nmol) and ethylene (3.71 nmol) were observed when linolenic acid was their precursor. The isomer of linolenic acid was found to be best for propane (0.21 nmol) and pentene (4.3 nmol) production, and arachidonic acid for pentane (28.64 nmol).

Olefins are the most biologically active in comparison with other released hydrocarbons. Their ability to influence biochemical processes is due to the double bonds in their molecules. Vlasyuk and Porutskii (1964) showed that the activity of unsaturated hydrocarbons depends on the amount of carbon atoms, diminishing as their number increases. Thus the, power of propylene and butylene action on isolated roots of leguminous plants is markedly lower than that of ethylene, while olefines, like sterol, show negligible activity. The influence of low-molecular mass hydrocarbons on the germination of seedlings of purslane, *Portulaca oleracea*, was estimated by Taylorson (1979). He showed that saturated hydrocarbons, even at a relatively high concentration, inhibited purslane seed germination by no more than 16–30%, unsaturated hydrocarbons, however such as ethylene, propylene, and propadiene by 55–77%. Unlike ethylene, propylene and propadiene were effective in 10–100-fold higher concentrations. Demonstrating the influence of low-molecular mass unsaturated hydrocarbons on the formation of apical hooks in pea seedlings grown in the dark and epinasty of leaves, Crocker (1948) showed that their physiological activity in the sequence: ethylene: propylene: butylene drops quickly with increased length of the chain, so that isoprene is inactive in the reactions.

One interesting feature of propylene, its ability to stimulate ethylene synthesis, should be noted (Sfakiotakis and Dilley 1973). The treatment of apples of cultivar Bad Delikons by propylene stimulated the formation of ethylene and this ability is increased during fruit ripening. At low concentration of O_2 (6.5% and lower), no similar induction of ethylene synthesis was observed.

Recently, the ability of propylene to change important physiological functions of plants was demonstrated (Lee et al. 1987). Treatment of nonripe fruits of tomato, *Lycopersicon esculentum*, by propylene (50 μl/l) for 2, 4, and 6 days increases both the rate of their respiration and the number of polysomes, and finally changes in protein synthesis occur. Moreover, protein with molecular mass 55 kdaltons which participates in the ripening of fruits can be synthesized only at a definite concentration of propylene. If the concentration is lower, the protein is not produced.

Hydrocarbons with triplet bonds are less physiologically active than ethylene and propylene. Seedlings of pea, *Pisum sativum*, are sensitive to the presence of acetylene in concentrations 1200-fold higher than those observed for ethylene, while epinasty of leaves of tomato, *Lycopersicon esculentum*, is by more than 500-fold susceptible to C_2H_4 than to C_2H_2 (Crocker 1948).

It should be noted that the character of physiological action of the emitted hydrocarbon depends not only on the features of its molecules but on the chosen test object or process as well. Hydrocarbons with high inhibitory power in one case, for a certain physiological process, can be absolutely inactive for another, or even for another species.

6.2.3 Terpenoids

Unlike short-chain hydrocarbons the significance of terpenoids as indicators of the stress state of the plant was demonstrated only 25 years ago. Wounding, infection, and other extreme factors often destroy the normal synthesis of terpenoids (see Chaps 3.6 and 4.3.2). Degradation of plant cells leads to a change in the qualitative and quantitative content of the compounds (Kuc and Lisker 1978). The substances can stimulate the healing of a wound or increase resistance to disease.

In the plant world, soft resin is one of the most significant origins of volatile terpenoids. Wounds, often observed under natural conditions after injury, but especially on tapping of trees, induce an abundant release of these substances. Under stress, the amount of pinene, kaurene, phellandrene, myrcene, and abietic acid in resin is changed (Kuc and Lisker 1978). According to widespread opinion, the amount and composition of the eliminated terpenes are indicators of plant resistance to stress. Spruce, *Picea abies*, sensitive to industrial smoke, contains only little camphene and limonene but much β-pinene, whereas the volatile oils of resistant plants are abundant in camphene. As a result of air pollution, the content of monoterpenes and sesquiterpene hydrocarbons, alcohols, and ketones is changed in the resin of resin ducts of damaged needles of *Picea abies*, with especially sharp shifts in the ratio α-pinene/β-pinene. In damaged needles, the amount of β-pinene is increased approximately four fold (Jutner 1988). The content of volatile terpenes may be an indicator of the tree's reaction to air pollution. Thus, resistant and nonresistant species differ in their amount of monoterpenes. In the 1980s, changes in monoterpene composition on fumigation of plants by sulfurous gas were shown. In balsam of fir, *Abies balsamea*, higher amounts of β-pinene, camphene, and α-pinene were found on some days under high concentrations of SO_2 than were observed in untreated plants (Penwic and Potter 1981 in Bucher 1982). Their concentration of borneol acetate and some other terpenoids is increased to a lesser degree. In experiments by Bucher (1982), fumigation of needles of pine, *Pinus sylvestris*, by SO_2 induced a five fold increase in α-pinene and a ten fold increase in camphene in comparison with the control. The amounts of myrcene, limonene, and some other volatile compounds also decreased. The total content of monoterpenes did not change. The elimination of the terpenes kaurene and isokaurene in response to fungal infection are also known (Melon and West 1979). Resinous monoterpenes and diterpene acids are formed in lodgepole pine, *Pinus contorta* var. *latifolia*, on wounding or inoculation by the fungus *Ceratocystis clavigera* (Croteau et al. 1987). On viral infection, compounds usually nonsynthesized by the plant can arise. For instance, besides 16 known sesquiterpenoids, three new ones were isolated from leaves of tobacco, *Nicotiana rustica*, infected by tobacco mosaic virus and identified as occidenol, occidentalol, and trans-occidentalol (Fig. 81) (Uegaki et al. 1985).

The biological activity of many plants is connected with the production of terpenes. Lakhno and Kozlova (1967) studied the composition of coniferous plants and concluded that their antimicrobial features are caused by the elimination of turpentine, the main part of which is terpenes. In the investigations of Muller and Muller (1964), camphor and cineole influence markedly the root growth of cucumber, *Cucumis sativus*, being more active than pinenes. Terpenoid derivatives can also be physiologically active and induce inhibition either of plant growth and/or growth of insects, thus acting as allelochemicals or

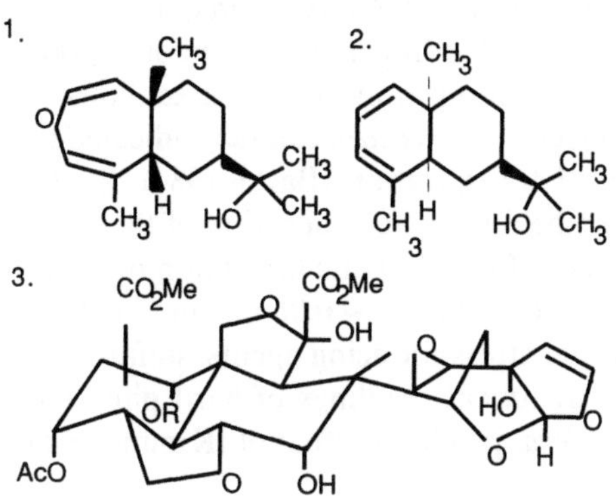

Fig. 81, 1,2. Defensive terpenoid compounds in plants. Sesquiterpenoids in leaves of tobacco *Nicotiana rustica* infected by tobacco mosaic virus, occidenol (**1**) and occidentalol (**2**). Azadirachtin from *Azadirachta indica* (**3**) R = tigloyl

kairomones, by some kind of plant resistance mechanism. For instance, tetranotriterpenoid azadirachtin (Fig. 81) is isolated from seeds of the tropical neem tree, *Azadirachta indica*, from the family Meliaceae, and on mechanical stress or feeding by insects has an inhibitory effect on the desert locust, *Schistocerca gregaria* (Rembold 1989). This Indian economic plant also shows strong allelopathic activity. Kairomones are found in volatile and nonvolatile excretions of leaf and pod hairs of chickpea, *Cicer arietinum*, and pigeonpea, *Cajanus cayan*. Their volatile aroma contains more than 200 peaks of terpenoids and four of the substances are active against the *Heliothis* moth. Their nonvolatile exudates include 12–75% of acids (pH $\sim$ 1), among them malic acid amounts to more than 70–100% dry matter.

Synthesis of cardiac glycosides or cardenolides by plants is also found to increase under anaerobiosis. Stuhlfauth et al. (1987) showed that a three fold increase of carbon dioxide in air leads to noticeable (ca. 60%) stimulation of their formation in *Digitalis lanata* (family Scrophulariaceae). The amount of digoxin, a derivative of digitoxigenin, is especially increased.

The study of the mechanisms of some terpenes' action, in particular of the bicyclic terpene, α-pinene, has shown that biological activity of terpenes is connected with their ability to block SH-groups of enzymes. The effects of terpenes are possibly due to their solution in lipophylic layers of the cuticle and protoplasma of plant cells.

6.2.4 Alcohols

Production of ethanol and other alcohols occurs under aerobic conditions, but under oxygen deficit, the processes are enhanced (Grineva 1975). The difficult aeration connected with the anatomical features of plant species is observed in the root systems, especially in the apical part. The oxygen deficit is characteristic for deep-lying tissues of woody plants. In summer, on hot days, traces of ethanol can be found in the internal air of balsam poplar, *Populus balsamifera*, common birch, *Betula verrucosa*, and other plants (Roshchina 1975). The concentration of the alcohol can reach 15 μM in the seed endosperm of castor bean, *Ricinus communis*, on 15 h of anoxia (Donaldson et al. 1985). As has been demonstrated for many plants, ethanol formation occurs on flooding due to the poor supply of oxygen to the tissue. Experiments with flooding done in the greenhouse on alfalfa, *Medicago sativa*, and birds'-foot trefoil, *Lotus corniculatus*, indicated that ethanol was released with the stream of transpiration (Barta 1984). In these cases, formation of the alcohol may also be due to weak aeration of the tissues. In seedlings of rice, *Oryza sativa*, the ability for enzymatic formation of ethanol increased, especially at a critically low concentration $\sim 0.06\ \mathrm{mol\,m^{-3}}$ (5%) O_2 (Atwell and Greenway 1987). Among other stress-inducing agents, sulfurous gas is of most interest, because under this treatment seedlings of hard pine, *Pinus resinosa*, and paper birch, *Betula papyrifera*, produced ethanol (Kimmerer and Kozlowski 1982).

To free themselves from an excess of alcohols, cells release substances into the environment. This process is most intensive in root systems and weakly aerated tissues, from which ethanol can be transported with the transpiration stream. The increase in exudation of ethanol, amino acids, and sugars by roots of pea, *Pisum sativum*, has been observed in artificial gaseous media containing nitrogen and 30% CO_2. Under these conditions, the amount of ethanol reached more than 250 mg g^{-1} dry weight within 96 h. Amino acids, mainly alanine, leucine, asparagine, and glutamine, compose approximately 1.34 mg g^{-1} dry weight. Traces of ethanol and other alcohols were found in the transpiration water of woody plants by gas chromatography (Roshchina 1973 a, b, c, 1974 a, c).

Ethanol is formed during glycolysis from pyruvate, which is then used predominantly in the Krebs cycle on aeration (see Scheme 9). Sometimes, acetaldehyde, known as an intermediatory metabolite in the process, is not completely transformed into ethanol. Part of the substance can be used for the production of other either lower or even higher alcohols (Glinka and Reinhold 1962). Indeed, methanol was found among the volatile compounds released by radish leaves (Dadykin et al. 1967), methanol and propanol in the transpiration water of woody plants (Roshchina 1974a, c), and butanol in the composition of volatile components of soybean (Nunomura et al. 1976). In roots of maize, *Zea mays*, and sunflower, *Helianthus annuus*, as well as in the solution surrounding the roots, isobutanol, 1-pentanol, and 1-hexanol were found. Higher amounts of alcohols, showing the strength of the glycolytic processes, can perfectly compensate the energy required for the life of the organism. As a result of mechanical damage and pathogen infection, peroxidation of lipids with participation of lipooxygenase takes place, leading to the formation of ketoalcohols, such as α-ketol and δ-ketol, as well as of hexanol and hexenal (Zimmerman and Vick 1988). The amount of "damaging"α-ketols increases three to five fold within 10 min of lipid oxidation. Like aldehydes, they can be important factors in the response on the wound, defending plant tissues against pathogenous fungi and morbific microorganisms.

Among other stress conditions which induced the appearance of alcohols in plant excreta are mechanical damages and changes in the gaseous media. Wallbank and Wheatley (1976) studied the chemical composition of vacuum extracts from intact plants of cabbage *Brassica oleracea*, turnip, *Brassica rapa*, and radish, *Raphanus sativus*. They identified small amounts of only one component, the alcohol derivative helen acetate. In ruptured leaf tissues of the same species, *cis*-hexenyl acetate prevailed among other excreted substances, such as acetone, methanol, hexyl acetate, etc.

The physiological action of ethanol has been better studied than that of other alcohols. Its low concentrations stimulate the growth processes. For example, enhanced growth of the mesocotyl was observed for seedlings of oat, *Avena sativa*, treated by this alcohol (Mer 1961). At a concentration of 0.02 M, ethanol stimulates the division of *Chlorella* cells. For other alcohols, the same effects were characteristic only for methanol. However, unlike methanol, which stimulates cell division only by 3–4%, ethanol does this by 75% (Bach and

Felling 1958). This stimulation by ethanol is explained mainly by its use as the origin of carbon groups, since at sufficient CO_2-uptake the effect on growth is lower (Bach and Felling 1958). When the concentration of ethanol is increased, the stimulation is changed by the inhibition of the process. Grineva (1975) also observed depression of seed germination of barley, *Hordeum vulgare*, and growth of callus tissue from core of tobacco, *Nicotiana tabacum*, which were induced by ethanol. An increase in ethanol concentration to 0.1% led to plant death.

Inhibition of growth is also caused by disturbance in the metabolism under chemical stress. The introduction of ethanol solutions into leaves is known to lead to both a change in utilization of various sugars and stimulation of respiration. Thus, ethanol influences the most important aspects of cell life.

Unlike CH_3CH_2OH, other alcohols are not so physiologically active. Moreover, their role in metabolism has not been well studied. The localization of their biosynthesis in the cell, and the pathways for their subsequent inclusion into the metabolic exchange are not yet clear.

6.2.5 Aldehydes and Ketones

Among the components of intact plant excreta, precursors of alcohols, volatile aldehydes, are often found (see Chap. 4.3.3). Under stress, the emission of aldehydes is usually increased. Mechanical trauma promotes the formation of the substances, because homogenates of leaves and other plant tissue differ from intact plants by their high aldehyde content. Indeed, volatile aldehydes have been found in vapors of disrupted buds of bird tree, *Padus racemosa*, common poplar, *Populus balsamifera*, and in infusions from leaves of the same plants (Roshchina 1973a, 1974b). In distilled infusions from apple, pear, grape, and currant, Pribella and Vasatko (1967) identified acetaldehyde, acetone, propionaldehyde, bytyraldehyde, methyl propylketone, n-valeraldehyde, methylbutylketone, and other products. Acetaldehyde is often found in homogenates of leaves and fruits (Masquelier and Vitte 1967).

Aldehydes and ketones were identified in excretions of injured cowberry, *Vaccinium vitis idaea*, birch, *Betula verrucosa*, bird tree, *Padus racemosa*, and oak, *Quercus* (Skvortsov and Smirnova 1972). Chubinidze (1966) found formaldehyde, acetaldehyde, and isovaleraldehyde in the vapor of injured leaves of false acacia, *Robinia pseudoacacia*, black poplar, *Populus nigra*, and other species by the polarographic method. The release of aldehydes and alcohols with C_6-atoms was also observed on wounding of tomatoes. Among these 2-hexanal, trans-2-hexanol, and n-hexanol, which possess antifungal activity, have been identified.

Many plants with damaged leaves form hexanol. A noticeable amount of the aldehyde was found in homogenates of false acacia, *Robinia pseudoacacia* (Schauenstein et al. 1977), and maidenhair-tree, *Ginkgo biloba* (Major et al. 1963). In a condensate from leaves of this woody plant, hexanal is present in concentrations higher than 3.78×10^{-7} mg per 3 μl. Oxygen is needed for the formation of the aldehyde. If leaves of *Ginkgo biloba*, collected during the growth period, are put into an atmosphere lacking oxygen, hexanal is not formed. Perhaps, the enzyme, requiring O_2, participates in this process. The release of hexanal has been considered in connection with plant resistance to fungal infection (Major et al. 1963) (see Chap. 7).

Many unsaturated aldehydes released by plants possess fungicidic characteristics, as described in the review of Lyr et al. (1987). They are formed from unsaturated fatty acids (linolenic and linoleic) with the participation of lipoxygenase. This enzyme is thought to participate in the production of hexanal and hexenal in plants (Zimmerman and Vick 1988). It can be illustrated by the following scheme:

Unsaturated fatty acid $\xrightarrow{O_2}$ peroxidized products $\longrightarrow$

peroxidized products $\xrightarrow[\text{hydroperoxide isomerase}]{\text{lipoxygenase}}$ α-ketol + γ-ketol $\longrightarrow$

↓

hydroperoxide lyase

↓

12-oxo-*cis*-9-dodecenoic acid + hexanol $\longrightarrow$ hexenal

↓

12-oxo-*trans*-10-dodecenoic acid

↓

traumatic acid

After homogenization, within 10 min, 100 g of apple fruit tissues released up to 2310 μg of hexenal (Lyr et al. 1987). Alk-2-en-al was found in the transpiration stream of *Robinia psuedoacacia* (Lyr et al. 1987), and E-hexenal (3 μg m^{-3}) in the air surrounding *Thea* plants (Hatanaka et al. 1978a). 2-Hexenal is formed in the presence of O_2 on stress induced by any damage – mechanical, insect attack, etc. (Major 1967).

The release of aldehydes increased under stress conditions. Indeed, the elimination of acetaldehyde is enhanced on treatment of seedlings of *Pinus resinosa* and *Betula papyrifera* by SO_2 in concentrations up to 3 μl l^{-1} (Kimmerer and Kozlowski 1982). There is also information that excretion of the same aldehyde by pea, *Pisum sativum*, marigold, *Tagetes*, and ribbon gum, *Eucalyptus viminalis*, is increased on water deficit, freezing, and exposure in liquid nitrogen (Kimmerer and Kozlowski 1982). Thus, the release of aldehydes on stress is high and can achieve biologically active concentrations.

The biological effects of aldehydes depend on the number of carbon atoms, their concentration, and the object on which their action is tested. Low concentrations of aldehydes have a weak influence or sometimes stimulate the tested process. The treatment by 100 ppm solutions of aldehydes can stimulate seed germination of purslane, *Portulaca oleracea* (Taylorson 1979). In the sequence acetaldede – propionaldehyde – butyraldehyde, the two latter were more than (20-fold effective, while acetaldehyde had a weak effect. A higher concentration of aldehydes induces negative changes in living cells. On the whole, aldehydes can shift the state of the water-absorbing capability of the cell and the permeability of the cytoplasmic membranes (Roshchina and Roshchina 1970, 1983; Roshchina 1974c). The toxic action of aldehydes is connected with their ability to react with amino- and sulfhydril groups of protein, inducing conformational changes of the latter (Schauenstein et al. 1977). However, the strength and character of their action depend on the structure, hydrophility, lipophility, and polar features of the molecule. Among saturated alkanals, formaldehyde is the most reactive and when the number of carbon atoms increases, the inhibitory ability of aldehydes, on the contrary, decreases (Schauenstein et al. 1977).

Saturated aldehydes have an inhibitory action on many physiological processes in plants and microorganisms. Formaldehyde or acetaldehyde (1 mM) irreversely inhibits cell division in *Escherichia coli*, and retards or stops the germination of seeds and growth of seedlings (Schauenstein et al. 1977). Acetaldehyde can depress photosynthesis (Kolesnichenko 1976). The action of alkanals on cell respiration has also been studied (Schauenstein et al. 1977). Formaldehyde inhibits the oxidation of succinate, glutamate, α-oxyglutamate, and pyruvate coupling with phosphorylation (Schauenstein et al. 1977). Moreover, it depresses ATPase and the reversion of electron transport by ATP in mitochondrial particles. Other aldehydes such as acetaldehyde, propionaldehyde, isobutyraldehyde, and isopentanaldehyde induce changes in the oxidative processes of mitochondria by acting on appropriate dehydrogenases and the metabolism of pyruvate. These effects are partly reversible. One example of irreversible inactivation can be depression of nitrogenase activity (Slatyer et al. 1983).

All these facts show that, as physiologically active substances, low-molecular mass aldehydes may participate in allelochemical interactions. The plant-donor of aldehydes acting on the protein of the plant-acceptor is proved to be competitive. On the other hand, the plant-acceptor can have defensive mechanisms against similar influences. For instance, it can include absorbed aldehydes in its own metabolism by way of redox conversions, forming organic acids and alcohols (Durmishidze 1977).

6.2.6 Hydrogen Cyanide

With improved methods, in addition to aldehydes and alcohols, other non-oxidized products of damage have been found. Among them is hydrocyanic acid or hydrogen cyanide, which has been identified in the composition of volatile compounds of injured plant tissues. For the first time Drabkin (1954) reported that vapors of disrupted leaves of the bird tree, *Padus racemosa*, released this substance. Hydrogen cyanide is also found in excretions of crumbled seeds of flax, *Linum linearis* (Ermakov 1960), being emitted to the air simultaneously with acetone.

The main biological precursors of HCN are the cyanogenic glycosides, which are contained at least in 800 plant species belonging to 70–80 families (Miller and Conn 1980). Cyanogenic glycosides has been identified but only two of them, linamarin and lotaustralin, are widespread. Usually they are found in different amounts in such species as lima bean, *Phaseolus limensis*, cassava, *Manihot utilissima*, flax, *Linum .linearis*, and clovers, *Trifolium newseeland*, *T. repens*, etc. (Kakes 1990). Cyanogenic glycosides with aromatic rings are often found, in particular amygdalin, which is present in the almond, *Amygdalus communis* (in a high quantity up to 1.8%); in peach, *Persica prunuca*, mountain ash, *Sorbus aucuparia*, sweet cherry, *Cerasus avium*, plum, *Prunus domestica*, as well as dhurrin, found in cereal sorghum, *Sorghum vulgare* (Miller and Conn 1980).

Cyanogenous glycosides are synthesized from amino acids, as is shown for linamarin (Fig. 82). The first steps of the process include the formation of hydroxyamine acid, which is converted to aldoximes by oxidative decarboxylation. Then these intermediators are transformed into cyanogenic glycoside via glycosidation of α-hydroxynitrile.

Cyanogenous glycosides are not distributed evenly within the plant. Most of them are concentrated in the seeds, or in other cases in the tips of stems and seedlings. In intact plants containing cyanogenous glycosides, formation of HCN does not occur because the substrates and glycosidases, enzymes which participate in the process, are localized in different compartments (see Sect. 6.1). When the tissue has been disrupted, specific glycosidases chip off sugar and the forming intermediate cyanohydrine is spontaneously decomposed to ketone or aldehyde and HCN. The amount of hydrocyanic acid which can be released from food plants after destruction varies among different species: 10 mg/100 g from seeds of some sorts of lima bean, 2–60 mg/100 g from young leaves of bitter seeds of apricot, linen flax, and common vetch, 90–400 mg/100 g from leaves of bitter cassava, wild cherry, and peach, seeds of bitter almond, peach, and lima bean (Poulton 1983). The highest emission (800–910 mg/100 g) of HCN is observed for disrupted tips of unripe sprouts of bamboo and seedling tips of linen flax, and the smallest ones (up to 10–20 mg/100 g) for mature seeds of sorghum and some sorts of lima bean (Poulton 1983). Sometimes, seeds, in particular of *Sorghum vulgare*, do not eliminate HCN; however, the etiolated tips of stems and young leaves which are formed from the same seeds, produce hydrocyanic acid in noticeable quantities. Single organs of *Hevea brasiliensis*–petioles of leaves, apical stem buds, endosperm, and embryo of seeds and even roots–release HCN after mechanical damage (Poulton 1983). The elimination of HCN can occur not only in similar cases but also as a response to chemical action and invasion of pathogenic fungi (Lieberei 1986).

The physiological meaning of the accumulation of cyanogenous glycosides is still unclear. Some of them are thought to be used in the synthesis of non-cyanogenous substances, as is observed during seed germination of *Hevea* (Lieberei et al. 1985). Cyanides of leguminous seeds appear to be a source of

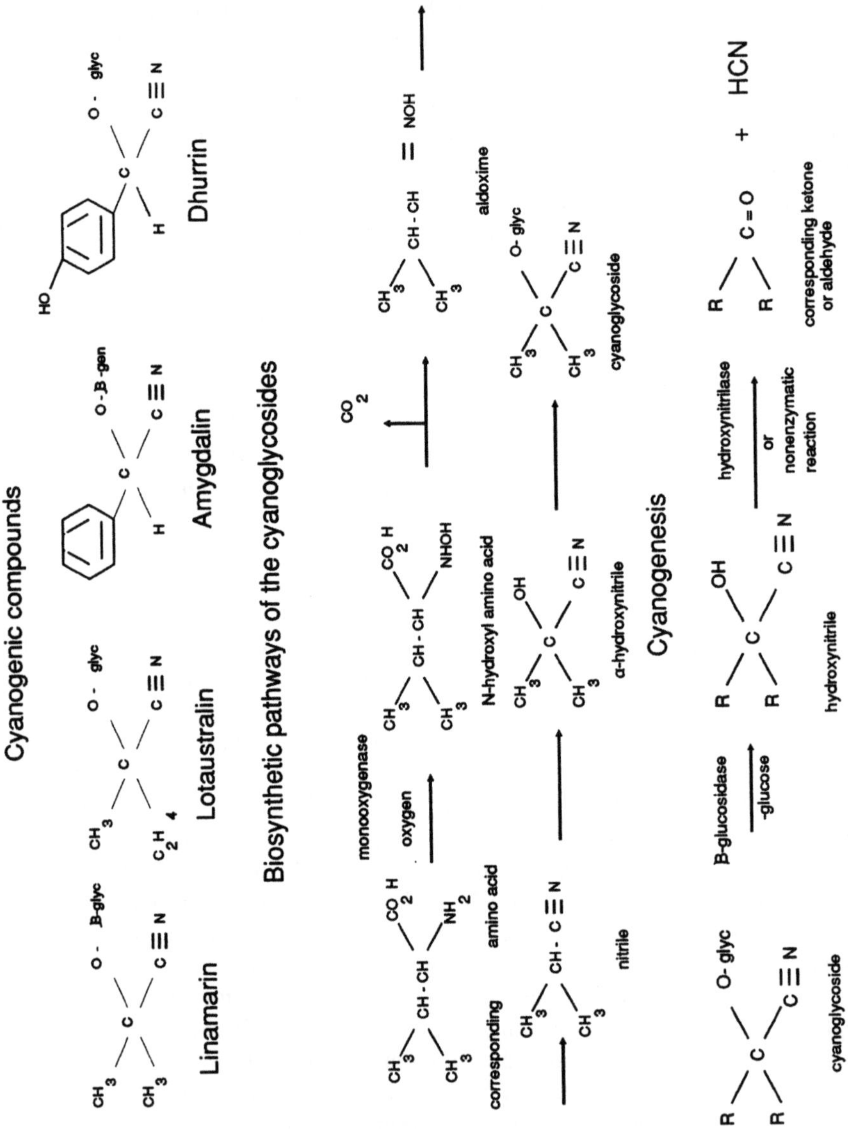

Fig. 82. Cyanogenic glycosides. *glc* glucose; *gen* gentiobiose

nitrogen for the embryo. However, the most evident biological role of cyanogenous glycosides of higher plants is as a defence against destruction by animals (Nahrstedt 1985). One of the reasons for the cultivation of manioc, *Manihot esculenta*, in India, Africa, and South America, where this plant is a main food product, is thought to be its resistance to locusts (Jones 1981). However, humans who use the leaves and stems of this plant for food often suffer from poisoning by the cyanides contained in manioc. There have been cases of poisoning of sheep and other herbivorous animals by cyanogenous plant glycosides (Vennesland et al. 1982). The lethal dose for insects is about 100–250 ng HCN on 50 mg of larvas, which is less than a young leaf of clover, *Trifolium repens*, releases on being damaged (Nahrstedt 1985). The mechanism of the action of the compound on animal cells appears to consist in the inhibition of cytochrome oxidase, which participates in the final stage of respiration.

6.2.7 Phenols

The phenolic compounds in living tissues of higher plants are located in the vacuoles in the form of glycosides. If the plant cell undergoes stress, the breaking or increasing of tonoplast permeability leads to an efflux of phenolic substances in the cytoplasm, where they are oxidized to the appropriate quinones with the participation of phenolase. This copper–containing enzyme oxidizes, in particular, chlorogenic, caffeic, p-coumaric acids, tyrosine, and catechol. The quinones formed are very strong oxidative agents which react with the α-amino- and thiolic groups of proteins. They can also be polymerized, transforming to black and brown products. This reaction of polymerization occurs also nonenzymatically.

Since these processes of phenolic transformation are not observed in intact plants, the enzyme oxidizing polyphenols and substrates is thought to be located in different cell compartments which are destroyed under stress (Rhodes and Wooltorton 1978). The site of the synthesis of phenolic compounds is the chloroplasts (Zaprometov and Kolonkova 1967), which are also the primary compartments of the phenols. Most water–soluble phenols are concentrated in the vacuole (see also Chaps. 1.4 and 3.8). As has been shown in experiments with extraction of the enzyme, phenolase is mainly in the soluble part of the cytoplasm, although its activity is also associated with both chloroplasts and mitochondria (Rhodes and Wooltorton 1978).

Stresses not only promotes the liberation and transformation of phenols already present in cell, but also stimulate phenolic metabolism. In the latter case, the synthesis of derivatives of caffeic and quinic acids is increased, in particular cinnamic, chlorogenic, and isochlorogenic acids (Fig. 83). The compounds are usually present in intact cells, but their accumulation is stimulated by stresses. The synthesis of caffeic, cinnamic, chlorogenic, and isochlorogenic acids occurs via the shikimate metabolic pathway, where aromatic acids are formed from phosphoenolpyruvate and erythrose-4-phosphate (Fig. 84).

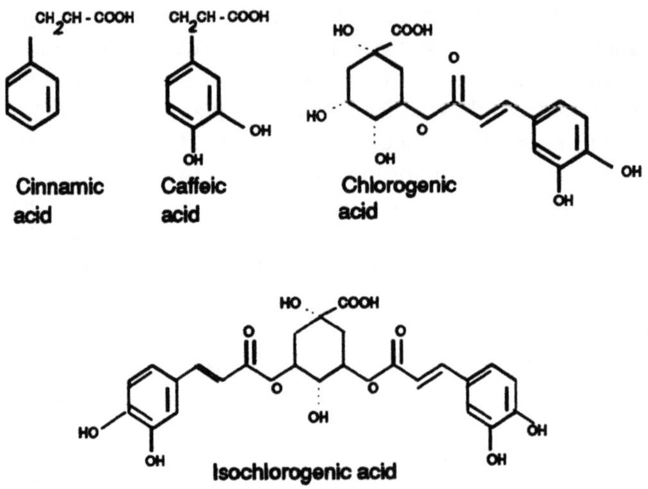

Fig. 83. Aromatic acids forming under stress

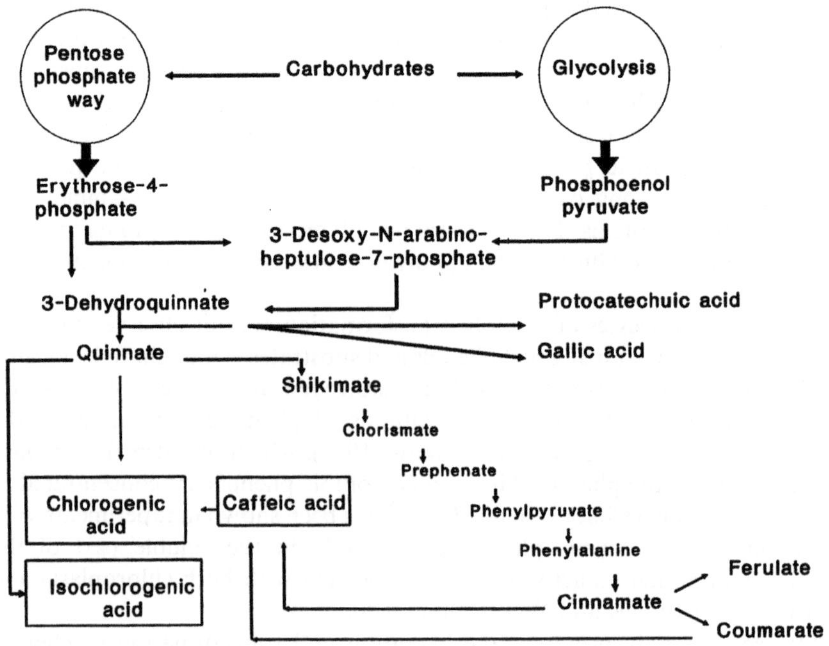

Fig. 84. Synthesis of aromatic acids

The amount of phenolic compounds becomes larger on various damage such as mechanical, chemical (in particular, by heavy metals), or infectional (fungal, bacterial, or viral). The formation of phenols is promoted by a deficiency in mineral nutrient elements (boron, calcium, magnesium, nitrogen, phosphorus, potassium, and sulfur). For example, the concentration of chlorogenic acid is increased (Putnam 1983). In combination with other stresses, drought can significantly increase concentrations of chlorogenic and isochlorogenic acids in plants. Indeed, the combination of drought and nitrogen starvation induces a 15-fold increase in the substances in sunflower. Hot and cold also influence the formation and elimination of plant phenols (Putnam 1983). Many other factors can act on the synthesis and release of plant phenolic compounds. Herbicides and natural inhibitors are able to stimulate their accumulation. On stress,

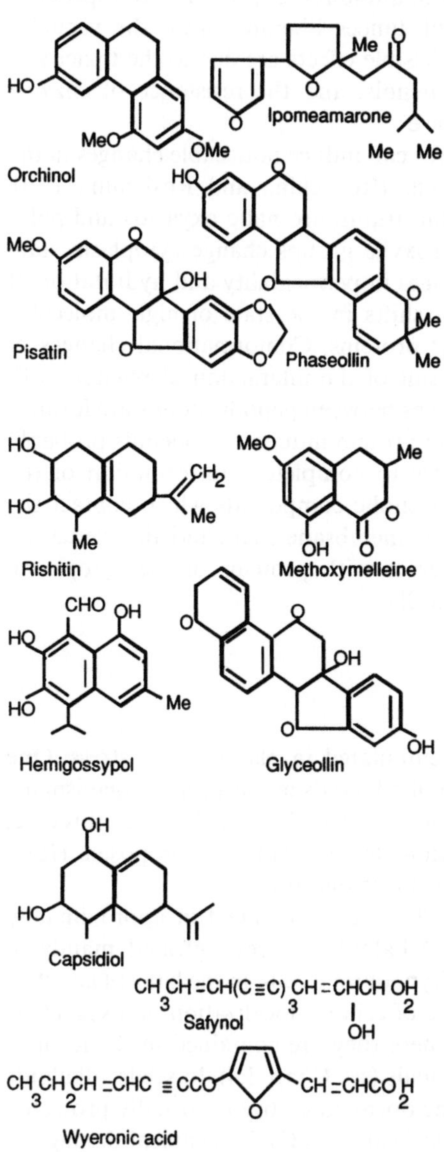

Fig. 85. Chemical structures of some phytoalexins of higher plants

mainly on fungal infection, new phenolic substances, the phytoalexins–phaseollin, pisatin, sativin, medicarpin, and others arise (Fig. 85) (see Sect. 6.2.11).

The phenomenology of the biological effects of phenols has been studied better than of other secondary stress metabolites. Their action is connected with the phenomena of allelopathy, the interaction between plants and insect-phytophages, and resistance to infection by pathogenic microorganisms. Chlorogenic acid, appearing as a result of stress, has an antimicrobial ability but weaker than the products of its oxidation, quinones (Stom 1979). It appears to be mainly products of oxidation of substances liberated from the vacuoles influence plant physiological processes. Visible effects are due to the toxicity of quinones, the high concentration of phenols, and the resistance of enzyme oxidizing phenols-o-dinitro-phenoloxidase.

Phenols and their aromatic derivatives can induce noticeable changes in the permeability and water exchange of cells (Roshchina and Roshchina 1970; Roshchina 1974a,c). At the same concentrations, aromatic oxyacids and polyphenols with *ortho*-position of the hydroxylic groups change cytoplasm state more strongly than other phenols. Changes in permeability and hydratation of cells are connected, in some way, with shifts in the state of high–molecular components of cytoplasm, in particular proteins. Conformational changes of macromolecules can take place as a result of the interaction of phenols with proteins. In the process, hydrogenic bridges between peptidic bonds are formed. The degree of conformational changes in protein induced by phenols probably varies from negligible structural changes to complete transformation of the position of the peptidic chains. Thus, phenolic compounds which contact the plasmalemma, influence the plasmalemma membrane characteristics mediating permeability. Simultaneously, they interact with proteins of the cytoplasm, acting on the water-retaining power of cells.

6.2.8 Alkaloids

Alkaloids are often supposed to be accumulated in plants under stress. One hypothesis holds endogenous formation of alkaloids in the animal organism for a possible cause of stress inducing psychic disorders (Metzler 1977). In this case, parallels between the biochemistry of animals and plants are proposed. However, there are few experimental data on these questions.

Alkaloids are widespread in plants. They are contained in approximately 20% of the families of higher plants. Alkaloids are concentrated mainly in actively growing tissues, epidermal and hypodermal cells, in the latex of laticifers and the vascular bundle sheath. The site of cellular localization of alkaloids is considered to be mainly the vacuole, where they are contained in the form of salts of malic, tartaric, citric, and other acids (see Chap. 1.4). In seeds, alkaloids are stored within the endosperm, and the dormant embryo is usually protected from their toxic influence (Friedman and Waller 1985). Mechanical damage of the plant tissue liberates alkaloids (Fig. 86) which can act very strongly on any living organism animal, plant, or microorganism.

Alkaloids belong to the nonhomogeneous class of chemical compounds and differ significantly in their structure. Depending on the nature of the nitrogenous heterocycle which is in a molecule, alkaloids are subdivided into the main groups: derivatives of piperidine (anabasine, coniine), pyridine (nicotine), tropane (atropine) quinoline (quinine), isoquinoline (morphine, berberine), prince (caffeine), and terpenoid (solanine) indole (strychnine) (Goodwin and Mercer 1983; Münz 1984).

Usually, in a plant there is a whole complex of alkaloids, relative to the chemical nature, such as, for example, in tobacco, *Nicotiana tabacum*, poppy, *Papaver somniferum*, and cinchona, *Cinchona succirubra*. It is shown for many alkaloids that their amount in plants varies and depends on the changing conditions of the external media. The strongest influencing factors are light, temperature, salting, drought, etc. The action of stress-inducing factors on the accumulation and release of alkaloids has been examined in experiments with suspensional culture of *Coffea arabica* in the stationary phase (Frischknecht and Baumann 1985). Salting (0.75 NaCl) induced an increase in the synthesis of the purine alkaloids theobromine and caffeine at high irradiation intensity. When

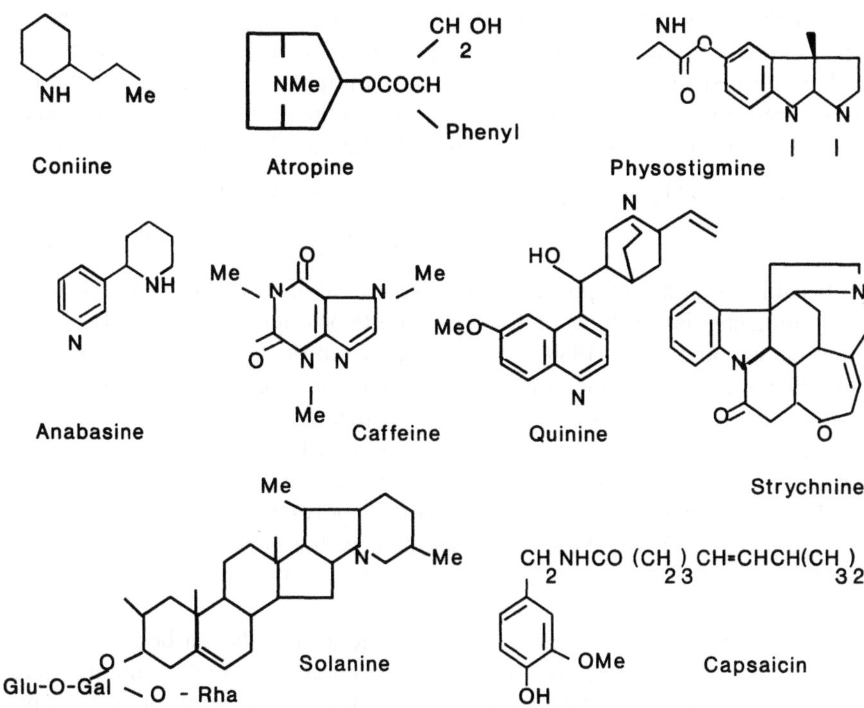

Fig. 86. Some alkaloids which are liberated on destruction of plant tissue

the salinity reaches 153.8 mol/m^3, it leads to the accumulation of tropane alkaloids in the leaves of stramony, *Datura innoxia*. In analogous experiments, deficiency in K^+-ions induced an increase in the level of diamine putrescine, which is known as a derivative of pyrrolidine (Brachet and Cosson 1986). Mechanical damage of leaves of *Nicotiana sylvestris* (wild type) led to a four fold increase of alkaloids in comparison with nondamaged ones (Baldwin 1989). After 19 h of treatment, the concentration of nicotine and nornicotine increased approximately ten fold. The formation of alkaloids occurs in roots by the condensation of pyridine and N-methylpyrrolidine. The process is catalyzed by phosphoribosyl transferase (EC 2.4.2.19) and putrescine methyltransferase (EC 2.1.1.53). Thus, stress changes the chemical composition of plant alkaloids, which is directly correlated with the appropriate shifts in enzymatic systems. Therefore, alkaloids are not now considered as a waste product of plants, and the existence of a close connection between the exchange of proteins and amino acids on one hand and formation of alkaloids on the other is proposed. The variability in the amount of alkaloids on stresses can be an indicator of changes in the nitrogen metabolism of the plant.

6.2.9 Polyacetylenes, Thiophenes, and Traumatic Acids

Among the stress metabolites liberated from cellular compartments or secretory ducts and cavities when they are damaged, a special role belongs to acetylene and its derivatives, the polyacetylenes and thiophenes (Bohlman 1988). The substances are synthesized from fatty acids with the participation of lipogenase (EC 1.13.11.12) linoleate: oxygen oxydoreductase as shown on the scheme:

$$\text{oleic acid} \rightarrow H_3C(CH_2)_4\ C{\equiv}CC_2CH_2{=}CH(CH_2)_7COO \rightarrow$$

$$H_3C\ (C{\equiv}C)_3CH_2CH_2{=}CH(CH_2)_7CO(O).$$

Noticeable amounts of polyethylenes are stored in secretory polyacetylene-containing cavities or ducts of flowering plants belonging to the family Asteraceae (Lersten and Curtis 1989). Secretory spaces in the whole plant of *Artemisia borealis* contain hennuinic polyacetylenes and their acetates. One of the substances possesses a high antilarval activity against the mosquito, *Aedes aegyptii*. This is heptadeca-1,9(Z),16-trien-4,6-diyn-3,8-diol. In secretory ducts of *Cicuta virosa* L. (family Umbelliferae), polyacetylene cicutotoxin is found (Anet et al. 1953; Roshchina et al. 1980). Taxon-specific polyacetylenes and their sulfur derivatives, named thiophenes, are identified in the green parts of plants and heads of flowers of *Dahlia tubulata* (family Asteraceae) and in plants of the family Cynnareae (Christensen and Lam 1990).

Besides the acetylene derivatives, stress-induced acids can be formed from fatty acids. These are 12-oxo-trans-10-dodecenoic acid and traumatic acids (Zimmerman and Vick 1988), which can be released as a result of any trauma, and can defend deep layers of cells from pathogen invasion. The substances are formed from fatty acids with the participation of lipooxygenase during the peroxidation of lipids.

6.2.10 Other Nitrogen- and Sulfur-Containing Compounds

Among other stress metabolites eliminated by plants are nitrogen- and sulfur-containing substances, proteins, amino acids, amines, etc. Excretion of some enzymes has been shown when the plant is infected or mechanically damaged. This is not a result of cell lysis, but a part of the secretory mechanism. Indeed, the hydrolase activity of glucanases and chitinases excreted by suspensional tissue culture of chick pea, *Cicer arietinum*, in response to the pathogen *Ascochyta rabiei* plays an important role in the defense of the plant cell (Vogelsang and Barz 1990). In the plant line resistant to this fungi, chitinase activity is five fold higher than in the sensitive line. In contrast, β-1,3-glucanase is three fold more active in the sensitive line than in the resistant one. Thus, fungal growth in plant cell cultures is inhibited because plant hydrolase breaks the cell wall of the pathogen. Using antibodies, it has been demonstrated that β-1,3-glucanase and chitinase are secreted into the intercellular spaces of tobacco, *Nicotiana tabacum*, inoculated with tobacco mosaic virus (Hosokava and Ohashi 1988). They appear to play an important role in plant immunity. Elimination of catalases and peroxidases by plant roots may be another example of the secretion of enzymes.

Besides enzymes, the glycoprotein extension/is also secreted by cell culture of *Nicotiana tabacum* into the culture medium, as a result of damage to the cell wall on wounding or infection, (Kawasaki 1991). This protein appears to participate in cell wall regeneration and elongation.

In angiosperms, plants grown on sandy dunes or pebbles of coastal zones accumulate osmotically active nitrogen-containing metabolites such as proline and betaine, which can also be considered as stress-induced compounds (Smirnoff and Stewart 1985). Since these soils have low saturation by water, and plants inhabiting them undergo by drought and overheating, some authors (Smirnoff and Stewart 1985) suggest that the compounds may be indicators of drought- and heat resistance. Indeed, the thermostability of certain enzymes, in particular glutamine synthetase and glutamate oxalacetate aminotransferase, is increased in the presence of these stress metabolites. Correlation between salt resistance and the amount of betaine in cultivated grasses is also possible. The transformation of amino acids to betaine $(CH_3)_3N^+CH_2COO^-$ by methylation is characteristic for nitrogen exchange in many plants, especially for alfalfa, *Medicago sativa*.

Nonprotein amino acids should also be considered as metabolites liberated from cells under stress (Fig. 87). In higher plants, more than 400 of these acids were found (Friedman and Waller 1985). Many of them are analogs of proteinogenous amino acids. Nonproteinogenous amino acids are rather widespread. Many of the compounds are toxic and taxon-specific, differing not only

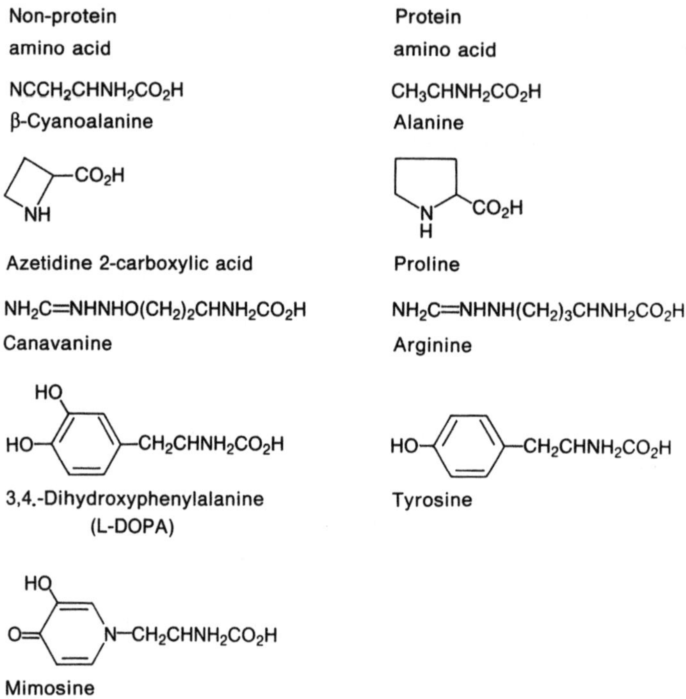

Fig. 87. Toxic nonproteinogenous amino acids and their analogs among proteinogenous ones

among families but even within genera. Indeed, azetidine-2-carboxylic acid is found in species belonging to the family Liliaceae, as well as in some from the family Amaryllidaceae, which is close to liliaceus plants. The amount of nonprotein amino acids may differ in plant organs (Fowden 1965); however, most of them appear to be in the seeds. In particular, 4–6% fresh weight of seeds of *Canavalia ensiformis* are attributed to canavanine, in seeds of *Dioclea megacarpa* from 7 to 10% (Harborne 1988). The presence of 6–9% L-DOPA (3,4-dihydroxyphenylalanine) in seeds is described (Harborne 1982). The role of nonproteinogenous amino acids in the metabolism of plants is still unclear, but it is evident that they have no universal significance in metabolic processes, unlike the protein amino acids. Only few nonprotein amino acids are often met in nature (Fowden 1965). Since these substances contain nitrogen, they can be considered as store compounds of seeds, where they serve as a source of easy-mobilized nitrogen. Nonproteinogenous amino acids possibly play a protective role in plants. They are formed in order to improve the plant's chance of survival. Such plant amino acids, in particular canaranine, djencolic acid, and β-cyanoalanine, are toxic for other living organisms. Azetidine-2-carbolic acid, which was isolated first from *Convallaria majalis*, is the most studied toxin known for leguminous species. Its toxicity is due to the inhibition of synthesis and the utilization of proline. Species which form azetidine-2-carboxylic acid are autoresistant to the substance, because their protein-synthesizing apparatus, in particular proline-tRNA-synthetase, is able to recognize this acid and prevent its inclusion into molecules of protein. Nonadopted organisms erronoeusly use azetidine-2-carboxylic acid instead of proline. They include such nonproteinogenous acid into protein and as a result can perish. Thus, the presence of nonprotein amino acids in plants, mainly in seeds, secures their defense against both insects and herbivorous animals.

Amines accumulated as stressory metabolites, are also physiologically active. When fruits of lemon, *Citrus limon*, grapefruit, *Citrus paradisi*, and red pepper, *Capsicum annuum*, are stored at low temperatures which induce damage to tissues, putrescine is formed (McDonald and Kushand 1986). The pericarp of pepper fruits is most sensitive to chilling. Storage at 1 °C for 21 days leads to a 2.5-fold greater increase in concentration of putrescine than at 7.2 °C. The amount of the substance is higher by 39–40% in pulp and sap of grapefruit at 4.4 °C for 60 days in comparison with 15.5 °C. There is a noticeable linear ratio between the concentration of putrescine and the degree of damage by chilling in lemon fruits which are stored at different temperatures for 21 days. The accumulation of polyamines under chilling stress is confirmed by the work of Wang (1987) on seedlings of cucumber, *Cucumis sativus*, grown in the greenhouse. Chilling treatment (5 °C for 2 days) led to an increase in the level of polyamines, in particular of spermidine.

Putrescine can be accumulated in leaves of plants such as barley, *Hordeum*, on potassium starvation as well as on poisoning by chloride (Kretovich 1971). An activation of decarboxylases is presumed in these processes because putrescine arises as a result of the decarboxilation of the amino acid, ornithine, Amines produced by decarboxylation of amino acids can be re-involved in the metabolism. They are used as a building material in many biosynthetic pathways, especially easy for synthesis of alkaloids.

Some sulfur-containing compounds in the sharp smells of disrupted tissues of species such as onion, garlic or horse-radish should also be considered as stress (Fig. 88). The specific smell of onion consists of the hydrolysates of four compounds, known as alkyl cystein sulfoxides: S-methyl, S-propyl, S-ethyl-L-cysteine sulfoxides and major component, S-trans-prop-1-enyl-L-cysteine sulfoxides and major component, S-trans-prop-1-enyl-L-cysteine sulfoxide. These components are precursors of the onion odor which arises after injury of the tissue (Collin et al.1986). Enzyme alliinase or alliin lyase (EC. 4.4.1.4) hydrolyzes them so that volatile lachrymatory substances, ammonia and pyruvate, are formed. The first three components of the onion smell are synthesized from serine and the fourth, S-trans-prop-L-enyl-L-cysteine sulfoxide, from valine and cysteine. Alliinase and precursors of the odor are localized in different cell compartments: the first in the vacuole and the others in the cytoplasm. In plant tissue culture, synthesis of these precursors does not occur, perhaps due to the absence of the necessary structures. As soon as these structures arise in the cytoplasm at differentiation, the compounds appear as well. The sites of their

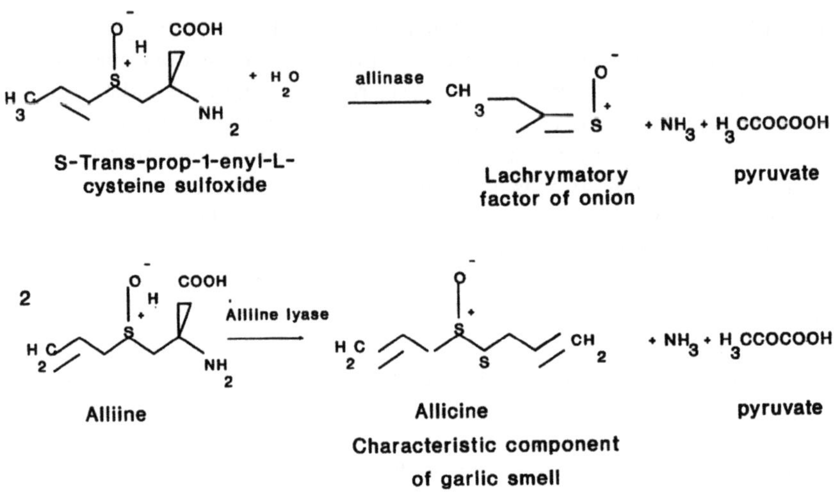

Fig. 88. Formation of odorous substances in onion and garlic

synthesis are supposed to be small vesicles and perhaps plastids (Collin et al. 1986).

6.2.11 Phytoalexins

Phytoalexins are believed to belong to the group of stress-induced metabolites which are accumulated in response to various traumas of plant tissue, and play a leading role in plant resistance to diseases (Haard 1983). Most often, accumulation of phytoalexins takes place on infection of the plant by fungi or bacteria, as well as a result of treatment by physical and chemical factors (Bailey 1982). It is the opinion of Mansfield and Bailey (1982) that traces of phytoalexins are also usually present in healthy tissues. However, it is difficult to verify exactly whether the plant is in a stress state or not. As a rule, the synthesis of phytoalexins is induced due either to the activation of enzymes which catalyze the process, or the synthesis of these enzymes de novo. This differentiates phytoalexins from other antibiotics which, although they also play a protective role, are found in both intact and damaged tissues or are formed from nonactive precursors by the way of the simple reactions of oxidation or hydrolysis. Phytoalexins are easily released from tissues and can be constituents of plant eliminations, as can be demonstrated in model experiments. If fungal spores in a water drop are placed on a leaf, after some hours of exposure, a noticeable amount of phytoalexins is found in the liquid. This principle is at the basis of the method of drop diffusates for the determination of similar compounds. Phytoalexins themselves are synthesized within the leaf, but part of them is eliminated on the plant surface as well. The substances are found in root excretions of pea,

Pisum sativum, on hydroponic growth of plants infected with microorganisms (Novak and Stanek 1986; Werner and Hohl 1987) and in suspension callus tissue cultures of pepper, *Capsicum annuum* (Brooks et al. 1986).

Phytoalexins are low-molecular mass compounds (250–600 daltans), which vary in their chemical nature. The type of compound produced depends on the plant taxon. To date, phytoalexins have been isolated from species belonging to more than 20 families. The chemical formulas of some taxon-specific phytoalexins are shown in Fig. 85. Plants belonging to the family Orchidaceae form phenanthrenes (orchinol), the Convolvulaceae furanoterpenoids (ipomeamarone), the Leguminosae (Fabaceae) isoflavonoids (phaseollin and pisatin), the Solanaceae sesquiterpenoids (rishitin), the Compositae polyacetylenes (safynol), the Umbelliferae isocoumarins (methoxymellein), and the Malvaceae terpenoid polyphenols (hemigossypol), and others. Year by year the number of similar compounds identified increases. Under the influence of infection, coumarin scopoletin was synthesized in leaves of *Hevea brasiliensis* (Giesemann et al. 1986), and glyceolline (hydroxyphaseollin) was identified in root exudates of soybean, *Glycine soja* (Werner and Hohl 1987). There is information on phytoalexins as stress-induced metabolites of sap-wood (alburnum), which arise if cells of the parenchyma have died (Kemp and Burden 1986). Stilbenes such as pinosylvin and its monoethyl ester are formed in pines, *Pinus*, infected by many fungi, and resveratrol, oxyveratrol in mulberry, *Morus*, damaged by fusariosis, etc. In soybean, *Glycine max*, isoflavonoid phytoalexin glyceolline is synthesized (Keen et al. 1986).

Phytoalexins are not always represented as individual substances but also as a mixture of closely related compounds to make the adaptation of parasitic organisms to them more difficult. For example, some ten sesquiterpenoid compounds, including rishitin, were isolated from potato, *Solanum tuberosum*. In tissues of cotton, *Gossypium*, seven terpenoids were identified (Metlitskii 1987). The ratio between the individual components varies depending on the nature of the infection, the physiological state of the plants, the vegetation period, and other factors. It is known that compounds belonging to different classes of chemicals are synthesized in response to one and the same stress. Indeed, in fruits and leaves of tomato, *Lycopersicon esculentum*, infected by *Cladosporium fulvum*, beside sesquiterpenes, polyacetylenes are formed (Watson et al. 1985). In the family Solanaceae, the main class of stress metabolites are diterpenes, although steroid alkaloids are synthesized as well (Haard 1983).

The accumulation of the above-mentioned phytoalexins can be induced not only by pathogen invasion, but also by various physical and chemical factors (Haard 1983). The physical factors stimulating formation of stress metabolites are believed to be chilling, visible light, anoxia, and rarefied atmospheric pressure. For example, short exposure of tissue sections to ultraviolet light induces the accumulation of furanoterpenes in sweet potato, *Ipomoea batatas*, diterpenes in slides of potato, *Solanum tuberosum*, and isoflavonoids in alfalfa, *Medicago sativa* and pods of kidney bean, *Phaseolus vulgaris* (Bailey 1982). There is conclusive evidence that a wide range of chemical agents can induce increase in stress metabolites. Salts of heavy metals (copper or mercury) stimulate the accumulation of ipomeamarone in *Ipomoea batatas*, rishitin in *Solanum tuberosum*, and phaseollin in *Phaseolus vulgaris* (Bailey 1982). Actinomycin D and other inhibitors of protein synthesis can also be effective stimuli for the formation of stress metabolites under experimentally controlled conditions. Metabolic inhibitors such as acetate and sodium fluoride, detergents, various sulfhydril reagents, and salts are additional examples of chemicals which are able to induce synthesis of stress metabolites (Stoessl 1982). Recent studies show that various pesticides also stimulate phytoalexin accumulation (Stoessl 1982). Formation of phytoalexins is stimulated by water stress (moisture deficit) when tissues are damaged. In this case, pisatin was identified in necrotic parts of leaf of pea, *Pisum sativum*, and several fungitoxic components, including wyerone and wyerone acid, were found in the same spots on foliage of horse bean, *Vicia faba*. Thus, phytoalexins can be accumulated in leaves strongly damaged by stress (Piumbe and Willmer 1985).

The chemical composition and quantity of phytoalexins can vary depending on the active extreme factor. This has been demonstrated in experiments by Vaverka (1986) on leaves of soybean, *Glycine soja*. He showed that the amount of phaseollin and 6, α-oxyphaseollin changed from 2 to 420 $\mu g\, g^{-1}$, depending on the active factor. The highest accumulation of these phytoalexins was induced by *Pseudomonas glycinea* and lowest by mechanical damage.

The stimulus for synthesis of phytoalexins is biochemical agents named inductors or elicitors, which are contained both on the surface (mainly in the cell walls) of intact cells of bacteria, fungi, and in tissues of the plant host itself. They are eliminated after cell death, and migrate to living cells where they induce synthesis of phytoalexins (Darvill and Albersheim 1984; Ozeretskovskaya and Chalova 1985). Perhaps elicitors influence the genetic systems of plants, activating the expression of those genes which code synthesis of enzymes participating in the production of phytoalexins (Darvill and Albersheim 1984). Elicitors are represented as fragments of pectin: oligosaccharides or polysaccharides. Enzymes which take part in the synthesis of the elements of the cell wall may play an analogous role. This has been confirmed in model experiments on tissue culture of tobacco, *Nicotiana tabacum*, treated with commercial cellulase (Watson et al. 1985) and on suspensional cell cultures of *Capsicum annuum* treated by microbial cellulase and pectinase (Brooks et al. 1986).

Precursors of stress-induced metabolites are ordinary compounds whose accumulation favors the synthesis of phytoalexins (Bailey 1982). The scheme of the latter process depends on the nature of the precursor and phytoalexin. Plants use or activate already functioning pathways of normal metabolism for phytoalexin production, especially those that can be connected with formation of enzymes de novo. This has been shown for diterpene hydrocarbon phytoalexin casbene, which is formed by castor-bean seedlings infected by fungi (Dudley et al. 1986). In this case, casbene is synthesized from mevalonate by the common metabolic pathway of isoprenoid formation to isopentenyl pyrophosphate, and then through dimethyl-allyl-PP, geranyl-PP, and farnesyl-PP. It is catalyzed by the corresponding enzymes. The activity of some of the enzymes of formation of isopentenyl-PP has not changed, whereas the amount of farnesyl-PP-synthase, geranylgeranyl-PP casbene synthase noticeably increases. Dudley et al. (1986) showed that isopentenyl pyrophosphate isomerase and farnesyl pyrophosphate synthase are localized in proplastids of both sterile and infected tissues, unlike geranyl pyrophosphate synthase and casben synthase which are exclusively in the same organelles of infected tissues. Therefore, synthesis of the enzymes is induced only by infection.

The phytoalexins released by plants are highly toxic. Their effective dose reaches $10^{-5} - 10^{-4}$ M (Smith 1982) and is enough to destroy the pathogenic protoplast. The action of such stress-induced metabolites can be specific, that is connected with their lipophyllity which permits them to penetrate the membrane and damage it (Darvill and Albersheim 1984).

CHAPTER 7

Biological Effects of Plant Excreta

Biological effects of plant excretions first became the subject of scientific research after the works of the embryologist B. P. Tokin, who showed that vapors of homogenated tissues (phytoncides) have a bactericidal effect (Tokin 1980a). Phytoncides as antimicrobial agents were then effectively studied for their therapeutic effects in medicine (Tokin 1980a; Aizenman et al. 1984). Other directions in the study of the biological activity were the influence on the growth processes in allelopathic interactions in phytocenosis (Grodzinskii 1965, 1991). The mechanisms of action of plant excreta on cellular and subcellular levels were investigated after the 1960s, and there is little information about them as yet. It should be noted that in most cases the experiments were performed on homogenated tissues on their extracts, where large amount of stress metabolites are liberated or formed during the destruction of cellular compartments. Thus, such plant preparations from damaged tissues are mixtures of products of normal metabolism (all cells cannot be destroyed simultaneously), products of their hydrolysis or oxidation, and partly stress agents synthesized de novo. The resulting data have hardly been interpreted because similar multicomponent preparations of pounded cells did not yield information as to the real contribution of each individual substance to the observed effect. Taking this into consideration in this chapter the main emphasis is placed on the mechanisms of action not only of natural mixtures but of the individual chemicals included in them.

7.1 Growth Processes and Cell Destruction

There is an extensive literature reviewing data dealing with the influence of plant excreta on the growth processes (Grodzinskii 1965, 1991; Ivanov 1973; Rice 1974, 1984; Kolesnichenko 1976; Rao 1990). In these publications, the stimulatory and inhibitory effects of water extracts and vapors from plants on seed germination, growth of seedlings, and other growth reactions, usually determined as biological tests, have been observed. However, any noticeable growth reaction occurs in response to relatively high concentrations of metabolites, which are seldom found in the excretions of intact plants. Therefore in spite of the simplicity and accessibility of the biotest method, it has certain limitations when studying the mechanisms of biological activity of plant excreta. The results of experiments with extracts and infusions should be observed as necessary phenomenological data indicating sensitive species for subsequent study. In this section we consider the information dealing with mechanisms of the influence of plant excreta and their individual components on the stages of plant growth – division and elongation of cells, as well as their destruction.

7.1.1. Division and Elongation of Cells

Water-soluble metabolites of plant tissue influence the stages of cell mitosis. Thus, after treatment of pea, *Pisum sativum*, roots by extracts from peel of black walnut, *Juglans nigra*, the number of cells entering the phase of mitosis is increased. The substances in the extracts retard the process or prevent its initiation (Jensen and Welbourne 1962). A similar effect on mitotic activity in roots of wheat, *Triticum vulgare*, rye, *Secale cereale*, and garden cress, *Lepidium sativum*, is observed for toxins from excretions of weeds-sow thistle, *Sonchus arvensis*, goosefoot, *Chenopodium album*, and creeping thistle, *Cirsium arvense* (Bukolova 1971). Recently, the number of such examples has increased. it has been shown (Shehab and Adam 1983) that water extract of *Anastatica hierohuntica* (1–3%) depresses mitosis in root meristems of onion, *Allium cepa*. In this case, DNA despiralization, chromosome sticking or forming of bridges between them, and retardation of the divergence of chromosomes to poles have been observed. Analogous changes are seen in roots of broad bean, *Vicia faba*, treated with extracts of *Ammi majus* (Adam and Rashad 1984). The extract from leaves of tobacco, *Nicotiana tabacum*, at low concentrations (0.75–0.25%) also induced various types of chromosome injuries in roots of onion, *Allium cepa*, and depressed mitosis at higher concentrations (1.5–1.0%) (Patnaik et al. 1984) The action of plant extracts on mitosis has been described in other works as well (Sarkar and Amiya Kumar 1974; Abraham and Dileep 1976; Shehab and Adam 1983). In all these studies the chemical nature of the active matter was unknown or only presumed. There are few data where the influence of individual components of plant excreta on mitosis have been studied. Ethylene (mainly stress-induced) is able to decrease the mitotic index of root meristems of 5-day-old seedlings of pea, *Pisum sativum*, and to block cell division (Rost et al. 1986). Another volatile compound, acetaldehyde, at concentrations of 0.05–0.2% induced duplication of chromosomes, stimulating polyploidy (Cortes et al. 1987).

Among substances extracted from plant tissue by water, phenols are the most active. The character of their action depends on the position of the hydroxylic groups in the benzoic ring. m-Phenols are able to stimulate, whereas o- and p-phenols inhibit the growth processes (Meskhi and Dzhokhadze 1973). The target of phenolic action on the transcription level is enzyme RNA-polymerase at the transcription level is enzyme RNA-polymerase at the transcription level (Meskhi and Dzhokhadze 1973). It has been shown that resorcine stimulates and pyrocatechin inhibits RNA synthesis. The action of phenols on the genetic apparatus is also illustrated by the following example. Djordjevic et al. (1987) showed that phenols excreted from roots of clover, *Trifolium repens*, sharply change the activity of the gene responsible for the formation of nodules induced as a result of symbiosis with *Rhizobium trifolii*. Hydroxylated flavones in low concentrations, 5×10^{-7}M (most active is 7,4-dihydroxyflavone), contained in the extracts stimulate the transcription of this gene. In contrast, coumarin, umbelliferone, and isoflavone formononetin at concentrations more than ten fold higher prevent induction.

Another class of biologically active compounds, alkaloids, are able to delay the phase of cell division. The following example illustrates the mechanism of inhibition. Alkaloid colchicine isolated from *Crocus automnalis*, binding with molecules of the protein tubulin, responsible for one mode of cytoplasm motility, suppresses formation of the mitotic spindle, and in this way blocks cell division.

Besides their action on the embryonic phase of growth, volatile and water-soluble metabolites of plant excreta retard or stimulate cell elongation. Iwanami and Hyodo (1983) showed that volatile compounds of lemon, *Citrus limon*, and lemon oil inhibit not only cell division but elongation of seedlings of mung bean, *Phaseolus aureus*. Earlier, it was supposed that this reaction could be induced by ethylene released from the plant, which plays a significant role as a plant hormone. However, volatile lemon excretions are demonstrated to have no similar effect, unlike ethylene. Essential oils of lemon induced the etiolation of leaves and depressed the formation of lamellar structures of chloroplasts.

According to the data of some authors, the inhibition of seedling elongation was observed under treatment by volatile compound from cut segments of oleander, *Nerium oleander*, cedar, *Cedrus japonica*, peel of sweet orange, *Citrus sinensis*, and roots of *Vasabia japonica* as well. The character of this inhibition was not similar to those induced by ethylene. The chemical nature of the active matter has not been identified.

In experiments of Heisey and Delwiche (1984), volatile compounds of *Trichostema lanceolatum* contained in 2.9 mg volatile oils collected from 100 g fresh leaves are able to depress hypocotile elongation in barley, *Hordeum vulgare*, and smooth brome grass, *Bromus inermis*. The main inhibitor was identified as the monoterpene alcohol, terpinene-4-o1, whose activity was 0.3 of the same parameter of camphor and 1.9 of cineole (Heisey and Delwiche 1984). Examples of the stimulatory action of some other volatile organic compounds are also known. Among them is the induction of elongation of barley roots by the isoprenoid alcohols, geraniol and farnesol (Wardle and Short 1982). Stimulation of radial stretching of lettuce hypocotils by ethylene is thought to be the basis of acceleration of seed germination (Abeles 1986). Only 10 μg/l of this gas was needed to stimulate seed germination in *Lactuca sativa*. Other hydrocarbons such as propylene and ethane had no similar effect (Abeles 1986). At low concentrations, $10^{-5} - 10^{-6}$M, volatile terpenes and their derivatives inhibit seed germination in radish, *Raphanus sativus* (Asplund 1968). Among the

Table 13. Effects of some individual components of plant excreta on the growth reactions

Class	Compound	Origin	Influence	Reference
Sesquiterpene lactones	Strigol	*Gossipium*	Stimulation of germination of parasite *Striga asiatica*	Fischer et al. (1989a)
Terpenes of essential oils	Products of oxidation of tujone	*Cyperus*	Inhibition of growth of some test species	Komai and Tang (1989)
Monoterpenes	Piquerol and their derivatives	*Piqueria*	Inhibition of growth of some test species	Ortega et al. (1990)
Terpenoid compounds	Abscisic acid	*Cyperus*	Inhibition of growth of some test species	Komai and Tang (1989)
	Parthenin	*Parthenium*	Inhibition of growth of some test species	Kelsey et al. (1984)
Phenols Hydroquinones	Juglone	*Juglans*	Retardation of root elongation	Neave and Dawson (1989)
Phenolic acids	Benzoic, p-coumaric, 2,4-dihydrobenzoic, vanillic	*Brachia*, *Digitaria*, *Imperata*, *Panicum*	Inhibition of growth	Chou (1989)
Biogenic amines	Serotonin	*Urtica* *Utrica*	Stimulation of root growth Stimulation of seed germination	Regula et al. (1989) Roshchina (1991a)
	Histamine	*Urtica* *Jatropha*	Stimulation of seed germination	Roshchina (1991a)
Chrolinic ester	Acetylcholine	*Urtica* and others	Stimulation of seed germination	Tretyn et al. (1988)

substances studied, pulegone was more toxic (four fold) than the others, including HCN.I_{50} for these terpenes were (in M): pulegone 1.5; (−)-camphor 3.1; (+)-camphor 3.3; (−)-borneol 21; 1,8- cineole 78; limonene 45; α-phellandrene 57; α-cymene 51; α-pinene 30; β-pinene 170, but for HCN 5.8. Terpenoid partenin also inhibited root growth of some test species (Gasiduenas et al. 1972 in Kelsey et al. 1984).

The effect on growth reactions depends on the chemical nature (Table 13) and concentration of the active matter. The stimulation of growth by low concentrations ($< 10^{-5}$ M) is characteristic for sesquiterpene lactones such as strigol and biogenic amines, for instance, serotonin. other classes of components in plant excreta often inhibit growth processes, mainly at higher concentrations.

7.1.2 Pollen Germination

The sole object of the study of the effect of plant excreta on the growth processes is the germination of pollen in favorable medium. The pollen tube grows fast and the rate of the process is often some millimeters per hour (Esau 1977). Basic works by Kovalenko (1972) and Golubinski (1974) showed that vapors released by many plants can stimulate or depress pollen germination. Long-term pollen of higher plants is thought to possess resistance to various factors of the environment. Attention is given to this question in Iwanami (1981), who demonstrated that the pollen does not lose its ability to germinate even under treatment of a high dose (> 100 kR) of x-and γ-rays or various solvents such as diethyl ether, chloroform, and phenol. Inhibitors such as actinomycin D, bromodeoxyuridine, and cycloheximide also had no effect on pollen germination. On the contrary, volatile compounds from plant tissue cuttings have been shown to inhibit the process for pollen grains of *Camelia sinensis* (Iwanami 1981). Vapors from the pericarp of lemon, *Citrus limon*, citron, *C. medica*, and orange, *C. unshiu* had a stronger effect than those from their pulp (~10-16-fold), from bulbs of onion, *Allium cepa*, tuber of garlic, *Allium sativum*, leaf of *Allium tuberosum*, rhizome of *Eutrema wasabi* (~6-10-fold). Acting on pre-germinating pollen tubes, volatiles from pieces of *Allium cepa* stopped their elongation within 1–3 s and induced abnormal swelling of their tips within 4–6 s. The level and character of the inhibition by plant excreta were similar to those observed under a high dose (200 kR) of γ-rays (Iwanami and Hyodo 1983).

As is seen from comparison of the amount of substances extracted from leaf lamina, seeds, pollen grains, stigma, and pistils of various clones of cabbage, *Brassica oleracea*, the highest quantity of the inhibitors of pollen germination is found in seeds. Extracts from the seeds were most active because they completely blocked pollen germination in *Petunia hybrida* and *Lilium lancougens*. The chemical nature of the inhibitors was not identified.

Our knowledge was enriched by the observations of Viswanathan and Lakshmanan (1984), who showed inhibition of pollen tube growth in *Calotropis gigantea* by extracts from stems of dodder, *Cuscuta reflexa*, leaves of the weed

Parthenium hysterophorus, the bulb of onion, *Allium cepa*, and seeds of maize, *Zea mays*. The character of the action depended on the plant species and the tincture concentration. Extracts from *Cuscuta reflexa* and *Zea mays* induced branching of pollen tubes in the test object which was not observed under natural conditions. Some solutions decreased pollen growth at high concentrations, in particular for maize (100 and 10%), and stimulated it at lower ones.

Interesting experiments of Kovalenko (1972) show that volatile excreta of flowers belonging to one species influenced pollen germination in other species which are blooming simultaneously. Thirty woody shrub species were studied in the Dendrarium of the Botanical Garden of Odessa State University. Two examples from this work are below:

Flowers	Pollen	(+) Stimulation or (−) inhibition
Robinia pseudoacacia	*Gleditschia triacanthos*	(−)
–	*Cytisus laburnum*	(+)
Wistaria sinensis	*Robinia pseudoacacia*	(+)
Paulownia tomentosus	–	(−)

In the first of these, the flower vapors of the same plant (*Robinia*) induced the opposite effects on pollen germination of different species. In another experiment, flower fragrances of two different species acted on the pollen from one test species (*Robinia*), stimulating or inhibiting its germination. Therefore, pollen has a highly specific sensitivity to the flower excreta of different species.

Among individual components of plant excreta which influence pollen germination, flavonoids and certain organic acids are known. According to Kudryavtseva and Volinets (1981), flavonoids from *Tulipa* excretions themselves acted in an analogous way. Benzoic, vanillic, and gallic acids inhibited pollen germination of pine, *Pinus*. Pollen, usually sensitive to various plant excretions, can itself contain highly active compounds. In experiments by Viswanatahan and Lakshmanan (1984), pollen of *Parthenium hysterophorum*, which was brought onto the surface of stigma of various species, inhibited ovary setting in seven species studied, such as *Lycopersicum esculentum*, *Phaseolus vulgaris*, and *Capsicum annuum*. Seven to ten days after treatment, the flowers fell.

7.1.3 Destructive Changes in Cells

If biologically active substances are present in plant excreta at concentrations higher than a certain level or if they are toxic, treatment by them not only stops the growth of the treated cell, but inhibits many of its important physiological functions, which leads to destructive changes.

Alteration in the motility of the cytoplasm and other cell organelles is a well-known sign of functional disintegration. Some plant excreta are shown to retard or facilitate chloroplast movement in leaves of *Elodea*, which is induced by light (photodinesis) (Roshchina 1965). The effect depended on the chemical nature of the active substances, their concentration and the duration of treatment. Visible changes in motility of organelles can be a result of blocking or stimulation of processes which generate energy (Roshchina and Roshchina 1977). Another cause of retardation of movement appears to be the induction of unreversed increase in viscosity by the steams of the disrupted tissues, but sometimes the phenomenon precedes a short the decrease in viscosity (Roshchina 1974c).

Besides those mentioned above, one of the first signs of pathological damage in the cell is the increase in membrane permeability. Disturbance of the barrier function leads to the efflux of many substances from their compartments surrounded by the membrane. In model experiments, the effects of water-soluble leaf extracts from 32 species of woody plants on the process have been studied (Roshchina and Roshchina 1970; Roshchina 1973c, 1974c). As model, living tissue of red beet, *Beta vulgaris* f. *rubra*, which releases the cell sap anthocyan on injury, was used. Only 14 of the species studied (Table 14) induced pigment exoosmosis, which was highest for seven of them. The rate of anthocyan efflux depended on the strength of the injuring action, which offers the possibility of estimating the level of the extract's toxicity. The damage to membranes was possibly induced by substances of a polyphenolic nature because their toxic effect correlated with the content of the compounds belonging to the phenols (Roshchina 1973c, 1974c). Disturbance of the membrane structure leads to change of the cell water-absorbing capability. These infusions from woody plant leaves have been shown to decrease in to a greater or lesser degree the possibility of plant tissue absorbing water (Roshchina 1974a, c). The active matter of the solutions was phenols. The influence of individual phenols of varying chemical structure differed, depending on the number and position of the hydroxy-groups in the benzoic ring. *o*-Phenols are found to be most toxic, while *m*-phenols showed no noticeable activity. The degree of inhibition of membrane permeability for anthocyan and the water-absorbing ability of cells changed depending on the addition of any functional group to the phenol molecule, the nature of the group, and its position in the benzoic ring. Toxic action was most increased after the introduction of the carboxylic group in the phenolic molecule.

Table 14. Exoosmosis of anthocyan from disks of roots of *Beta vulgaris* f. *rubra* induced by water-soluble excreta from leaves of woody plants, in units of optical density. (Roshchina and Roshchina 1970; Roshchina 1974c)

Plant	Anthocyan efflux	Plant	Anthocyane efflux
Betula verrucosa	0.440	*Fraxinus pubescens*	0.235
Sorbus aucuparia	0.249	*Quercus robur*	0.259
Cotinus coggygria	0.219	*Tilia cordata*	0.254
Rhus typhina	0.186	*Ulmus leavis*	0.113
Padus racemosa	0.420	*Sambucus racemosa*	0.239
Acer campestre	0.187	*Syringa vulgaris*	0.147
Populus nigra	0.609	Control (water)	0.000
Corylus avellana	0.324		

Important criteria of the pathological state of the cell are visible changes in the protoplast surface. Similar morphological damage to plasmolyzed cells of the epidermis in blue onion scale treated by steams of disrupted leaves of bird tree, *Padus racemosa*, mountain ash, *Sorbus aucuparia*, and bulb of garlic, *Allium sativum*, was seen, by using a modification to the microscope (Roshchina 1972c, 1974c). Volatile compounds of the plants not only induced ruptures in the plasmalemma, but more often soluted it gradually. Sometimes, the cell membrane formed vacuoles that also show on injury to the cell surface. The interaction of the cytoplasm of onion cell with volatile leaf substances resulted in profound changes, accompanied by swelling. Pathological damages were spread to the nucleus, which also swelled, and various inclusions arose in it. The nucleolus was clearly outlined. Gradually changes spread to the tonoplast, which was ruptured so that vacuolar sap leached out. When the cell died, the protoplast was broken into small grains and vesicles so that it was dispersed throughout the whole cell volume, or sometimes the coagulated cytoplasm reserved its form. The latter case looks like fixation and may be explained by the inactivation of hydrolytic enzymes, especially proteases, which destroyed the cytoplasm after its atrophy. Most of the intracellular damages and ruptured membranes surrounding the nucleus, mitochondria, and dictyosomes were observed in root sections of various plants treated by steams of injured leaves of *Salvia leucophylla* (Lorber and Muller 1976). In some works, gelatinization of the nucleolus and cytoplasm of herbaceous plants on treatment by steams of disrupted bulbs of *Allium cepa* and *Allium sativum* was noted. Destruction and decoloration of chloroplasts, cytoplasm, and other cell organelles after treatment by volatiles from homogenated leaves of cow-parsnip, *Heracleum*, were shown by Chasovennaya (1961). Isolated chloroplasts were wrinkled, changed their form, and agglutinated in infusions from leaves containing phenols. Visible changes in the structural cell organization are induced due to the ability of phenols and their derivatives to interact with proteins. They inactivated also the enzymatic systems of chloroplasts, which, occurring simultaneously with the destruction of the structure, leads to the depression of photosynthesis (Roshchina 1973b). Based on these results, new biological tests for the determination of plant excreta activity were proposed (Roshchina 1965, 1974c). These are changes in the rate of chloroplast movement in the cytoplasm stream, injury to the barrier function of cytoplasmic membranes (enhanced efflux of some cellular metabolites), and the water-holding power of cells.

7.2 Cellular Membranes as Targets for Action of Plant Excreta

The mechanism of action of plant excreta at the subcellular and molecular levels has only begun to be studied. Few experimental works are devoted to this question, so that the material of this section represents rather a scheme of main directions for future investigations than concrete facts, most of which have been worked out by animal physiologists for solving other tasks, but can be enlisted to the analyzed problem as well.

One of the main targets of action of normal and stress components of plant excreta in the cells of animals, plants, and microorganisms are the cellular membranes, composed basically of lipids and protein complexes (Houslay and Stanley 1982). Among the latter there are complexes of electron carriers in the mitochondria and chloroplasts, such as the cytochrome f(c)/b_6 complex, pigment-protein complexes of proteins which form ion channels and (or) receptors, the transport K^+-Na^+-ATPases (Changeux et al. 1984), ATPases of coupling membranes, and other enzymatic systems. The noncontinuousness of the membraneous structure makes it possible to indicate sites where one or the other chemical compound is specifically binding. Possible targets for similar substances are shown in Fig. 89. Based on this scheme, the literature data are interesting. There is little information as to how plant excreta influence the lipid phase of cellular membranes. It could be connected with their lipophylic components such as alkaloids with unsaturated binding, in particular capsaicin, which is contained in the secretory cavities of fruits of red pepper, *Capsicum annuum* (Zamski et al. 1987). According to data for animal cells, capsaicin induces the efflux of some proteins from the membranes, for instance an important neurotransmitter, peptide P (Dun and Kiraly 1983). Exudation of trichomes in the genus *Solanum* (*Solanum tuberosum, S. herthaultii, S. berthaultii*) contains sesquiterpenes, to which insects are very sensitive (Gregory et al. 1986). Among them E-β-farnesene, β-caryophyllene, β-cubebene, and Δ-cadinene are most active. Perhaps, all of them act on the lipids of membranes, inducing a block age in the digestive function of insects.

The protein components of membranes are sensitive to chemicals. Some of them function mainly as catalysts, i.e., as enzymes participating in metabolic and energic reactions. For material on this influence of plant excreta see Sections 7.3 and 7.4. Other proteins have the function of "recognition", i.e., are receptors of low molecular hormones and mediators. The binding of the released substances with similar receptors serves as a chemical signal to change either the enzymatic processes or membrane permeability for certain ions. Ion channels are mediated by receptors or can form receptors themselves, in particular the Na^+-channel (Changeux et al. 1984). Protein subunits of ion channels can have an enzymatic function as well (Eisenberg 1990). The conception of membrane receptors is well known for animals and can be applied to the theory of the action of neuro-

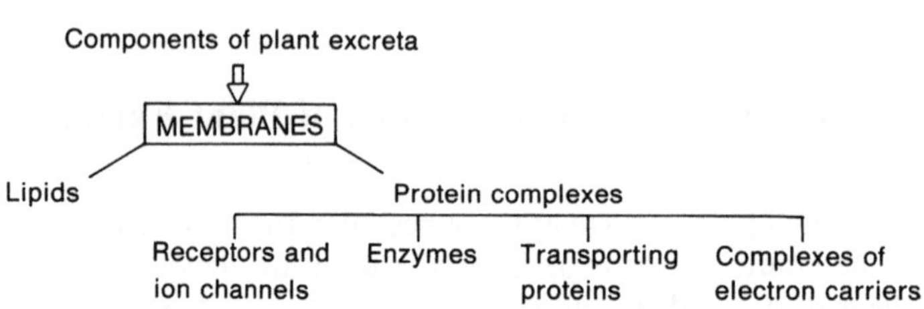

Fig. 89. Targets of action of plant excreta on membranes

transmitters and hormones (Mikhelson and Zeimal 1973, Rozen 1986). All similar receptors are proteins, which can have a high affinity in specific binding (noncovalently) with some low-molecular weight agents. Recently, this idea has been applied to explaining the action of plant hormones (Venis 1985) and neurotransmitters such as acetylcholine, dopamine, noradrenaline, adrenaline, serotonin, and histamine, which are also found in plants (Roshchina 1991a,b). The conception could be useful for understanding the mechanisms of the action of plant excreta.

Protein receptors for some known phytohormones, including ethylene, a component of many volatile excretions, have now been isolated (Sisler and Yang 1984, Venis 1985). Receptors of ethylene can be located in membranes of the endoplasmatic reticulum and the plasmalemma. The receptor which is bound with ethylene is protein with molecular mass 52 000–60 000 daltons (Venis 1985). The perception by animals of many volatile plant compounds takes place by means of special olfactory receptors which can be very sensitive to smells and odors, in particular for camphor, even at concentrations of $10^{-14} - 10^{-16}$ M. Most effective progress in evolution is the "recognition" system in animals due an internal regulatory system, such as the cholinergic system (McQueen 1987). The cholinergic system includes the mediator, acetylcholine, the enzyme of its synthesis, choline acetyltransferase, the enzyme of its hydrolysis, cholinesterase, and the cholinoreceptor sensitive to acetylcholine. The mechanism of such a system is as follows. Acetylcholine, which is synthesized with the participation of choline transferase, binds with the cholinoreceptor that changes the conformation of the receptor. As a result, the ion channel opens and ion permeability is increased, which induces the appearance of an electric potential known as the action potential which moves along the membrane as the impulse of irritation. The ion channel is not open for long, since destructive changes can appear. To prevent these, cholinesterase hydrolyzes acetylcholine and the cholinoreceptor is liberated from the irritating agent; the ion channel then closes.

Acetylcholine has been found in many plant excreta as well (see reviews of Fluck and Jaffe 1976; Roshchina and Mukhin 1986; Hartmann and Gupta 1989; Tretyn and Kendrick 1991, and the monograph of Roshchina 1991a). At a high concentration, it is found in the stinging glandular trichomes of some plants from the family Urticaceae (see Chap. 3.10). Choline acetyltranseferase was found in plant tissues (Smallman and Maneckjee 1981) as well as cholinesterase (Fluck and Jaffe 1976; Roshchina and Semenova 1990). Acetylcholinesterase is also widespread in plant tissues. This is a protein with a molecular mass of subunits from 60 to 80 kdaltons, localized in the plasmalemma (Fluck and Jaffe 1976; Tretyn and Kendrick 1991), chloroplasts (Roshchina 1986, 1988a,b, 1990a), and the nucleus (Maheshwari et al. 1982). It has been shown that acetylcholinesterase can be a component of plant excreta since the enzyme with molecular mass 70 kdaltons has been found in latex from *Synadenium grantii* belonging to the family Euphorbiaceae (Govindappa et al. 1987). Bednarska and Tretyn (1989) demonstrated the acetylcholinesterase activity in pelliculae of excreta covering the stigma surface in the pistil of *Pharbitis nil* flower (family

Convolvulaceae). Similar enzymatic activity has also been found in the pollen grains and tubes of the same species. The enzyme appears to be needed for interaction between the pistil stigma and pollen. Fluck and Jaffe (1976) proposed that acetylcholinesterase and acetylcholine take part in interactions within phytocenosis.

Some experiments have shown that blockators of animal cholinoreceptors influence acetylcholine-dependent reactions such as C^{14}-acetate uptake by roots (Fluck and Jaffe 1976), emission of ethylene by the leaf disks (Jones and Stutte 1986), and water uptake by the roots (Zolkevitch et al. 1979). As observed in animals, exogenous acetylcholine can change the permeability of plant cells for K^+ (Yurin et al. 1979; Hartmann and Gupta 1989, Tretyn and Kendrick 1991) and Na^+ (Roshchina 1987, 1989a) and membrane potential (Yurin et al. 1979). Now, new, more direct data indicate the existence of ion channels (Tester 1990) and functional analogs of animal cholinoreceptors which can control the opening of the Na^+ and K^+-ion channels (Roshchina 1987; Tretyn and Kendrick 1991) or the rate of photophosphorylation in chloroplasts (Roshchina 1989a, 1990cc, 1991a). Acetylcholine, being transported in secretory vesicles, may play a role in the transfer of the information from the plasmalemma to the organelles or between organelles (Roshchina 1989a, 1990b,c, 1991a). The presence of some components of the cholinergic system in plants shows a similarity to "recognition" sites in biological membranes.

The targets of action of plant excreta are two components of the cholinergic system. The first is the cholinoreceptor, a complicated protein with molecular weight > 200 000–250 000 daltons, which controls the work of ion channels, its subunits forming ion channels themselves in some cases (Changeux et al. 1984). The second is the enzyme cholinesterase (mainly acetylcholinesterase). Table 15 gives data dealing with the inhibition or stimulation of these membrane components by plant excreta, including mainly alkaloids and terpenoids. The modes of action differ from each other. One group of substances, in particular alkaloids binding with cholinoreceptor, act as either blockators-antagonists of acetylcholine (atropine, d-tubocurarine) or agonists of acetylcholine (nicotine, arecoline). Other substances, such as the alkaloids physostigmine and capsaicin, terpenoid permethrine, many monoterpenes, and the polyacetylene cicutotoxin, inhibit cholinesterase. In both cases the functioning of the cholinergic system is changed.

Besides the above-mentioned system, animals also have dopaminergic, adrenergic, serotoninergic, histaminergic, and other regulatory systems which include low-molecular mass effectors, mainly neuromediators (dopamine, noradrenaline, serotonin, histamine), receptors sensitive to them, and enzymes participating in the synthesis and catabolism of the substances. Dopamine, noradrenaline, serotonin, and histamine are found in plant excreta: the two first substances mainly in latex and the others in the stinging trichomes (see Chap. 3.10). The activity of enzyme in their synthesis and catabolism is observed also in plants (Roshchina 1991). The targets of low-molecular mass effectors are thought to be functional analogs of the corresponding animal receptors. At least

Table 15. Effect of some secondary metabolites included in plant excreta on the cholinergic system of plants and animals

Plant	Excreted substance	Effect on	
		Cholinoreceptor or its analog	Cholinesterase
Amanita muscaria	Muscarine	*An (1), *P (6)	0 (1)
Areca catechu	Arecoline	*An (1), *P (5)	–
Atropa belladonna	Atropine	+ + + An (1), + P (5)	+ An (1), + P (2)
Capsicum annuum	Capsaicin	0 (4)	+ P (4)
Chondodendron tomentosum	d-Tubocurarine	+ + + An (1), + P (6, 7)	0 (4)
Chrysanthemum cinerariaefolium	Permethrine	–	+ + + An (3)
Cicuta virosa	Cicutotoxin	–	+ + P (4)
Juglans regia	Juglone	–	+ + An (9)
Nicotiana tabacum	Nicotine	*An (1)	0 (1)
Odor of flowers of many species (fam. Labiatae, Orchidaceae, etc.)	1,8-Cineole	–	+ + + An (8)
	Carvacrol	–	–
	Carvone	–	–
	Fenchone	–	–
	Thujone	–	–
	Thymol	–	–
Physostigma venenosum	Physostigmine	–	+ + + An (1), + P(10)
Rosa sp.	Limonene and its derivatives	–	+ P (10)

* –stimulation; (+) – inhibition (+ weak, + + medium, + + + strong); (0) – no effect; (–) – no data. An. – animal, P – plant.
1. Mikhelson and Zeimal (1973), 2. Fluck and Jaffe (1976), 3. Bandyopadhyay (1982), 4. Roshchina (1987), 5. Roshchina (1990b,c), 6. Roshchina (1987), 7. Tretyn and Kendrick (1991), 8. Gracza (1985), 9. Sastry and Sadavongvivad (1979), 10. Riov and Jaffe (1973).

there are some experimental data based on reactions sensitive to noradrenaline and dopamine (Roshchina 1990a,b,c, 1991a,b). Thus, components of plant excreta can participate in signaling processes based on the interaction between the low-molecular effector and the membrane-bound receptor. This conception could be used to explain the toxic action of some alkaloids, terpenes, and neurotransmitters found in plant excreta, on insects and herbivorous Vertebrata. An analogous mechanism is proposed for the membrane "recognition" system of plants.

Recognition in plants, animals, and microorganisms can be connected with protein lectins, sensitive to glucans and some other sugars as well. Lectins have been detected in plant cells of more than 1000 species, mainly on the surface of the plasmalemma and chloroplasts (Sharon and Lis 1989). The proteins specifically bind mannose, fucose, and N-acetylgalactosamines, which are found in

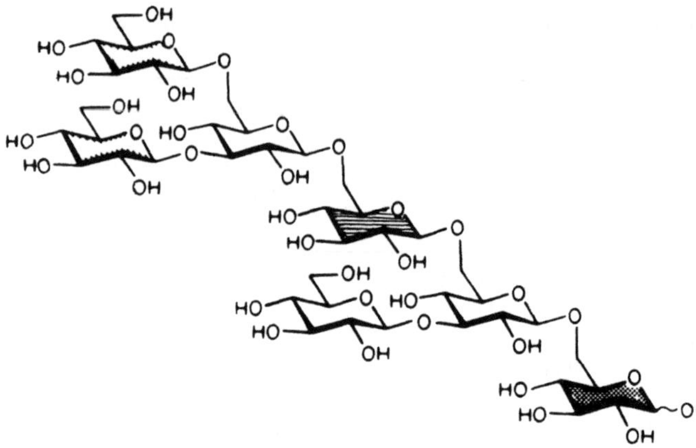

Fig. 90. Structural elements of hepta-β-glucoside elicitor, which, as found by Cheong et al. (1991), are important for its biological activity

excreta of plants and microorganisms. Their main function is as a defense against foreign invasion by agglutination erythrocytes (in animals) or precipitation of polysaccharides and glycoprotein of strangers (in plants and microorganisms). Another function of lectin is participation in the recognition either of the partner in formation of pea, *Rhizobium*, symbiosis (Kijne et al. 1990) or of the elicitor in signal transduction in host and nonhost plant/fungi pathogenic interactions (Scheel et al. 1990). Another aspect of the problem of membranous targets for plant excreta is a possible response to stress metabolites. As reply to parasite invasion, phytoalexins arise in a complex of releasing substances. In model experiments, the phytoalexin glyceollin induced the leakage of protons and other ions from the plasmatic membranes of the fungus *Phytophthora megasperma* and from the tonoplast membranes of red beet (Giannini et al. 1990). It is believed to be a nonspecific damage of the membranes.

Molecular recognition in plants is also connected with phytoalexins (Cheong and Hahn 1991). Signal transduction by oligoglucoside elicitors induces the synthesis of phytoalexins. A specific, high-affinity binding site, in particular for the hepta-β-glucoside elicitor (Fig. 90), has been found in soybean membranes (Cheong and Hahn 1991). Released cell wall components play the role of elicitors in plant-pathogen interactions (Hahn et al. 1989).

7.3 Energetic Reactions

The molecular mechanism of the action of plant metabolites released normally and under stress is carried out via membranous energetic processes such as electron transport in photosynthesis and respiration. In the 1970s, it was shown

(Roshchina 1973c, 1974a,c) that water infusions from dried and fresh leaves of woody plants inhibit the Hill reaction with 2,6-dichlorophenolindophenol (DCPIP). The aetive matter of the solutions was polyphenols. The strength of inhibition of the process correlated with the amount of phenolic groups in the compounds. From 12 woody plants studied (Table 16), the most active as inhibitors were infusions from leaves of stags' horn sumach, *Rhus typhina*, common oak *Quercus robur*, silver birch, *Betula verrucosa*, and smoke tree, *Cotinus coggygria*. After the addition of the infusions into a chloroplast suspension, a whole chain of reactions occurs. This is possibly a dark reaction of some part of DCPIP accompanied by oxidation of polyphenols to quinones. The latter compete with DCPIP for hydrogen ions released in the light as a result of water photolysis. Polymeric tannins and the accompanying polyphenols inhibited enzymatic systems due to the interaction with proteins and the changed structural organization of chloroplasts. Finally, these events led to a decrease in the ability of chloroplasts to reduce DCPIP. Reduction of quinones in the light to appropriate hydroquinones was shown in experiments by Zweig et al. (1969).

Recently, the action of individual components of plant excreta on electron transport and photophosphorylation in chloroplasts and mitochondria has been studied, as is shown in the scheme below:

The components of plant excretions influenced the coupling of electron transport and photophosphorylation in chloroplasts and mitochondria (given from Akulova 1977; V.V. Roshchina and V.D. Roshchina 1983; Knobloch et al. 1986; Moreland and Novitzky 1987a,b).

Terpenes		Phenols			
Monoterpenes their alcohols	Naphto-quinones	Phenolic acids	Phenolic aldehydes	Flavones	Flavonones
Anetol	Juglone	Benzoic	Benzaldehyde	Quercetine	Naringin
Borneole	p-Benzo-quinone	Cinnamic	Vanillic	Kaempferol	Naringenin and
Carvacrol	2,5-dimethyl-	Chlorogenic	Cinnamic		their glycosides
Curvacrone	-benzoquinone	p-Coumaric			and acyl dervatives
Eugenol		Caffeic			
Geraniol		Ferulic			
Linalool		Gallic			
Methylchavicol		Salicylic			
Methyleugenol		Syringic			
Nerol		Vanillic			
α-Pinene					
β-Pinene					
Piperitone					
Thymol					

The active matter of most samples was identified as phenols and terpenes. The group of phenols included naphthoquinones, flavonones phenolic acids, and others, the group of terpenes mainly monoterpenes.

Table 16. Changes in rate of Hill reaction with DCPIP on treatment of pea chloroplasts by infusions from fresh leaves (water/leaves ratio = 10: 1; temperature 0–3 °C, exposure in the solution 24 h. (Roshchina 1974c)

Leaves	Photochemical activity, % of control	Confidence coefficient of difference in comparison with control
Control	100	–
Rhus typhina	0	4.4
Salix alba	86	0.5
Phellodendron amurense	95	0.0
Populus balsamifera	68	1.7
Syringa vulgaris	83	0.2
Quercus robur	31	8.0
Betula verrucosa	26	5.1
padus racemosa	74	1.5
Acer campestre	91	0.2
Sorbus aucuparia	89	0.0
Populus nigra	77	1.4
Cotinus coggygria	59	3.0

For phenols isolated from common walnut, *Juglans regia*, juglone (hydroxy-1,4-naphthoquinone) at a concentration of 10^{-4} M inhibited O_2-uptake by 50% in disks from leaves of tomato, *Solanum lycopersicum*, and haricot bean, *Phaseolus vulgaris* (Perry 1967). Approximately the same concentrations of p-benzoquinone and 2,5-dimethylbenzoquinone depressed respiration in mitochondria from tubers of potato, *Solanum tuberosum*, which was accompanied by a drop in the rate of phosphorylation and in the activities of the enzymes malate and succinate dehydrogenases (Makovec and Sindelar 1984). Regulators of electron transfer and phosphorylation in respiratory and photosynthetic electron transport chains are found among flavonoids (Stenlid 1970; Prokhorchik and Volinets 1973; Akulova 1977). Kefeli and Turetskaya (1964) showed the marked uncoupling effect of some flavonoids. Compounds such as quercetine, kaempferol, and their glycosidic and acetylic derivatives in mitochondria and chloroplasts have characteristics of inhibitors of energy transfer and uncouplers (Lang and Racker 1974; Akulova 1977; Cojocaru et al. 1977; Muzafarov and Zaletskaya 1977). Phloridzin, a well-known component of apple-tree root excreta, is also an inhibitor of energy transfer (Winget et al. 1969). An important role is attributed to the hydroxylic group in the α-position of the phenolic ring. Certainly it regulates the activity of ATP-ase and oxidative phosphorylation of mitochondria (Stenlid 1970). The possibility of the direct participation of flavonoids in the electron transport chain of chloroplasts as oxidants or reductants was shown in experiments of Muzafarov et al. (1983) and Takahama (1983). Glycosidic derivatives of quercetine, quercetine-glucoside-p-coumarate, and others possess protonophorous characteristics, decrease the value of ΔpH (Cojocaru et al. 1977), and interact with ATP-synthetase as allosteric regulators (Malyan et al. 1977). Phloridzin inhibits ATP-synthesis in chloroplasts on the reductive side of cytochrome f and plastocyanin (Roshchina and Akulova 1978). Phloridzin and esculetin are thought to act directly on ATP-synthetase of chloroplasts, and their affinity to this enzyme is closer than that of quercetine-glucoside-p-coumarate.

Comparing the influence of different groups of flavonoids found in plant excreta on the reactions of chloroplasts and mitochondria, Moreland and Novitzky (1987a, b) showed phosphorylation to be more sensitive than other responses of the organelles. Thus, ATPase of coupling membranes should be considered as a target for the action of flavonoids. Besides, flavones are more effective inhibitors of ATP synthesis than flavonols and flavonones. Electron transport is less sensitive to flavonoids. The possibility of the participation of flavonoids in regulation is very high because glycosides of the compounds are contained in the chloroplasts (1%). In cells of the leaf epidermis, as much as 10 mM is accumulated (Moreland and Novitzky 1987a). Flavonoids, coumarins, and benzoic and cinnamic acids inhibited the O_2 release by intact chloroplasts ($I_{50} < 1$ and 10^{-3} M) and ATP synthesis in thylakoids, but at higher concentrations electron transport as well (Moreland and Novitzky 1987b). Their effectivity was decreased in the following sequence; flavonoids > coumarins > cinnamates = benzoates. Generally, the substances had no action as uncouplers. The former acted as inhibitors of electron transport and sometimes as inhibitors of ATP synthesis in the mitochondria of mung bean. Most active were flavonoids (10–80 μM). Oxidation of malate was more sensitive than oxidation of succinate or NADH. Compounds such as naringenin, flavones, and cinnamic acid inhibited ATP-hydrolysis catalyzing by mitochondrial Mg^{2+}-ATPase. The derivatives of benzoic, cinnamic, and chlorogenic acids strongly inhemical reactions of isolated chloroplasts. However, it has been shown that substances such as salicylic acid (25×10^{-6} M), caffeic acid (25×10^{-6} M), and gallic acid (50×10^{-6} M) increase the rate of the Hill reaction in flag leaf of rice, *Oryza sativa* (Sharma and Singh 1987).

Alkaloids with phenolic groups can influence energetic reactions as well, but their role in the processes has not been studied sufficiently. The action of capsaicin, the pungent principle of the red pepper, *Capsicum annuum*, fruit secretory tissues, on the light reactions of photosynthesis has been shown (Roshchina et al. 1986). Alkaloid capsaicin (the product of condensation of vanillilamide and decenic acid) inhibits both electron transfer and coupled photophosphorylation. Inhibition of electron transport was presumed at the site of photosystem 1 because a decrease in the rates of the $NADP^+$- or ferricyanide photoreduction was demonstrated. The inhibitory effects of capsaicin were been concerned with its action on proteins such as electron carriers. Capsaicin is thought to act on isolated chloroplasts as a membrane-active agent, which changes the membrane state. Besides capsaicin, physostigmine (eserine), an alkaloid from the calabaric bean, *Physostigma venenosum*, influenced energetic reactions of chloroplasts (Roshchina and Mukhin 1985a, b). Like many phenolic compounds, alkaloid eserine inhibited photophosphorylation by 50%, but had no effect on electron transport.

Among the metabolites leaching from plants, polyacetylenes and cyanogenic glycosides are found. The best known polyacetylene is a strong toxin for animals, cicutotoxin, from *Cicuta virosa* secretory tissues. It has been shown (Roshchina et al. 1980; V.V. Roshchina and V.D. Roshchina 1983) to inhibit significantly electron transport and coupling photophosphorylation in chloroplasts (Table 17). The active concentrations are very small (1.5×10^{-10} to 10^{-9} M), but it is enough to cause coagulation of the important proteins of the electron transport chain, ferredoxin and plastocyanin. The action of extracts from leaves and rhizomes of *Cicuta virosa* was almost analogous (Table 17).

Synthesis of cyanogenous glycosides by plants as well as formation of volatile HCN under stress may lead to the supposition that these compounds play an important role in the regulation of energetic processes. Known cyanides such as sodium cyanide and potassium cyanide are often used for the study of electron transport chains of chloroplasts and mitochondria. Moreover, cyanides and HCN have been found to switch over electron transport in mitochondria from the basic (with participation of cytochrome oxidase) to the alternative (without cytochrome oxidase) pathway, which has an important adaptive significance (Laties 1982). Cyanides can also bind with plastocyanin in chloroplast blocking electron transport on the side of photosystem 1 (Izawa et al. 1973).

In some cases, the volatile components of excreta such as hydrocarbons, including terpenes, act on the energetic reactions of plants. Indeed, gaseous compounds of leaves of sage, *Salvia*, have been shown to depress O_2 uptake by mitochondria (Muller et al. 1969). Pure preparations of cinneole act like vapors from leaves. Plhak (1972) studied the action of volatile excreta from basket willow, *Salix viminalis*, on the oxidative phosphorylation in seedlings of rye, *Secale cereale*. He found stimulation of ATP synthesis in the samples, but in roots of the same plant depression of the process was observed. The author suggested that the active matter of the volatile excreta was ethylene. The effects of other volatile compounds, in particular steams of garlic, *Allium sativum*, needle of pine, *Pinus*, and α-pinene on ATPase have also been studied. The

Table 17. Action of extracts from leaves and rhizomes of *Cicuta virosa* and pure cicutotoxin on electron transport and ATP synthesis in pea chloroplasts, % of control. (Roshchina et al. 1980)

Experimental condition (additions)	The $NADP^+$ photoreduction	Cytochrome f		Plastocyanin		ATP synthesis
		PO	DR	PR	DO	
Extract from leaves 1:20 (0.03 ml)	39	80	80	75	80	50
Extract from rhizomes 1:20 (0.03 ml)	37	85	83	78	77	50
Cicutotoxin 3×10^{-9} M	62	70	87	75	80	30

PO and DR photooxidation and dark reduction of cytochrome f; PR and DO photoreduction and dark oxidation of plastocyanin.

inactivation of the enzyme is explained as the interaction of volatile compounds with sulfohydric from essential oils on chloroplasts and mitochondria in nonpathogenic photosynthetic purple bacteria, *Rhodopseudomonas sphaeroides* and *Rhodospirillum rubrum.* At concentrations of 0.02–5 mM, they could inhibit both O_2 uptake and coupling oxidative phosphorylation in isolated membranous preparations and whole bacterial cells. This shows that essential oils penetrate through the cell wall. The most active inhibitors of O_2 uptake were carvacrol and thymol, which blocked it completely, whereas piperitone, vanillin, and cymene demonstrated less activity. Therefore terpenoids are the main inhibitors of electron transport. In some cases, coupled oxidative phosphorylation was inhibited more strongly than electron transport, which indicates a mechanism of action of terpenoids as inhibitors of energy transfer. This is characteristic, in particular for anethol, linalool, methylchavicol, and nerol, which act on NAD-dependent electron transfer, and carvacrol, borneol, carbine, eugenol, geraniol, methylchavicol, methyleugenol, nerol, piperitone, thymol, and cinnamic aldehyde acting on succinate-dependent electron transport. Allelopathically active monoterpenes β-pinene and limonene (1 mM) are toxic for cells of *Saccharomyces cerevisiae.* They inhibit O_2 uptake by 80% (Uribe and Pena 1990).

One of their mechanisms of action is connected with their interaction with the phospholipidic double layer of biological membranes, mainly on the quinone level, where transport of protons and electrons occurs and the coupling factor is located. Some allelopathically active inhibitors of growth monoterpenes, piquerol and diacylpiquerol (Fig. 91), released by the roots of the Mexican genus *Piqueria* (family Eupatoriaceae), inhibit H^+-ATPase activity in microsomes and plasmalemma (Ortega et al. 1990).

The oxygenated hydrocarbons, alcohols, and ketones (methanol, ethanol, propanol, isopropanol, butanol, isobutanol, pentanol, hexanol, and acetone), in millimolar concentrations, stimulate the ATPase activity of the coupling factor isolated from liver mitochondria and pea chloroplasts (Toxobaeva et al. 1987).

Thus, volatile and nonvolatile excreta can be able, often under stress, to influence on energetic processes of chloroplasts and mitochondria. The targets of their action are at different sites of the membranes of the organelles. As is seen from analysis of the above data, phenolic compounds, alkaloids, alcohols, and ketones act mainly on the coupling factor, ATPase-synthetase, but other compounds on electron transfer. The targets for some polyacetylenes and

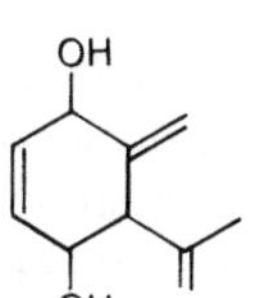

1

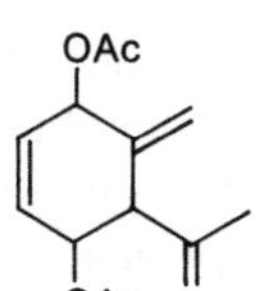

2

Fig. 91, 1,2. Piquerol A (**1**) and diacyl piquerol (**2**). *Ac* acyl terpenoids may be certain sites of electron transport chains, where the compounds are bound and, as a result, induce conformational changes of the proteins or even their coagulation.

7.4 Metabolic Processes

Changes in metabolism are important biological effects induced by plant excreta. The mechanism of this influence can be realized either via the regulation of enzymic activity (Horlacher and Poskuta 1986) or via the changes of direction in the metabolic process, mainly due to the involvement of exometabolites themselves in the intracellular metabolism (Benz and Rudiger 1981). There is little information on the action of plant excreta on carbon and nitrogen metabolism. Some flavonoids, such as quercetine, rutin, phloridzin, and epigallocatechin, which influence energetic processes as well (see Sect. 7.3.) can regulate the rate of CO_2-assimilation in isolated chloroplasts of spinach, *Spinacia oleracea* (Lyubimov et al. 1986). Rutin activates the process by 40%, while other compounds depress it by 60–80%, lowering the degree of carbon involvement in the metabolic chain of regeneration of ribulose-1,5-bisphosphate. Flavonoids can influence the enzyme activity as well. Infusions from roots of cherry plum, *Prunus divaricata*, sour cherry, *Cerasus vulgaris*, and quince, *Cydonia* sp. stimulated the activity of ribulose bisphosphate carboxylase, whereas infusions from leaves of apple tree, *Malus domestica* (active matter is phloridzin), decreased it (Kossakovskaya et al. 1983). There is information that the activity of the malic enzyme, which takes part in the main pathways of photorespiration and respiration, is inhibited by flavonoids (Oparina and Ruzieva 1986). Other groups of phenols such as β-naphthol, caffeic acid, and salicylic acid stimulate the activity of nitrate reductase in chick pea, *Cicer arietinum* (Sharma et al. 1984), or, like the naphthoquinone, juglone, inhibit nitrogenase (Neave and Dawson 1989).

Volatile oxides of carbon and nitrogen, as well as some hydrocarbons, are actively involved in the cellular carbon metabolism, mainly by activation or inhibition of the appropriate enzymes (Table 18).

The hydrocarbon ethylene can significantly depress (by 20–50%) CO_2 fixation at photosynthesis in 13 species from 7 families (Taylor and Gunderson 1986). The maximum inhibition by the compound (2.1×10^{-4} M for 4 h) was observed for peanut, *Arachis hypogaea*, cotton, *Gossypium hirsutum*, soybean, *Glycine max*, and pumpkin, *Cucurbita pepo*.

Similar effects were also shown under treatment by terpenoids soluble in organic solvents. Nonvolatile components of resin of pine, *Pinus sylvestris*, such as abietic and dihydroabietic acids, inhibited CO_2 uptake and some reactions of the Calvin cycle in chloroplasts of spinach, *Spinacia sativa* (Martin et al. 1984).

There are some few examples of the influence of some compounds from plant excreta on nitrogen exchange. The first of these are mainly phenolcarbonic acids, which are widespread in plants. Indeed, cinnamic acid (10^{-5} M for 3.5 h)

Table 18. Some examples of the action of individual components of plant excretions on cellular metabolism

Compound	Effect	Reference
Carbon monooxide	Activation of guanilate cyclase	Brüne et al. (1990)
Carbon dioxide	Stimulation of enzymes of the ethylene biosynthesis	Philosoph-Hadas et al. (1986)
Acetylene and N_2O	Inhibition of enzymes of the biosynthesis of bacterio-chlorophyll d	Ormerod et al. (1990)
Ethylene	–	–
	Inhibition of CO_2 fixation	Taylor and Gunderson (1986)
	Activation of chitinase	Broglie et al. (1986)
	Involvement in cyanide metabolism	Goudey et al. (1989)

is shown to induce a decrease in ^{14}C-glucose involvement in proteins of rose cell suspension, and simultaneously an increase in the binding of the marker with the amino acids (Croak 1972). Under the same conditions, ferulic acid blocked the synthesis of proteins and also diminished ^{14}C-radioactivity in amino acids. Perhaps the depression of the protein synthesis is one of the modes of action of the released substances on plant metabolism. Information on the inhibition of certain specific enzymes, including pectolitic, (in particular cellulases), catalases, peroxidases, amylases, and others by phenolcarbonic acids and tannins is summarized in monographs Rice (1974, 1984).

It is very important to understand how components of plant excreta are involved in the common metabolism of other plants. This method in particular for alkanals, is thought (Durmishidze 1977) to be used for the utilization of similar toxic substances. Terpene alcohols such as geraniol and farnesol are also included in chlorophyll synthesis in etiolated membranes of oat plastids (Benz and Rüdiger 1981). Some elicitors released from the cell under pathogen invasion change the direction of plant metabolism more sharply. They are shown (Di Cosmo and Misawa 1985) to induce synthesis of some secondary metabolites in tissue culture of the plant. A similar phenomenon takes place due to the three fold increase of CO_2-concentration under water stress in comparison with normal carbon nutrition. It intensifies the synthesis of such secondary products as cardiotonic steroid glycosides, mainly digoxin and digoxin-monodigitoxoside, by 60% (Stuhlfauth et al. 1987).

7.5 Problems and Perspectives in the Use of Plant Excreta

Recently, knowledge of the mechanisms of the biological action of plant excretions have become extremely useful in practice. Future perspectives will be further considered. Since many of these are reviewed in special publications

(Grodzinskii 1965, 1991; Rice 1974, 1984; Tokin 1980a, b; Bailey 1982; Aizenman et al. 1984 and others), we shall concentrate only on the main aspects of the problems.

7.5.1 Plant Resistance to Pathogens

Mechanisms of chemical resistance to diseases are at the basis of the problem of the fight against damage to agricultural and forest plants by microorganisms and other parasites. They include synthesis and release of substances which are present in the plant before infection and/or those which are formed in response to the invasion of a foreigner. For first group of defensive compounds, most of them are alkaloids and terpenoids. They differ in their antifungal effect and are toxic for spores (Harborne 1982, 1988; Bailey 1982, and others). The chemical nature of post-infectional substances, named phytoalexins, varies. Each of them is specific to the plant species.

The selection of economic plants based on either a high content of phyto alexins or high rate of their synthesis will permit the creation of certain cultivars resistant to disease. This aspect is especially interesting for industrial production and the use of analogous compounds which are synthesized by the plant in response to infection. Polyacetylenes, so-called safflowrins, from safflower, *Carthamus tinctorius* (fam. Compositae) formed in plant cell culture after treatment of the fungal pathogen are able to depress fungal mycelium growth on nutritious medium (Tietjen and Matern 1984). The synthesis of analogous polyacetylenes is readily available to the chemical industry, and is preferred because the compounds are easily decomposed under natural conditions, which distinguishes them from other pest-killers. Table 19 shows some examples of the chemical nature of phytoalexins produced by certain families, as a perspective for artificial synthesis by the chemical industry.

Another mode of external regulation of plant resistance to diseases can be the stimulation of phytoalexin accumulation in the organism after treatment of biotic and abiotic elicitors (Table 20). The use of the particular localization of plants in artificial phytocenosis, where complexes of volatile compounds acting

Table 19. Chemical nature and origins of phytoalexins available for artificial synthesis

Chemical	Class	Family of plants
Capsidiol	Sesquiterpene alcohol	Solanaceae
Casbene	Diterpene	Fabaceae (Leguminosae)
Rishitin	Sesquiterpene	Solanaceae
3-Methylbutanol	Volatile alcohol	Fabaceae
Pentanal	Volatile aldehyde	Many families
"Safflowrins"	Polyacetylenes from safflower	Compositae (Asteraceae)

Table 20. Some abiotic and biotic elicitors which can be used for stimulation of synthesis of phytoalexins in agriculture. (After Bailey 1982; Brindle and Threifall 1983)

Abiotic elicitor	Condition for use	Biotic elicitor	Condition for use
Ultraviolet radiation	Special films and lamps of UV-light in greenhouse	Killed culture of bacteria or fungi (like vaccine)	Sprinkling in greenhouse or open plantings
Partial freezing	Refrigerating machines in greenhouse or climatic cameras (phytotron)	–	–
Fungicides	Open plantings or greenhouse Treatment of seeds and seedlings	Nonpathogenic lines of fungi	Sprinkling
Detergents	Sprinkling of open plantings or the greenhouse cultivars	–	–

as phytoalexins depress pathogens, is also promising. Indeed, volatile compounds released by roots of *Acacia pulchella* are demonstrated to inhibit strongly the growth of the fungus *Phytophtora cinnamomi*, which affects cultivated fruit species (Whitefield et al. 1981). The planting of *Acacia* among the other cultivated plants could significantly contribute to a defense against these pathogens.

7.5.2 Chemical Interactions: Plant–Insect and Plant–Plant

Chemical interactions between organisms in biocenosis have still not been successfully explained. The substances released by plants can act on other plant species and animals at a distance and have a noticeable effects.

The most interesting example of this is the recognition of the odor of a flower containing nectar, which the insect needs for nutrition. The signaling smell may belong to a component of the nectar itself or the essential oils of the secretory glands. Insects are sensitive to a negligible amount of odorous substances, and therefore the flower flavors are effective at very low concentrations (Harborne 1988). Recently, volatile compounds of plants were shown to function not only as attractants for some insects but as repellents as well. These remarkable features of some plant excreta could well be used either in a complex of activities in biological defense of plants against pests, or for the attraction of insect pollinators. The industrial production of attractants and repellents or their analogs is very promising. Individual compounds, attractive for some insect pollinators, were found in volatile plant excreta (Table 21). Most of them include

Table 21. Main attractants for insects found in plant excreta

Attractant	Class of chemical	Plant species and families	Reference
Methyleugenol	Terpenoid alcohol	*Palea anisata* (Rutaceae)	Tuthil (1953, cited in Lewis et al. 1988)
Methyleugenol	Terpenoid alcohol	*Cassia fistula* (Caesalpinaceae)	Kawano et al. (1968, cited in Lewis et al. 1988)
Methyleugenol	Terpenoid alcohol	*Spathiphyllum cannaefolium* (Araceae)	Lewis et al. (1988)
Methylchavicol	Terpenoid alcohol	*Zieria smithii* (Araceae)	Fletcher et al. (1975)
Methylchavicol	Terpenoid alcohol	*Ocimum basilicum* (Araceae)	Chen et al. (1970, cited in Lewis et al. 1988)
1,8-Cineole	Terpenoid alcohol	*Spathiphyllum cannaefolium* (Araceae)	Lewis et al. (1988)
1,8-Cineole	Terpenoid alcohol	*Stanhopea ecornuta* (Orchidaceae)	Williams and Whitten (1983)
Ethylacetate	Ester	*Dendrobium superbum* (Orchidaceae)	Flath and Ohinata (1982)
2-Tridecanone	Hydrocarbon	*Ocimum basilicum* (Araceae)	Chen et al. (1970, cited in Lewis et al. 1988)
Myrcene	Terpene	*Stanhopea ecornuta* (Orchidaceae)	Williams and Whitten (1983)
Camphene	Terpene	*Stanhopea ecornuta* (Orchidaceae)	Williams and Whitten (1983)
p-Cymene	Terpene	*Stanhopea ecornuta* (Orchidaceae)	Williams and Whitten (1983)
Methylcinnamate	Ester	*Stanhopea ecornuta* (Orchidaceae)	Williams and Whitten (1983)
Anysilacetate	Ester	*Spathiphyllum floribundum* (Araceae)	Williams and Whitten (1983)
Benzylacetate	Ester	*Spathiphyllum floribundum* (Araceae)	Williams and Whitten (1983)
Benzylbenzoate	Ester	*Stanhopea panamensis* (Orchidaceae)	Williams and Whitten (1983)

terpenes, terpenoid alcohols, and their esters. The greatest variability is characteristic for the flower odors of plants from the families *Orchidaceae* and *Araceae*. The number of attractants may achieve high values, from 20% (myrcene) to 60% (cineole) flavor. For details see the review of Williams and Whitten (1983) and the monograph of Harborne (1988), which are devoted to this question.

For insect-repelling or insect-damaging compounds (Fig. 92) found in plant excreta, some examples are:

2-Hexenal	from *Ginkgo biloba*	Repellent
Strigol	from *Striga asiatica*	Insecticide
β-Asarone	from *Acorus calamus*	Chemi-sterilant

There are also kairomones among the plant excreta, in particular either common volatile aliphatic aldehydes and monoterpenes, as in chick pea and pigeonpea, or characteristic for species, like azadirachtin from neem tree, *Azadirachta indica* (Rembold 1989). Natural photosensitizers are released from plants. They include polyacetylenes, furanocoumarins, β-carbolines, and extended quinones, which play a role in plant resistance to insects (Arnason et al. 1989). The mechanism of their action consists either in interaction with DNA, which causes chromosomal abnormalities, or in the production of singlet oxygen and free radicals. It should be especially noted that natural products, such as terthienyl, are more toxic to mosquito larvae than DDT.

On the other hand, the same compounds show cytotoxic (for many animals), insecticidic, molluskocidic, and fungicidic effects (Fischer et al. 1989a, b). Many

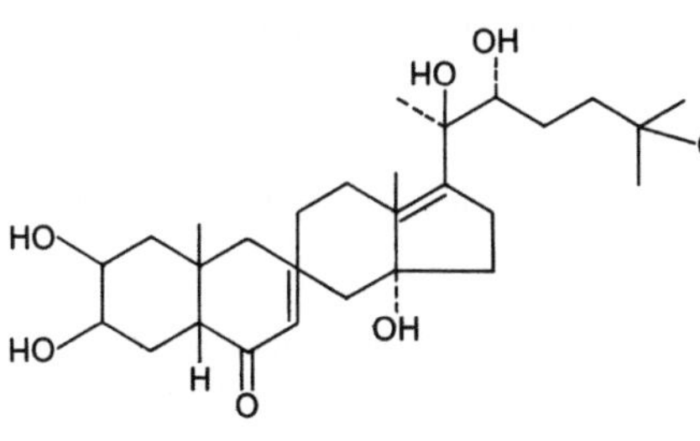

1

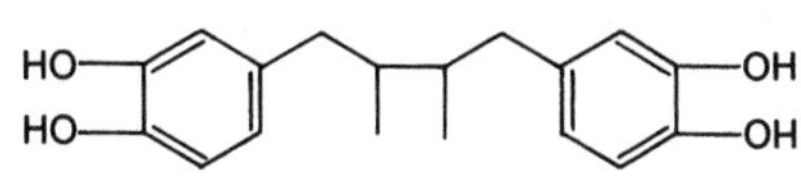

2

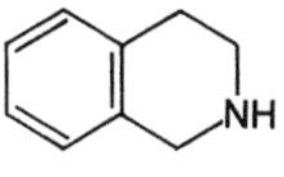

3

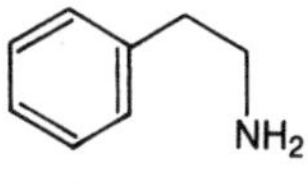

4

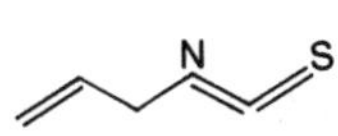

5

Fig. 92, 1–5. Plant excreta compounds toxic for insects. **1** β-Ecdysone from *Ajuga renata* (Labiatae), **2** Nordihydroquairetic acid from *Larrea* or *Acacia*, **3** Isoquinolines from Cactaceae, **4** Phenylethylamines from Cactaceae, **5** Allylisothiocyanate forming from sinigrin (many species of Brassicaceae) plant excreta have insecticidic features, in particular water extracts from water-arum, *Calla palustris*, and false hellebore, *Veratrum vulgaris*, are shown to kill larvas and chrysalisces of gypsy moth, *Osneria dispar* (Gur'ev 1981). Insecticidic effects are also known for extracts of red pepper, *Capcicum annuum*, and many other plant species. Synthesis of analogs of plant insecticides opens new possibilities for defense against parasites.

A number of works have dealt with allelopathy, the chemical interaction between plants in phytocenosis, and future perspectives for the practical use of the phenomenon were studied, most important Grodzinskii (1965, 1981, 1991 and Rice 1974, 1984), who coordinated investigations in this field in the former USSR and in the USA respectively. Their analysis of world literature showed that there are few examples of verifiable allelopathic influence of intact plants on the growth of other plant species. They have been described in detail in the monograph of Rice (1984), who claimed this phenomenon to be peculiar to arid or semiarid regions. Perhaps in some cases, direct allelopathic influence takes place in forest plantings (Kolesnichenko 1976; Matveev 1985). However, products released under stress or accumulated in soils and forest litter as a result of the action of heterotrophs play a far more important role in chemical interactions. However, consideration of this question is not included here, having no direct relation to the normal excretory function of plant.

The participation of vital plant excreta in allelopathic interactions needs special discussion. Products of plant excretory function influence growth mainly under stress, as is shown in Table 22, based on information from previous chapters. Normally, plants excrete such a negligible amount of the substances that they cannot influence the growth and development of other plant species to any great extent. However, due to the relatively high concentration and toxicity of the stress metabolites released, they induce more or less notable changes in

Table 22. Substances released by plants under normal and stress conditions and their allelopathic activity

Class of chemicals	The released substance, μg/g of dry or fresh mass		Concentrations, inducing effect on growth, μg/g of fresh mass
	Norma	Stress	
Aldehydes	6.2	23.1–126	1–60 (+)
Monoterpenes	0.125[a]	0.13[a]	100 (−)
Sesquiterpene lactones	0.0018	0.01	0.0186 (+)
Unsaturated hydrocarbons	0.005–0.003[a]	0.56[a]	1 (+)
Phenols	2.0	2–420	1.46–4.4 (±)

(+) or (−) stimulation or inhibition of growth.
[a] Dry mass.

the structural and functional state (including the growth process) of the whole organism sensitive to them. However, a quantity of excretions of intact plants are usually enough to signal or trigger functions which are included in the problem of "recognition". Recognition is considered as a common biological phenomenon embracing different levels of processes from intermolecular to biocenotic (Grodzinskii 1985). It is also thought to take place in interactions such as plant-plant or plant-lower organisms (Grodzinskii 1981). The supposition is based on the following facts. During pollination, viable pollen falls from the plants to a stigma of the pistil. However, only pollen of the same or compatible species can germinate. Pollen of incompatible species does not germinate, due to the presence of certain chemicals in the excreta of the stigma itself, which act negatively. Seeds of root parasitic plants such as broom-rape, *Orobanche*, striga, *Striga*, or dodder, *Cuscuta campestris*, are preserved in the soil for decades, being incapable of beginning their growth until the root excreta of an appropriate plant host appears in their environment. No germination is observed also for seeds of many orchids until the appearance of metabolites of the appropriate mycosymbiont in the surrounding media (Grodzinskii 1981); 10^{-14} M of the active substance is enough for its function as trigger or stimulator of germination of the parasitic plant (Fischer et al. 1989a, b) or as attractant for microorganism symbionts (Stanier et al. 1975). The phenomenon of recognition is found in the study of plant infection by nodular bacteria (Stanier et al. 1976). Certain leguminous plants stimulate the germination of *Rhizobium* in the soil at a radius as great as 10–20 mm from the roots. This event is specific. The development of bacteria differing from *Rhizobium* is slightly stimulated or is not stimulated by the root excreta.

Besides, the growth of bacterial species able to infect an appropriate leguminous plant is facilitated more than is observed for other clones of nodular bacteria. Recently, stimulators of the process have been identified. In exudates of seeds and roots of leguminous plants, inductors of germination of the bacteria, *Rhizobium meliloti*, *R. trifolii*, and *R. leguminosarum*, mainly phenols, have been found. Of these, the flavones luteolin and apigenin-7-o-glycoside (from seeds) and naringenin (from roots) were the most active (Zaat et al. 1988). Their concentration in root exudates was equivalent to 50–200 nM naringenin. Bacterial growth was induced very quickly, within 5 min after treatment by pure naringenin or root exudates from a plant. 4′,7-Dihydroxyflavone; 4′,7-dihydroxyflavanone; 4,4′-dihydroxyflavonone and dihydroxy-2′-methoxy-chalcone of plant exudates are attributed to inducors of the transcription of *Rhizobium* genes (Maxwell et al. 1989).

There are even more interesting results of the study of similar recognition. At a CO_2 high concentration in air, volatile excretions of mint, *Mentha*, and rose, *Rosa*, as well as their components menthol, citronellol, and others, have been demonstrated to change their cellular electric potential in 26 plant species (Matsuoka 1987). The reaction to the individual components and their mixtures, characteristic for the odor, differed, which suggests the existence of special sensors in plants. These phenomena can be used for the diagnosis of various volatile substances which are contained in the excreta.

Cellular recognition systems in plants have some analogs in animal organisms, where they deal with the information transfer between cells in synapses in the form of chemical signals such as transmitters or mediators. In plant excreta, nitrogen-containing compounds which play the role of neurotransmitters in neural and muscle tissues are also found. Among them are the cholinic ester acetylcholine and the biogenic amines dopamine, noradrenaline, serotonin, and histamine (see Chap. 3.10), which participate in intercellular chemical signalization in animals.

Transmission of the information in the form of chemical signals has arisen in unicellular organisms at the beginning of evolution. Their selective chemosensibilization deals with chemoreception, which is based not only on food requirement but also on the participation in relationships such as attractivity or repellency, partnership or competition with other species or persons of the same species. The evolution of chemoreception has evidently been directed to the development of informative systems both in biocenosis and inside the individual organism, as is shown below:

Chemoreception of unicellulates →

Biocenosis

Chemoreception of multicellular animals (olfaction)
Chemoreception of multicellular plants (allelopathic recognition)

Chemoreception of unicellulates →

Individual multicellular organism

Transmission of information by mediators inside organism (cholinoreception, adrenoreception, etc.)

The peak of evolution in the animal organism is the appearance of mediator transmitters of information between cells of a unicellular organism such as acetylcholine and biogenic amines (see Chap. 3.10). This recalls the fact that in the bacterial cell "recognition" systems, for instance at chemotaxis, use compounds of plant excreta or their analogs as stimulus (Macnab 1979). Among these are attractants (sugars, amino acids, amines, cyclic AMP, sesquiterpenes) and repellents (short-chain and unsaturated fatty acids, aliphatic C_2-C_4- alcohols, hydrophobic amino acids leucine, isoleucine, valine, tryptophan, indole, skatole, aromatic acids). These groups of substances participate in plant-animal interaction (scheme below):

Examples of attractants and repellents for animals among the components of plant excreta (based on Major 1967; Rodriguez 1983; Harborne 1988; Goffreda et al. 1989; Roshchina and Roshchina 1989).* = food attractants for repellents (feeding deterrents).

Attractants

Monoterpenes and their alcohols

Limonene
Cadinene
Linalool
Geraniol

Monoterpene lactones

Nepetalactone*

Phenylpropanoid derivatives of terpenes (aromatic)

Eugenol
Methyleugenol
Methylisoeugenol*
E-Asarone*

Indoles

Indole
Skatole

Amines

C_1-C_6-Methylamine-hexylamine
Putrescine
Cadaverine

Alkaloids

Coniine

Nectar feeding components

Sugars*
Amino acids*
Lipids*
Essential oils

Repellents

Monoterpenes

β-Ocimene*
Caryophyllene

Diterpenes

Acyllandrometodol

Sesquiterpene lactones

Glaucolide-A*

Terpenoid derivatives

Gossypol
Kaurenoic acid

Hydrocarbons

Ethane*

Aldehydes

Hexenal

Alkaloids

Demissine

Phenols

Arbutin
Juglone
Tannins

Glucose esters with fatty acid

2,3,4-Thri-o-acylglucose ester of C_4-C_{12} fatty acids

Steroid derivatives

β-Ecdysone*

Isothiocyanates

Allylisothiocyanate*

Analysis of the scheme shows that the main classes of substances attractive or repellent to prokaryotic organisms play a similar role for eukaryotic ones (predominantly for insects). Monoterpenes and their phenylpropanoid derivatives, components of the odor of flowering plants, are mainly known as attractants for pollinators.

Many substances are simultaneously attractants and allelochemicals. The odor of *Artemisia ludoviciana* Nutt (fam. Asteraceae) includes a number of allelochemicals such as cineole, camphor, etc., drawing insects as well. Pheromones for some pollinators are also found among the flavor components (Blust and Hopkins 1987). Excreta of *Dalechampia magnoliifolia* (fam. Euphorbiaceae) include different attractants: 1,8-cineole, special for the bee, *Euglossa allosticta*, methyl salycilate for *E. imperialis*, eugenol for *Eulaema cingulata* (Armbruster et al. 1989).

The results of these studies can already be used in practice. Firstly, components of plant excreta may act as pesticides: insecticides, nematocides, fungicides, herbicides (see for details see works of Rice 1983, Harborne 1988). The use of purified compounds of the excreta, in particular coniine as insecticide or cadinene and copaene as pheromones, has become traditional as well as the synthesis of their artificial analogs, which are not able to pollute the environment. Extracts of many species of orchids, including female pheromones of fruit fly of the genus *Dacus* have been shown to be a good lure for male individuals (Flath and Ohinata 1982; Lewis et al. 1988). It could be promising to use special plantings of some species of the genus *Tagetes*, whose roots release nematocide, the -S-S-polyacetylenic derivative α-terthienyl, and its analogs (Gommers and Bakker 1988). α-Terthienyl also acts as an allelochemical and insecticide, being toxic for mosquitos and dark flies (Arnason et al. 1981; Wat et al. 1981) as well as larvas of various Lepidoptera (Iyengar et al. 1987). Knowledge of the composition of plant excreta could be useful for the determination of plant resistance to pests. Indeed, the amount of β-myrcene analyzed is the test for resistance to nematode invasion (Ishikawa et al. 1987). Estimating the amount of the substance (mg) in 10 g of bark and wood, very sensitive species, such as *Pinus sylvestris*, contain 0.33–4.00, less sensitive ones 0.1–9.23, resistant ones such as *P. resinosa*, *P. regida*, and *P. excelsa* 0.13–2.54, and very resistant, in particular *P. strobus* and *P. palustris*, only traces (0.00–1.79). In order to decrease pest number, the planting of woody species resistant to insects and fungi may be recommended. In this connection, the ancient genus *Ginkgo* is unique and multifunctional. This tree is found to be extremely resistant to bacteria, viruses, fungi, insects, and industrial and cigarette smoke. Resistance to insects is thought to be due to the formation of unsaturated aldehyde 2-hexenal in the presence of oxygen (Major 1967). It serves as a killer of fungi and beetles, but the concentration as a repellent for larvas is low. Other components of the resistance of *Ginkgo biloba* are thought to be the high acidification of the leaves and the presence of hydrolactones, two of which have been identified as ginkgolides.

The use of allelopathically active substances as natural herbicides is prom ising in practice (Putnam 1983; Rice 1983; Harborne 1988). The synthesis of

Table 23. Allelopathically active substances of plants which can be a base for industrial production of natural pesticides

Chemical	Class	Plant origin
Arbutin	Phenol	*Arctostaphylos*
Azadirachtin	Tetranotriterpenoid	*Azadirachta indica*
Gallic acid	Tannin	*Euphorbia virgata*
Dhurrin	Cyanogenic glycoside	*Sorghum vulgare*
Camphor	Monoterpene	*Salvia sclarea*
Capsaicin	Alkaloid	*Capsicum annuum*
Caffeic acid	Phenol	*Coffea arabica*
Psoralen	Furanocoumarin	*Psoralea*
Phloridzin	Flavonoid	*Malus domestica*
Cicutotoxin	Polyacetylene	*Cicuta virosa*
Cinnamic acid	Aromatic acid	*Parthenium argentatum*
Juglone	Quinone	*Juglans nigra*
α-Terthienyl	Polyacetylenic derivative	*Tagetes erecta*, *T. patula*

similar compounds or their active derivatives can be useful for the struggle against weeds. Some chemicals can well be put to use in this connection (Table 23). Such herbicides could be very effective for the control of weeds without polluting the environment, as they are either quickly decomposed in the soil or metabolized by plants. At the same time, plants with high allelopathic potential, such as *Polygonum orientale* L., most effective under the conditions in India, are proposed to control the growth of weeds (Datta and Ghatterjce 1980). The selection of allelopathically active or tolerant varieties could also be recommended for agricultural practice.

7.5.3 Use in Medicine

First reports about the ability of wounded higher plants to kill micro- and macroorganisms at a distance were given by B.P. Tokin in Saint Peterburg in the years 1928–1929 (Tokin 1980a). In the light of this discovery, the medicinal action of excreta of onion, *Allium cepa*, garlic, *Allium sativum*, horse-radish, *Armoracia lapathifolia*, bird tree, *Padus racemosa*, and other plants, as earlier described by many generations of physicians and naturalists, has been explained.

Plant excreta with antimicrobial effects were called "phytoncides" by B.P. Tokin. Phytoncidic activity has now been observed in 1365 species of plants in many taxons (Aizenman et al. 1984), which have become the basis for the production of highly effective medicinal preparations. Indeed, plant excretions, including polyacetylenes and thiophenes, act as antiviral agents in concentrations of 0.001–1.000 $\mu g\,ml^{-1}$ (Hudson and Towers 1988). This determines human resistance to the diseases. A similar antiviral and antimicrobial action is characteristic for terpenes (camphor, α- and β-pinenes, cineole, etc.) (Kelsey et al. 1984). Flavonoids and coumarins of nectars such as esculetin and umbelliferone have a noticeable spasmolytic and antibacterial activity (Andon and Denisova 1974). An analgesic effect is observed for terpene myrcene, which is an active part of essential oils of the medicinal Brazilian plant *Cymbopogon citratus* (DC) Starf (fam. Gramineae). A possible target of this substance are the α_2-adrenoreceptors responsible for the secretion of endogenous opioids. Concrete results of clinical investigation of phytoncidic medicinal action have been summarized in the Proceedings of All-Union Conferences on Phytoncides (Kiev) and the monograph of Aizenman et al. (1984). Phytoncides are at present successfully used for therapy against quinsy, influenza, tuberculosis, and intestinal, fungal, and staphyllococcus infections (Vichkanova 1981; Aizenman et al. 1984). The complex methods of phytoncidotherapy could be recommended to combat chronic pulmonary diseases. Researchers from the Ukraine Academy of Sciences, Zabolotny Institute of Microbiology, prepared new drugs such as imanin, novoimanin, and others from St. John's wort, *Hypericum perforatum* (Aizenman et al. 1984). Vichkanova (1981) composed a list of plant families, including species which are promising in the search for new natural antimicrobial preparations. An analysis of the list showed that in future concentration should be on the study of species belonging to the families Nymphaeaceae, Salicaceae, Betulaceae, Asclepiadaceae, Myrtaceae, and Polygonaceae because most of them possess biological activity.

In the sanitary organization of large urban complexes, plants with phytoncidic activity could be widely recommended for air disinfection as well as for medicinal purposes, especially in health resorts, as has been done in Yalta (Crimea). This occurs due to the release of a number of volatile substances with antimicrobial features, which have a good influence on vitally important functions of the human organism. Another possibility is to use decorative plants with phytoncidic characteristics for purposes of preventive medicine. Recently, publications devoted to volatile plant excreta as in hygiene in the environment have appeared (Ivanchenko et al. 1989; Makarchyuk et al. 1990). The mode of optimizing health, based on these issues, under Professor Grodzinskii, may be summarized as follows:

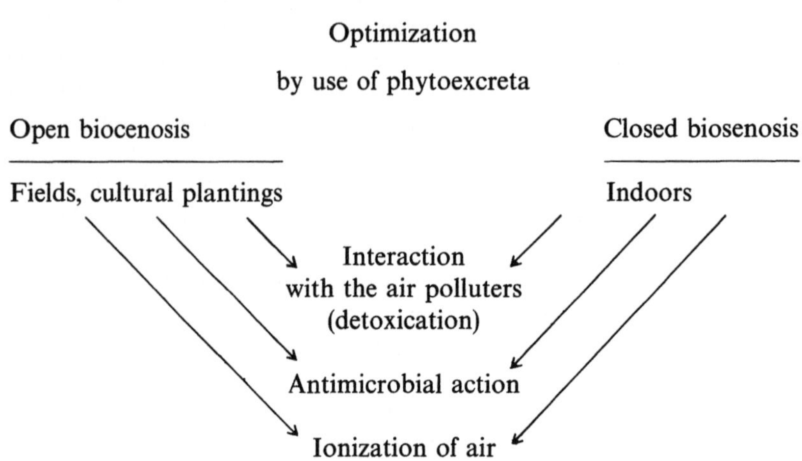

Volatile substances can react with the molecules of air pollutors promoting their precipitation and evacuation from the inhabited zone. The antimicrobial effects of many plants in the open air and indoors induce air sterilization. Aerotherapy in closed working rooms, so-called "phytodisain", is important in sanitary programs for workers in many industries. Phytoncides promote the increase of ions in the air and this, as a secondary factor, improves human resistance to diseases. However, "phytodisainers" have to consider also possible allergic reactions of individuals, also to the combination of various phytoexcreta (Makarchyuk et al. 1990). Sometimes, during an epidemic, the dispersion of volatile phytoncides can play an important role in stimulating immunity.

Conclusion

The secretory function is peculiar to all living organisms. On the cellular level, the release of metabolic products is induced by the need to eliminate compounds which are not necessary for the cytoplasm to function at a given moment. A summary of this book shows that secretion is a feature common to every living being because, in any cell, secretion occurs with the participation of the same cellular structures (Golgi apparatus, secretory vesicles, endoplasmic reticulum, contractile proteins) and the excretory mechanisms (active and passive) are also similar.

Unlike in other organisms, intracellular secretion in the plant cell deals with special organelle vacuole, and excretion from the cell needs to overcome not only the plasmatic membrane, but also the cellulose wall. Intracellular secretion occurs mainly into the vacuole and the free space of the cell. The release of substances is connected with the formation of constitutional parts of the cell (cell wall and its incrustation) and/or the preliminary preparation of the products of exchange for their evacuation into the environment or the intercellular spaces of multicellular organisms. The secretion into the vacuole has a different significance. In the vacuole, the compounds are stored and can be included in cellular metabolism again or, as secondary byproducts, should be excluded from the actively metabolizing part of cell.

Although synthesis and secretion of metabolites occur in nonspecialized cells, these processes seem to be rather weak in the cells, unlike cells such as idioblasts, where secretion is the prevailing function. Idioblasts are the most primitive secretory cells which begin to function as specialized ones. In the multicellular plant organism, there are specialized secretory organs consisting of the whole complex of cells organized in ducts and cavities covered within by epithelial cells, or in glands (Fig. 93). According to the origin and position on the plant secretory tissues, two main types can be distinguished – secretory tissues which are located within the plant and eliminate substances into the intracellular or intercellular spaces, and secretory structures which are located on the plant surface (glands and glandular trichomes) and excrete substances out of the plant. The intercellular spaces and other cavities of the plants are filled with

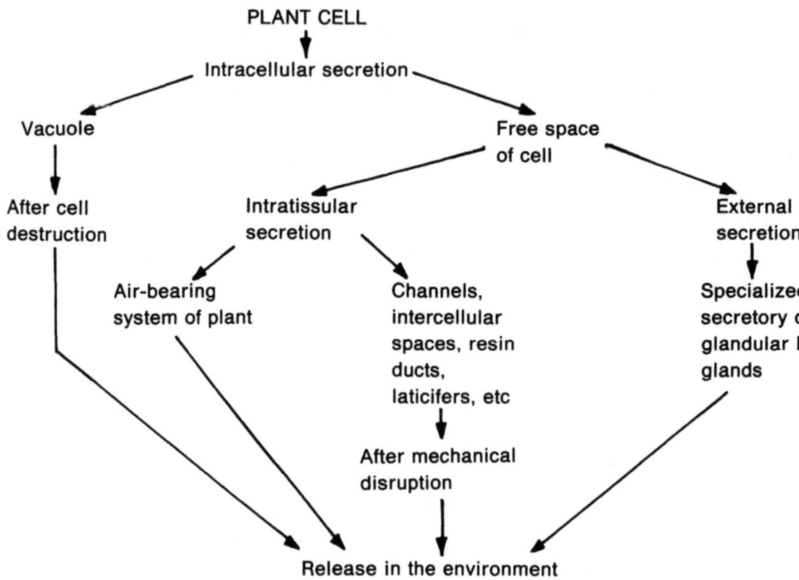

Fig. 93. Secretory system of plants

either atmospheric gases mixed with volatile organic excreta (the air-bearing system of plant) or water-insoluble secondary products (resins, gums, latex, essential oils). The development of the air-bearing system and the gas composition within the plant are mainly dependent on the surrounding medium. Well-developed intercellular spaces are found in wet or arid zones. Intratissular gas excretion varies depending on the season, especially in woody plants. The accumulation of internal gases, such as ethylene, aldehydes, and alcohols, can be an indicator of plant resistance to stress factors or the stressed state of the plant itself. Unlike the air-bearing system, the composition of the secretion of intratissular channels, ducts, and cavities depends not only on the surrounding medium but on the taxonomy of plants which have specialized secretory cells where water-insoluble products are synthesized. External secretion in the latter case occurs by glands and glandular hairs. Here, biosynthesis of the secretion takes place and the mechanism of their elimination functions. The amount and composition of the released secretion, including both the main products of primary metabolism (proteins, carbohydrates, lipids) and secondary metabolism (terpenes, essential oils, alkaloids, etc.), as well as water and salts, depend on the systematic position in the plant kingdom.

The modern concept of the excretion of metabolites by plants attempts to understand the chemical nature of the released substances and their mechanisms of action. This knowledge is necessary for a correct estimation of the role of excreta in the vital function of the plant organism itself and the formation of the surrounding media.

All plant cells are able to release metabolites into the environment. The excretion occurs either by the normal secretory function of the organism or under the action of a stress-inducing factor. The first type of excretory activity represents the usual physiological function. During the process of evolution, specialized secretory structures participating either in intratissular or external secretion have arisen in many plants, which has allowed them to release the products of secondary metabolism most effectively. The presence of stress-induced metabolites which are not peculiar to the usual excreta, and often the lack of specialized secretory structures, distinguish the second type of excretion from the first.

Normal excretory function is characteristic for any plant cell and is needed for the evacuation of secondary products, including stored, excessive, and so-called "defensive" metabolites, toxic for pathogens or herbivorous animals, from the cytoplasm, and then for their accumulation in the vacuole. The results of the secretion of toxic metabolites or products of specialized secretory cells, having various functions, are released from the cell. The position of intracellular secretion in the whole secretory system of the plant is shown in Fig. 93. Intracellular secretion is not only a usual and necessary function, found in any plant cell, but is also the initial stage of the secretory activity of a multicellular plant organism by means of a specialized system.

On the subcellular level, the production of any secretion is the result of metabolic processes, occurring in the plastid membranes, mitochondria, Golgi apparatus, and other organelles. The components of the excreta are products of either primary or secondary metabolism. In specialized secretory structures, products of secondary exchange such as aldehydes, terpenoids, flavonoids, and others predominate. Many of them are synthesized in the branches of the main metabolic pathways: shikimate (flavonoids and other phenols, alkaloids), mevalonate (terpenes), etc. The release of substances can have different meanings. In some cases these functions are evident, in particular the elimination of nectar for insect-pollinated plants, or trapping mucilage and digestive saps in glands of carnivorous plants. Excreta of salt glands and hydathodes have an adaptive meaning as the mode of liberation from excess salts and water. The role of the released secondary compounds is only beginning to be studied. It is supposed that they play an important role in interactions between plants and animals. They can be attractants or olfactory and gustatory repellents. Excreta, having bacteriostatic or bactericidal action, play a certain role in plant immunity and make the plant environment more healthy. Of course, the significance of the secretory function is not limited to the above aspects. The answers to these questions should be found in future studies.

Besides normal secretion by intact plants, substances can be released under the influence of stress factors such as sharp changes in temperature, irradiation, pollution, and others. The composition and quantity of "stress-induced" excreta differ noticeably from the usual ones. Damages to membranes leads to an increase in the permeability of many compounds, which were earlier separated in compartments. Due to hydrolysis and oxidation, the degradation of their membranes occurs, which results in a larger variety of metabolites being released. The synthesis of some metabolites, both usual and stress-induced, is increased. The differences between native and stress compounds are illustrated in Fig. 94.

Since under stress specific compounds are found, the excreta can be used as an indicator to analyze the physiological state of the plant. Modern methods of gas and gas–liquid chromatography are sensitive to trace amounts of substances and may be the main instruments for physiological diagnostics.

The quantity of stress-induced metabolites in plant excreta often exceeds the level of normal exchange products. Most of these are toxic for human and other mammals. Moreover, this appears to be the cause of serious poisonings and diseases, which must be kept in mind if the plants are used as food.

The products of excretion under normal and stress conditions are biologically active. However, intact plants release them in such small amounts that the substances cannot induce noticeable changes in the growth of other plants, although it is enough to be able to realize their signaling or "recognition" function. In the contrast, due to the relatively high concentration and toxicity of stress metabolites, they lead to stronger disintegration in the cells of animals, plants, and microorganisms, which extends to the structural and functional state of the cells and leads to negative changes in growth or perhaps even to destruction and death.

Stress metabolites may play the main role in biochemical competition in phytocenosis. Unlike cases when they are synthesized de novo under extremal factors and usually absent in unstressed conditions, it is difficult to find the quantitative threshold which distinguishes normal from stress metabolites. Stress develops step by step and does not extend to all cells simultaneously. The difference is usually found only in extremal conditions, for instance in arid or wet regions or in the tropical zone, where biochemical interactions between plants are more pronounced. In plants growing in middle temperate zones, the influence of their life time excreta, even stress-induced is less strong in comparison with the influence of products of decomposing plant biomass. However, the latter have no connection to the normal excretory activity of plants.

An important question is the evolution of the excretory function. Secretory and excretory activities are peculiar to any cell of the living organism. The facts

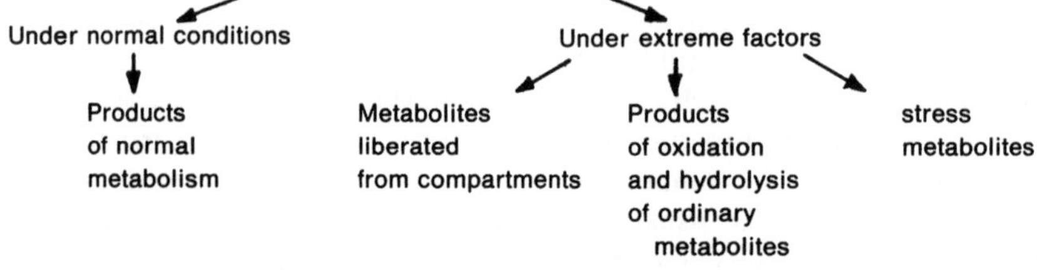

Fig. 94. Plant excreta under normal and stress conditions

in this book allow us to estimate its evolution of the excretory function in plants in comparison with other kingdoms. On the cellular level, the excretory function of any organism includes both the compartmentation and elimination of stored (reserved) or toxic substances and the transmission of chemical information or the transport of substances between organelles and the surrounding media. This occurs via special cellular structures: Golgi apparatus, secretory vesicles including clathrin-coated ones, endoplasmic reticulum, and proteins which are organized in microfilaments and microtubules. Common to all living organisms is the excretory mechanism of exocytosis and other forms of active or passive transport through the plasma membrane. However, unlike other organisms, in the plant cell the excreted products have to overcome the cellulose wall.

As living organisms develop in evolution, multicellular organisms with specialized excretory organs arise, as is illustrated in Fig. 95.

There is a similarity between plants and animals in the release of substances by nonspecialized and specialized organs. On the cellular level, all living organisms from microorganisms to higher plants and animals are capable of gas excretion and passive leaching, as well as the active elimination of substances from the cytoplasm. Besides, multicellular eukaryotes (both plants and animals) have specialized organs of excretion, including the secretory (excretory) epithelium and glands, including secretory trichomes. Although differing in anatomy, they have an analogous function. The most significant differences in comparison with animals are as follows: higher plants have single secretory cells, named idioblasts, but lack specialized respiratory organs. The significance of the release of substances by both unicellular and multicellular organisms is similar. It includes (1) the regulation of the exchange of ions and metabolites between the

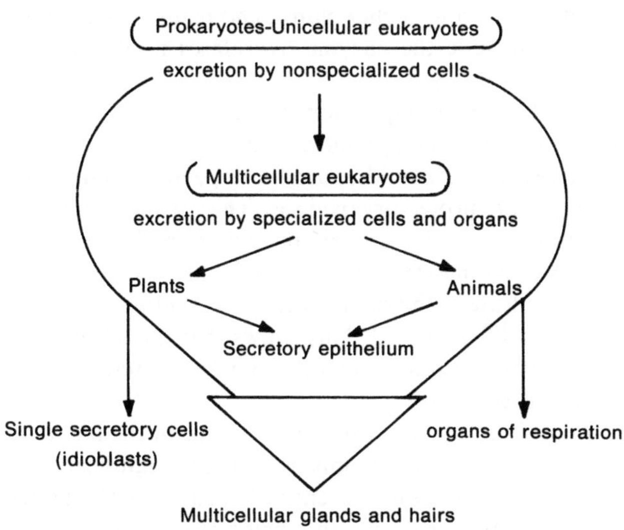

Fig. 95. Possible path of evolution of the excretory function

cytoplasm and the surrounding media; (2) the evacuation of waste toxic products; (3) competition or attraction in the relationships within a homogeneous population, or multipopulational biocenosis.

Another important question is the evolution of the chemical composition of the excreta which include sex or food attractants, repellents, allelochemicals, antiviral, antimicrobial, antifungal agents, etc. The excretory function as a factor of chemoisensibilization is known as early as in bacteria, where it plays a role in chemotaxis. With the appearance of multicellular organisms, special olfactory organs appeared in animals, whereas there is no information about analogous sensitive structures in plants connected with allelopathy. Allelopathic interactions exist in phytoplankton (Khailov 1971) and depend on the composition of the excreta. Their content also includes amino acids, proteins, hydrocarbons, and phenol terpenoids. The same classes of substances also participate in all kinds of interactions between higher plants (Grodzinskii 1965, 1991; Rice 1974, 1984; Roshchina and Roshchina 1989) as well as in competing or attracting relationships of animals and microorganisms (Harborne 1988). There is a convergence in the composition of defensive excreta in all living organisms. In particular, α-pinene is released both by higher plants and specialized glands of insects such as *Isoptera* and Termitidae (Everaerts et al. 1990). Sex attractants of unicellular green algae such as *Chlamydomonas* and *Volvox* are shown to be glycoproteins, amino acids, and even sesquiterpene lactones, which, for example in the algae *Ectocarpus siliculosus* (Saut and Whittick 1987) create an aroma similar to that of orange. The rather simple composition of sex attractants is known for mammals – volatile alcohols, aldehydes, and ketones, which are also characteristic of both plants and microorganisms like *Pseudomonas fragii* and *P. fluorescens*. There is a similarity between the mucilage of the prokaryote *Ciliata*, in particular *Tetrahymena thermophilla* (Florin-Christen et al. 1990), the digestive glands of mammals and carnivorous higher plants, and plant root exudates (see above). All these excreta include the enzyme hydrolases needed for the utilization of nutrients.

The main differences in chemical composition between excreta of higher plants and animals consist in specific compounds, characteristic only for the plant kingdom, such as taxon-specific alkaloids, phenols, terpenoids, and certain proteins, mainly enzymes which are protectory oxidases. The chemical composition within the plant community has also undergone evolution. For example, some complex terpenoids are synthesized and released only by specialized organs – the glands. In tissue culture of the same species, this synthesis does not occur, since the tissue becomes differentiated. This shows that the structure of the secretory organs and the composition of plant excreta are interconnected and arise at a higher stage of evolution.

Information on the composition of plant excreta, their role in the life of the plant itself, and in its interaction with other organisms is of great significance for the development of fundamental and applied sciences. It promotes the understanding of the life functions of plant organisms and their resistance to extremal factors, their defense against pathogens, and of their reproductive function.

These results can be valuable for botanical science, in particular chemosystematics, where the knowledge of the chemical composition of plant excreta is used as an additional systematic sign. At last, modern medicine has introduced a new concept called "medicinal botany", which forecasts the production of new natural drugs based on the study of the chemical composition and biological effects of plant excreta. It is greatly needed in our century of chemistry and stress. The development of biotechnology envisages new industrial products including those excreted by plants, mainly natural insecticides, herbicides, and fungicides.

References

Abeles FB (1972) Biosynthesis and mechanism of action of ethylene. Annu Rev Plant Physiol 23:259–292

Abeles FB (1986) Role of ethylene in *Lactuca sativa* cv Grand Rapids seed germination. Plant Physiol 81:780–787

Abell CA, Hursh CR (1931) Positive gas and water pressure in oaks. Science 73:449

Abraham S, Dileep (1976) Somatic cell abnormalities induced by water extracts of *Gloriosa superba*. Nucleus (India) 19:17–20

Adam ZM, Rashad Th (1984) Cytological effects of water extracts of medicinal plants. 1. Influence of *Ammi majus* on root tip of *Vicia faba*. Cytologia 49:265–271

Adams CM, Hutchinson TC (1987) Comparative abilities of leaf surfaces to neutralize acid raindrops. II. The influence of leaf wetability, leaf age and rain duration on changes in droplet pH and chemistry on leaf surfaces. New Phytol 106:437–456

Adams DO, Yang SF (1981) Ethylene the gaseous plant hormone: mechanism and regulation of biosynthesis. Trends Biochem Sci 6:161–164

Adkins SW, Shiraishi T, McComb JA (1990) Rice callus physiology–identification of volatile emissions and their effects on culture growth. Physiol Plant 78:526–531

Agnihotri VP, Vaartaja O (1969) Seed exudates from *Pinus resinosa* and their effect on growth and zoospore germination of *Pythium fertile*. Can J Bot 46:1135–1141

Aizenman BE, Smirnov VV, Bondarenko AS (1984) Phytoncides and antibiotics of higher plants. Naukova Dumka, Kiev

Akazawa G, Hara-Nishimura I (1985) Topographic aspects of biosynthesis extracellular secretion and intracellular storage of proteins in plant cells. Annu Rev Plant Physiol 36:441–472

Akhalkatsi MS, Vasil'ev AE (1984) The structure of stinging trichome of *Urtica dioica* (fam. Urticaceae). Abstracts of the 1st All-Union Conference on Plant Anatomy, Leningrad, October 1984. Nauka, Leningrad, p 10

Akulova EA (1977) Flavonoids–endogenous regulators of chloroplast energetic exchange. In: Akulova EA, Muzafarov EN (eds) Regulation of energy exchange of chloroplasts and mitochondria by endogenous phenolic inhibitors. Department of Scientific and Technical Information, Pushchino (USSR), pp 100–125

Alieva D, Kozlova NA, Sokolov PD (1976) Certain anatomical characteristics of the structure of vegetative organs of *Polygonium coriarium* substances in their tissues. Rastit Resur 12:30–41

Amoros M, Durand G (1964) Liberation de diverses substances par des graines de legumineuses au cours de leur imbibition. Note préliminaire. Ann Inst Pasteur 3 (Suppl):79–85

Amthor JS (1986) An estimate on the "cost" of nutrient leaching from forest canopies by rain. New Phytol 102:359–364

Anderson DMW, McDougall TJ (1985) The proteinaceous components of the gum exudates from some phyllodinous *Acacia* species. Phytochemistry 24:1237–1240

Anderson DMW, Hendrie A, Munro AC (1972) The amino acid and amino sugar composition of some plant gums. Phytochemistry 11:733–736

Andon TM, Denisova GA (1974) Localization of coumaric compounds in secretory reservours of *Ruta graveolens* L. Rastit Resur 10:528–540

Anet E, Lythgoe B, Silk MN, Trippett S (1953) Oenanthotoxin and cicutotoxin. Isolation and structure. J Chem Soc: 309–322

Apelbaum A, Vinkler C, Sfakiotakis E, Dilley DR (1984) Increased mitochondrial DNA and RNA polymerase activity in ethylene-treated potato tubers. Plant Physiol 76: 461–464

Archer BL (1980) Polyisoprene. In: Bell EA, Charlewood BV (eds) Secondary plant products. Encyclopedia of plant physiology New Ser Vol. 8, Springer, Berlin, Heidelberg. New York, pp 309–327

Arens K, Arens T (1970) Einflüsse der Temperatur auf die *kutikulare* Exkretion von Kalium Blöttern von *Coffea arabica.* Ber Dtsch Bot Ges 83: 501–504

Ariel AMR, Lemos-Pastrana A (1980a) Quimiotaxis de *Azospirillum lipoferum* y *Azospirillum brasiliensis* hacia exudados radiculares de gramineas. 1. Actividad de maiz trigo y sorgo. Rev Latinoam Microbiol 22: 131–135

Ariel AMR, Lemos-Pastrana A (1980b) Quimiotaxis de *Azospirillum lipoferum* y *Azospirillum brasiliensis* hacia exudados radiculares de gramineas. 11. Accion de los aminoacidos contenidos en exudados radiculares de gramineas. Rev Latinoam Microbiol 22: 137–142

Arisz WH, Gamphuis IJ, Hakens H, van Tooren AJ (1955) The secretion of the salt glands of *Limonium latifolium.* Acta Bot Neerl 4: 322–338

Armbruster WS, Keller S, Matsuki M, Clausen TP (1989) Pollination of *Dalechampia magnoliifolia* (Euphorbiaceae) by male Euglossine bees. Am J Bot 76: 1279–1285

Armstrong W (1979) Aeration in higher plants. In: Woolhouse HW (ed) Advances in botanical research, vol 7. Academic Press, New York, pp 225–332

Arnason JT, Swain T, Wat CK, Graham EA, Partington S, Towers GHN, Lam J (1981) Mocquito larvicidal activity of polyacetylenes from species in the Asteraceae. Biochem Syst Ecol 9: 63–68.

Arnason JT, Towers GHN, Philogene BJR, Lambert JDH (1989) Role of natural photosensitizers in plant resistance to insects. In: Hedin PA (ed) Plant resistance to insects. 183 rd Meet Am Chem Soc, Las Vegas, March 28–April 2, 1982. ACS Symp Ser vol. 208. Am Chem Soc, Washington, pp 139–151

Arny CJ, Pell EJ (1986) Ethylene production by potato, radish and soybean leaf tissue treated with simulated acid rain. Environ Exp Bot 26: 9–15

Arriaga-Giner FJ, Wollenweber E, Hradetzky D (1986) New flavonoids from the exudate of *Baccharis bigelovii* (Asteraceae). Z Naturforsch 41C: 946–948

Arriaga-Giner FJ, Rullkotter J, Peakman TM, Wollenweber E (1991) New triterpenes from the frond exudate of the fern *Notholaena rigida.* Z Naturforsch 46C: 507–512

Arthur JJ, Leone IA, Flower FB (1985) The response of tomato plants to simulated landfill gas mixtures. J Environ Sci Health Part A20: 913–925

Asakawa Y, Wollenweber E (1976) A novel phenolic acid derivative from buds of *Populus lasiocarpa.* Phytochemistry 15: 811–812

Asakawa Y, Takemoto T, Wollenweber E, Aratani T (1977) Lasiocarpin A, B and C, three novel phenolic triglycerides from *Populus lasiocarpa.* Phytochemistry 16: 1791–1795

Ascensao L, Pais MSS (1981) Ultrastructural aspects of secretory trichomes in *Cistus monspeliensis.* In: Margaris NS, Mooney HA (eds) Components of productivity of Mediterranean climate regions. Proc Int Symp, Kassandra, Greece, September 13–15, 1980. Junk, The Hague pp 27–38

Ascensao L, Pais MS (1985) Differentiation et processus secreteur des trichomes d'*Artemisia campestris* ssp. *maritima* (Compositae). Ann Sci Nat Bot Biol Veg 7: 149–171

Ascensao L, Pais MSS (1987) Glandular trichomes of *Artemisia campestris* (ssp. *maritima*): ontogeny and histochemistry of the secretory product. Bot Gaz 148: 221–227

Asplund RO (1968) Monoterpenes: relationship between structure and inhibition of germination. Phytochemistry 7: 1995–1997

Atwell BJ, Greenway H (1987) The relationship between growth and oxygen uptake in hypoxic rice seedlings. J Exp Bot 38: 454–465

Atwell BJ, Drew MC, Jackson MB (1988) The influence of oxygen deficiency on ethylene synthesis, 1-amino-cyclopropane-1-carboxylic acid levels and aerenchyma formation in roots of *Zea mays.* Physiol Plant 72: 15–22

Azarova TC (1983) Influence of volatile root excreta and its components on the formation of rhizospheric microflora. Trans All-Union Sci Res Inst Agric Microbiol (USSR) N 53:141–149

Baas P, Gregory M (1985) A survey of oil cells in the dicotyledons with comments on their replacement by and joint occurrence with mucilage cells. Isr J Bot 34:167–168

Babu AM, Menon ARS (1990) Distribution of gum and gum-resin ducts in plant body: certain familiar features and their significance. Flora 184:257–261

Babu AM, Shah JJ (1987) Unusual tissue complexes formed in association with traumatic gum cavities in the stem of *Bombax ceiba* L. Ann Bot 59:293–299

Babu AM, Nair GM, Shah JJ (1987) Traumatic gum-resin cavities in the stem of *Ailanthus excelsa* Roxb. IAWA Bull 8:167–174

Bach M, Felling J (1958) Effect of ethanol and auxins on the growth of unicellular algae. Nature 182:1359

Backhaus RA (1985) Rubber formation in plants – a mini–review. Isr J Bot 34:283–293

Bailey JA (1982) Mechanisms of phytoalexins accumulation. In: Bailey JA, Mansfield JW (eds) Phytoalexins. Blackie, Glasgow, pp 289–318

Baker DA, Kallarackal J, Milburn JA (1990) Water relations of the banana. II. Physicochemical aspects of the latex and other tissue fluids. Aust J Plant Physiol 17:69–77

Baker HG, Baker I (1975) Studies of nectar-constitution and pollinator-plant coevolution. In: Gilbert E, Raven PH (eds) Coevolution of animals and plants. Univ of Texas Press, Austin pp 100–140

Baker HG, Baker I (1982) Chemical constituents of nectar in relation to pollination mechanisms and phylogeny. In: Nitecki MH (ed) Biochemical aspects of evolutionary biology. University Chicago Press, Chicago pp 131–171

Baker HG, Baker I (1983) A brief historical review of the chemistry of floral nectar. In: Bentley B, Elias T (eds) The biology of nectaries. Columbia University Press, New York, pp 126–152

Bakker ME, Gerritsen AF (1990) Ultrastructure and development of oil idioblasts in *Annona muricata* L. Ann Bot 66:673–686

Baldwin IT (1989) Mechanism of damage-induced alkaloid production in wild tobacco. J Chem Ecol 15:1661–1680

Bandyopadhyay R (1982) Inhibition of acetylcholinesterase by permethrine and its reversion by acetylthiocholine. Indian J Exp Biol 20:488–491

Bandyukova VA, Oganesyan ET, Lisevitskaya LI, Sidel'nikova VI, Shankarenko AL (1968) Some results of the study of chemical content of the North Caucasus plants. In: Kursanov AL, Zaprometov MN (eds) Phenolic compounds and their biological functions. Nauka, Moscow, pp 95–100

Barber DA, Gunn KB (1974) The effect of mechanical forces on the exudation of organic substances by the roots of cereal plants grown under sterile conditions. New Phytol 73:39–45

Barber DA, Ebert M, Evans NTS (1962) The movement of ^{15}O through barley and rice plants. J Exp Bot 13:397–403

Barber J (1984) Further evidence for the common ancestry of cytochrome b–c complexes. Trends Biochem Sci 9:209–210

Barmicheva EM (1982) Cytochemical study of mucilage secretion by the root cap cells of *Raphanus sativus* (Cruciferae). Bot Zh (USSR) 67:1627–1635

Barmore CR, Briggs RH (1972) Ethylene diffusion through citrus leaf and fruit tissue. J Am Soc Hortic Sci 97:24–27

Barta AL (1984) Ethanol synthesis and loss from flooded roots of *Medicago sativa* L. and *Lotus corniculatus*. Plant Cell Environ 7:187–191

Barz W (1969) Stoffwechsel aromatischer Pflanzeninhaltstoffe .1. Uber den Umsatz von Isoflavonen und Cumostennen in *Cicer arietinum* L. und *Phaseolus aureus* Roxb. Z Naturforsch Teil B 24 b:234–239

Bauer K, Seiler W, Giehl H (1979) CO-Produktion höherer Pflanzen an naturlichen Standorten CO. Z Pflanzenphysiol 94:219–230

Beale MH (1991) The biosynthesis of C_5–C_{20} terpenoid compounds. J Agric Food Chem 39:387–407

Beardsell DV, Williams EG, Knox RB (1989) The structure and histochemistry of the nectary and another secretory tissue of the flowers of *Tryptomene calycina* (Lindl) Stapf (Myrtaceae). Aust J Bot 37:65–80

Beaumont J, Cutler DF, Reynolds T, Vaughan JG (1985) The secretory tissue of aloe and their allies. Isr J Bot 34:265–282

Becker M, Kerstiens G, Schönherr J (1986) Water permeability of plant cuticles: permeance, diffusion and partition coefficients. Trees 1:54–60

Beckman CN, Mueller WC, McHardy WE (1972) The localization of stored phenols in plant hairs. Physiol Plant Pathol 2:69–74

Bednarska E, Tretyn A (1989) Ultrastructural localization of acetylcholin-esterase activity in the stigma of *Pharbitis nil.* Cell Biol Int Rep 13:275–281

Belikov PS, Kirillova TV (1958) Intensity of the compound release as a test on a functional state of plant cell. Izv Timiryazev Agric Akad (USSR) N 2:21–38

Belin-Depoux M (1987) Biological aspects of folia nectaries and hydathodes in tropical plants. Abstr 14th Int Bot Congr. Berlin-Dahlem Bot Garden, West Berlin, 240

Belin-Depoux CC (1983) Aspects morphologiques et histologiques des hydathodes du *Pilea semidentata* (Juss) Wedd et du *P. microphylla* (L.) *Liebm* (Urticaceae). Rev Gen Bot 90:117–129

Belin–Depoux M, Clair Maszulajtys D (1975) Introduction àl' étude des glandes foliaires de l' *Aleurites moluccana* Willd (Euphorbiaceae). 11. Aspect histologiques et cytologiques de la glande petiolaire fonctionnelle. Rev gen Bot. 82:119–115

Bell EA (1972) Toxic amino acids in Leguminosae. In: Harborne JB (ed) Photochemical ecology. Academic Press, London, pp 163–178

Bell AA, Stipanovic RD, Elzen GW, Williams HJ Jr (1987) Structural and genetic variation of natural pesticides in pigment glands of cotton (*Gossypium*). In: Waller GR (ed) Allelochemical role in agriculture and forestry. ACS Symp Ser vol 330. Am Chem Soc, Washington, pp 477–490

Ben-David A, Bashan Y, Okon Y (1986) Ethylene production in pepper (*Capsicum annuum*) leaves infected with *Xanthomonas campestris* pv. *vesicatoria.* Physiol Mol Plant Pathol 29:305–316

Benedict CR (1983) Biosynthesis of rubber. In: Porter JW, Spurgeon SL (eds) Biosynthesis of isoprenoid compounds, vol 2. Wiley, New York, pp 355–369

Bentrup FW (1979) Reception and transduction of electrical and mechanical stimuli. In: Haupt W, Feinleib ME (eds) Physiology of movements. Encyclopedia of Plant Physiology, New Ser, vol 7. Springer, Berlin Heidelberg New York, pp 42–70

Benz J, Rüdiger W (1981) Incorporation of 1-^{14}C-isopentenyldiphosphate, geraniol and farnesol into chlorophyll in plastid membrane fractions of *Avena sativa* L. Z Pflanzenphysiol 102:95–100

Berenbaum M (1985) Brementown revisited: interactions among allelochemicals in plants. In: Cooper-Driver GA, Swain F, Conn EE (eds) Chemically mediated interactions between plants and other organisms. Recent Adv Phytochemistry, vol 19. Plenum Press, New York, pp 139–169

Berestetskii OA, Kravchenko LV (1980a) The release of free amino acids by seedlings of maize and lupine. In: Grodzinskii AM (ed) Chemical interactions of plants. Naukova Dumka, Kiev, pp 32–38

Berestetskii OA, Kravchenko LV (1980b) Release of free amino acids by seeds of wheat and pea. Fiziol Rast (USSR) 27:419–422

Berestetskii OA, Kravchenko LV, Fomicheva AP (1978) Quantitative analysis of volatile excrets of plants and microorganisms. Dokl VASKHNIL (USSR) N 1:17–19

Berger RG, Akkan Z, Drawert F (1990) The essential oil of *Coleonema album* (Rutaceae) and of a photomixotrophic cell culture derived thereof. Z Naturforsch 45C:187–195

Berlin J, Whitte L, Schubert W, Wray V (1984) Determination and quantification of monoterpenoids secreted into the medium of cell cultures of *Thuja occidentalis.* Phytochemistry 23:1277–1279

Bernard-Dagan C, Pauly G, Marpeau AL, Gleizes M, Carde JP, Barachat P (1982) Control and compartmentation of terpene biosynthesis in leaves of *Pinus pinaster*. Physiol Veg 20:775–795

Bernardello LM, Galetto L, Juliani HR (1991) Floral nectar, nectary structure and pollinators in some Argentinian Bromeliaceae. Ann Bot 67:401–411

Beyer EM Jr. Morgan PW (1970) A method for determining the concentration of ethylene in the gas phase of vegetative plant tissues. Plant Physiol 46:352–354

Bhaumic Ch (Mondal), Datta PC (1989) Hormonal effect of mentholic gland initiation. Indian J Biol 21:55–57

Bicchi C, D'Amato A, Frattini C, Nono GM, Cappelletti E, Caniato R (1985) Analysis of essential oils by direct sampling from plant secretory structures and capillary gas chromatography. J High Resolut Chromatogr Chromatogr Commun 8:431–435

Biedinger U, Schnabl H (1991) Ethane production as an indicator of the regeneration potential of electrically fused sunflower protoplasts (*Helianthus annuus* L). J Plant Physiol 138:417–420

Bieleski RL, Redgwell RJ (1980) Sorbitol metabolism in nectaries from flowers of Rosaceae. Aust J Plant Physiol 7:15–25

Billings WD, Godfrey PJ (1967) Photosynthetic utilization of internal carbondioxide by hollow-stemmed plants. Science 158:121–123

Bloudeau R (1966) Elimination de substances organiques au cours de germination du colza (*Brassica napus oleifera*). II. Vitamines du groupe B. Bull Soc Bot Nord Fr 19:72–78

Blust MH, Hopkins TL (1987) Olfactory responses of a specialist and a generalist grass hopper to volatiles of *Artemisia ludoviciana* Nutt (Asteraceae). J Chem Ecol 13:1893–1902

Bochler-Kohler BA, Lapple G, Hellman V, Boger P (1982) Paraquat-induced production of hydrocarbon gases. Pestic Sci 13:323–329

Bohlman F (1988) Naturally occurring acetylenes. In: Lam J, Breteler H, Arnason T, Hansen L (eds) Chemistry and biology of naturally occurring acetylenes and related compounds (NOARC). Elsevier, Amsterdam pp 1–19

Bohlman F, Bukhardt T, Zdero C (1973) Naturally occurring acetylenes. Academic Press, New York

Boichenko EA (1946) The hydrogen release by isolated chloroplasts. Dokl AN SSSR 52: 525–528

Boivin C, Malpica C, Rosenberg C, Denarie J (1989) Metabolic signals in the rhizosphere: catabolism of calystegins and trigonelline by *Rhizobium melilotti*. In: Lugtenberg BJJ (ed) Signal molecules in plants and plant-microbe interactions. NATO Ser vol H 36. Springer, Berlin Heidelberg New York, pp 401–407

Boldyrev AA (1985) Biological membranes and ion transport. Moscow University Press, Moscow (USSR)

Bollmark M, Eliasson L (1990) Ethylene accelerates the breakdown of cytokinins and thereby stimulates rooting in Norway spruce hypocotyl cuttings. Physiol Plant 80:493–496

Bolton EF, Erickson AE (1970) Ethanol concentration in tomato plants during soil flooding. Agron J 62:220–226

Bonner J (1965) The isoprenoids. In: Bonner J, Varner JE (eds) Plant biochemistry. Academic Press, New York, pp 665–692

Bones A, Iversen TH (1985) Myrosin cells and myrosinase. Isr J Bot 34:351–376

Bopaiah BM, Shetty HS, Nagaraja KV (1987) Biochemical characterization of the root exudates of coconut palm. Curr Sci 56:832–833

Borner H (1959) The apple replant problem. 1 The excretion of phlorizin from apple root residues. Contrib Boyce Thompson Inst 20:39–56

Borner H (1960) Liberation of organic substances from higher plants and the soil sickness problem. Bot Rev 26:393–424

Boulter D, Ieremy II, Wilding M (1966) Amino acids liberated into the culture medium by pea seedling root. Plant Soil 24:121–127

Bouzayen M, Latche A, Pech JC, Alibert G (1986) Involvement of vacuoles in ethylene metabolism in plant cells. In: Martin B (ed) Plant vacuoles. Plenum Press, New York, pp 449–454

Bovey RW, Diaz-Colon JD (1969) Occurrence of plant growth inhibitors in tropical and subtropical vegetation. Physiol Plant 22:253–259

Brachet J, Cosson L (1986) Changes in the total alkaloid content of *Datura innoxia* Mill. Subjected to salt stress. J Exp Bot 37:650–656

Breauv P, Joshi YC, Dwivedi R, Snehi BA, Bal AR, Qader A (1983) Salt excretion by glands in *Diplachne fusca* (Linn). Indian J Plant Physiol 26:203–208
Brindle PA, Threifall DR (1983) The metabolism of phytoalexins. Biochem Soc Trans 11:516–522
Broglie KE, Gaynor JJ, Broglie RM (1986) Ethylene-regulated gene expression: molecular cloning of the genes encoding an endochitinase from *Phaseolus vulgaris*. Proc Natl Acad Sci USA 83:6820–6824
Brooks CJW, Watson DG, Freer IM (1986) Elicitation of capsidiol accumulation in suspended callus cultures of *Capsicum annuum*. Phytochemistry 25:1089–1092
Brouillet L, Bertrand C, Cuerrier A, Barabe D (1987) Les hydathodes des generes *Begonia* et *Hillebrandia* (Begoniaceae). Can J Bot 65:34–52
Brown HT, Escombe F (1900) Static diffusion of gases and liquids in relation to the assimilation of carbon and translocation in plants. Philos Trans R Soc Lond 193B:223–291
Brown JT, Hegarty PK, Charlwood BV (1987) The toxicity of monoterpenes to plant cell cultures. Plant Sci 48:195–201
Brown PH, Brodl MR (1988) Hormonal and heat-stress regulation of protein synthesis in the aleurone layers of barley seeds. BioEssays 8:199–202
Brüne B, Schmidt KU, Ullrich V (1990) Activation of soluble guanylate cyclase by carbon monooxide and inhibition by superoxide anion. Eur J Biochem 192:683–688
Bryantseva ZN (1951) About excretion of volatile organic compounds to air by living plants. Uch Zap Gor'k Univ Gor'ky (USSR), pp 21
Bucher JB (1982) Einfluss von SO2 auf Terpen-Emissionen von Kiefern (Pinus silvestris L.). In: Materials of the XII Int Arbeitstagung forstlicher Rauchschadenssachverstandiger. IUFRO, Oulu, pp 1–4
Bukolova TP (1971) To the question about mechanism of water-soluble kolines of weed plants action on cultivated ones. In: Grodzinskii AM (ed) Physiologo-biochemical base of plant interactions in phytocenosis, vol 2. Naukova Dumka, Kiev, pp 66–69
Burg SP, Burg EA (1967) Molecular requirement for the biological activity of ethylene. Plant Physiol 42:144–152
Burström H (1959) Growth and formation of intercellularies in root meristems. Physiol Plant 12:371–385
Buta JG, Spaulding DW (1989) Allelochemicals in tall fescue-abscisic and phenolic acid. J Chem Ecol 15:1629–1636
Buttery RG, Kamm JA, Ling LC (1982) Volatile components of alfalfa flowers and pods. J Agric Food Chem 30:739–742
Buttery RG, Maddix DM, Light DM, Ling LC (1986) Volatile components of yellow starthistle. J Agric Food Chem 34:786–788
Buvat R (1989) Ontogeny, cell differentiation and structure of vascular plants. Springer, Berlin Heidelberg New York
Buzuk GN, Lovkova MYa (1986) Calcium, transmembranic transport and accumulation of alkaloids in plants. Dokl AN SSSR 289:749–750
Byers RE, Baker LR, Sell HM, Herner RC, Dilley DR (1972) Ethylene: a natural regulator of sex expression of *Cucumis melo* L. Proc Natl Acad Sci USA 69:717–720
Bykova OP, Yakovleva OV (1991) Mucilage cells of leaf of some representatives of the genus *Alcea* L. Rastit Resur 27:82–90
Caemmerer S von, Farquhar GD (1981) Some relationships between the biochemistry of photosynthesis and the gas exchange of leaves. Planta 153:376–387
Camara B (1984) Terpenoid metabolism in plastids. 1. Sites of phytoene synthetase activity and synthesis in plant cells. Plant Physiol 74:112–116
Caniato R, Cappelletti EM (1984) Localizazione dei lattoni sesquiterpenici e degliidroperossidi in *Liriodendron tulipifera* L. G Bot Ital 118 (Suppl 2):351–352
Cappelletti EM, Caniato R, Appendino G (1986) Localization of the cytotoxic hydroperoxyeudesmanolides in *Artemisia umbelliformis*. Biochem Syst Ecol 14:183–190
Carter CD, Gianfagna TJ, Sacalis JN (1989a) Sesquiterpenes in glandular trichomes of a wild tomato species and toxicity to the Colorado potato beetle. J Agric Food Chem 37:1425–1428

Carter CD, Sacalis JN, Gianfagna TJ (1989b) Zingiberene and resistance to Colorado potato beetle in *Lycopersicon hirsutum* f. *hirsutum*. J Agric Food Chem 37:206–209

Chaboud A, Rougier M (1984) Identification and localization of sugar components of rice (*Oryza sativa* L) roots cap mucilage. Plant Physiol 116:323–330

Chameides WL, Lindsay RW, Richardson J, Kiang CS (1988) The role of biogenic hydrocarbons in urban photochemical smog: Atlanta as a case study. Science 241:1473–1475

Chong W, Chang (1975) Carbon dioxide and senescense in cotton plants. Plant Physiol 55:515–519

Chang LA, Hammett LK, Pharr DM (1983) Internal gas atmospheres, ethanol and leakage of electrolytes from a flood-tolerant and a flood-susceptible sweet potato cultivar as influenced by anaerobiosis. Can J Bot 61:3399–3404

Changeux JP, Devillers-Thiery A, Chemouilli P (1984). Acetylcholine receptor: an allosteric protein. Science 225:1335–1345

Chapman DJ, Tocher RD (1966) Occurrence and production of carbon monooxide in some brown algae. Can J Bot 44:1438–1442

Chappelle EW, Krall AR (1961) Carbon monooxide fixation by cell-free extracts of green plants. Biochim Biophys Acta 49:578–580

Charest PM, Brisson L, Ibrahim RK (1986) Ultrastructural features of flavonoid accumulation in leaf cells of *Chrysosplenium americanum*. Protoplasma 134:95–101

Charlwood BV, Hegarty PK, Charlwood KA (1986) The synthesis and biotransformation of monoterpenes by plant cells in culture. In: Morris P, Scragg AH, Stafford A, Fowler MW (eds) Secondary Metabolism in plant cell cultures. Cambridge University Press, Cambridge, pp 15–34

Charriere-Ladreix Y (1977) La secretion flavonoique chez *Populus nigra* L signification, functionelle des remainiements ultrastructuraux glandulaires. Physiol Veg 15:619–640

Chase WW (1934) The composition, quantity and physiological significance of gases in tree stems. Minn Agric Exp Stn Tech Bull p 99

Chasovennaya AA (1961) To a question about mechanism of chemical interaction of plants. Vestn Leningr Univ Ser Biol N 1:54–66

Chavchavadze ES (1973) Classification of resinous ducts and cavities in conifers wood. Lesn Zh N 6:165–167

Chavchavadze ES (1979) The conifers' wood: morphological characteristics, diagnostic significance. Nauka, Leningrad

Chen SL, Chen CT. Kao CH (1991) Polyamines promote the biosynthesis of ethylene in detached rice leaves. Plant Cell Physiol 32:6, 813–817

Cheong JJ, Hahn MG (1991) A specific, high-affinity binding site for the hepta-β-glucoside elicitor exist in soybean membranes. Plant Cell 3:137–147

Cheong JJ, Birberg W, Fügedi P, Pilotti A, Garegg PJ, Hong N, Ogawa T, Hahn MG (1991) Structure-activity relationships of oligo-β-glucoside elicitors of phytoalexin accumulation in soybean. Plant Cell 3:127–136

Chernobrivenko SI (1956) Biological role of plant excreta in interspecies interactions in mixtured cultivars. Siviet Nauka, Moscow.

Chiba H, Takahachi N, Kitabatake N, Sasaki R (1979) Enzymatic improvement of food flavor. III. Oxidation of the soybean protein-bound aldehyde by aldehyde dehydrogenase. Agric Biol Chem 43:1891–1897

Chin CK, Frenkel C (1977) Ursurge in respiration and peroxide formation in potato tubers as influenced by ethylene, propylene and cyanide. Plant Physiol 59:515–518

Chirkova TV (1968) Characteristics of the oxygen supply of roots of some woody plants being put up at anaerobiosis. Fiziol Rast (USSR) 15:565–568

Chirkova TV (1988) About ways of plant adaptation to hypoxia and anoxia. Fiziol Rast (USSR) 35:393–411

Chirkova TV, Gutman TS (1972) About physiological role of the branch lenticels of willow and poplar at root anaerobiosis. Fiziol Rast (USSR) 19:352–359

Cholodny N (1932) Zur Kenntnis der durch das regnerische Wetter verursachten Ertragsabnahme bei Getreidearten. Ber Dtsch Bot Ges 50:562–570

Chou CH (1989) Allelopathic research of subtropical vegetation in Taiwan. IV. Comparative phytotoxic nature of leachate from four subtropical grasses. J Chem Ecol 15: 2149–2159
Chrestin H, Gidrol X, Pean M, Marin B (1986) DmH^+-controlled reversible fluxes of H^+ and Calcium at the tonoplast but quasi-total citrate sequestration within the intact vacuoles from the latex cells of *Hevea brasiliensis*. Implications in the production of natural rubber. In: Marin B (ed) Plant vacuoles. Their importance in solute compartmentation in cells and their applications in plant biotechnology. NATO ASi Ser A, vol 134. Plenum Press, New York, pp 325–338
Christensen LP, Lam J (1990) Acetylènes and related compounds in Cynareae. Phytochemistry 29:2753–2785
Chrominski A, Khan MA, Weber DJ, Smith BN (1986) Ethylene and ethane production in response to salinity stress. Plant Cell Environ 9:687–691
Chubinidze VV (1966) Season dynamics of plant volatile excretions and identification of some aldehydes in them. PhD Thesis, Tbilisi State University, USSR
Churms SC, Stephen AM, Steyn CB (1986) Analytical comparison of gums from *Acacia hebeclada* and other Gummiferae species. Phytochemistry 25:2807–2809
Cicerone RJ, Oremland RC (1988) Biogeochemical aspects of atmospheric methane. Global Biogeochem Cycles 2:299–327
Clair-Maczulajtys D, Bory G (1983) Les nectaries extrafloraux pedicelles chez l'*Ailanthus grandulosa*. Can J Bot 61:683–691
Clarkson DT (1974) Ion transport and cell structure in plants. McGraw-Hill, London
Cline SD, Coscia CJ (1989) Ultrastructural changes associated with the accumulation and secretion of sanguinarine in *Papaver bracteatum* suspension cultures treated with fungal elicitor. Planta 178:303–314
Cockburn W, Ting IP, Sternberg LO (1979) Relationships between stomatal behaviour and internal carbon dioxide concentration in Crassulaceae acid metabolism plants. Plant Physiol 63:1029–1032
Cody AM, Horner HT (1985) Analytical resolution of the crystalline sand pyramids. Am J Bot 72:1149–1158
Cojocaru AF, Ruzieva RKh, Topaly VP, Topaly EE (1977) Protonophoric activity of flavonoids and uncoupling of oxidative phosphorylation. In: Akulova EA, Muzafarov EN (eds) Regulation of energy exchange of chloroplasts and mitochondria by endogenous phenolic inhibitors. Department of Scientific and Technical Information, Pushchino USSR, pp 73–99
Coleman J, Evans D, Hawes C (1988) Plant coated vesicles. Plant Cell Environ 11:669–684
Coleman LW, Hodges CF (1987) Ethylene biosynthesis in *Poa pratensis* leaves in response to injury or infection by *Bipolaris sorokiniana*. Phytopathology 77:1280–1283
Collet GF (1975) Exsudation racinaire d'enzymes. Soc Bot Fr Coll Rhizosphere 122:61–75
Collier HOJ, Chesher GB (1956) Identification of 5-hydroxytryptamine in the sting of the nettle (*Urtica dioica*). Br J Pharmacol 11:186–189
Collin HA, Musker D, Britton G (1986) Flavor production in onion tissue culture. In: Morris P, Scragg AH, Stafford A, Fowler MW (eds) Secondary metabolism in plant cell cultures. Cambridge University Press, Cambridge, pp 54–62
Condon JM, Fineran BA (1989a) Distribution and organization of articulated laticifers in *Calystegia silvatica* (Convol vulaceae). Bot Gaz 150:289–302
Condon JM. Fineran BA (1989b) The effect of chemical fuxation and dehydration on the preservation of latex in *Calystegia silvatica* (Convol vulaceae). Examination of exudate and latex in situ by light and scanning electron microscopy. J. Exp Bot 40:925–939
Cook CE, Whichard LP, Turner B, Wall ME, Egley GH (1966) Germination of witchweed (*Striga lutea* Lour): isolation and properties of a potent stimulant. Science 154:1189–1190
Cook CE, Whichard LP, Wall ME, Egley GH, Coggan P, Luhan PA, McPhail AT (1972) Germination stimulants. II. The structure of strigol–a potent seed germination stimulant for witchweed (*Striga lutea* Lour). J Am Chem Soc 94:6198–6199
Corbineau F, Engelmann F, Come D (1990) Ethylene production as an indicator of chilling injury in oil palm (*Elaeis guineensis* Jacq) somatic embryos. Plant Science 71:N 1, pp 29–34

Cori OM (1983) Enzymic aspects of the biosynthesis of monoterpenes in plants. Phytochemistry 22:331–341

Cortes F, Mateos S, Escalza P (1987) Acetaldehyde induces mature endoreduplicated *Allium cepa* root cells to divide. Experientia 43:205–206

Cotton CM, Evans LV, Gramshaw JW (1991a) The accumulation of volatile oils in whole plants and cell cultures of tarragon (*Artemisia dracunculus*). J Exp Bot 42:365–375

Cotton CM, Gramshaw JW, Evans LV (1991b) The effect of α-naphtalene acetic acid (NAA) and benzylaminopurine (BAP) on the accumulation of volatile oil components in cell culteres of tarragon (*Artemisia dracunculus*). J Exp Bot 42:377–386

Couillerot JP, Byung YY(1985) Exsudation recinaire de gibberellines lors de laculture hydroponique de *Lycopersicon esculentum* Mill. C R Acad Sci D 301:99–102

Craker LE (1971) Ethylene production from ozone injured plants. Environ Pollut 1:299–304

Cresti M, Keijzer CJ, Tiezzi A, Ciampolini F, Focardi S (1986) Stigma of *Nicotiana*: ultrastructural and biochemical studies. Am J Bot 73:1713–1722

Criegee R (1959) Products of ozonization of some olefins. In: Ozone chemistry and technology. Washington, Amer Chem Soc pp. 133–135

Croak M (1972) Effects of phenolic inhibitors on growth, metabolism, mineral deletion, and ion uptake in Paul's scarlet rose cell suspension cultures, PhD Diss, University of Oklahoma, Norman.

Crocker W (1948) Growth of plants. Reinhold New York.

Croteau R, Johnson MA (1983) Biosynthesis of terpenoids in glandular trichomes. In: Rodriquez E, Hedley PL, Mehta I (eds) Biology and chemistry of plant trichomes. Plenum Press, New York, pp 133–185

Croteau R, Gurkewitz S, Johnson MA, Fisk HJ (1987) Biochemistry of oleoresinosis. Monoterpene and diterpene biosynthesis in lodgepole pine saplings infected with *Ceratocystis clavigera* or treated with carbohydrate elicitors. Plant Physiol 85:1123–1128

Croteau R, Miyazaki JH, Wheeler CJ (1989) Monoterpene biosynthesis: Mechanistic evaluation of the geranyl pyrophosphate: (−)-endo-fenchol cyclase from fennel (*Foeniculum vulgare*). Arch Biochem Biophys 269:507–516

Curry KJ (1987) Initiation of terpenoid synthesis in osmophores of *Stanhopea anfracta* (Orchidaceae): A cytochemical study. Am J Bot 74:1332–1338

Curry KJ, Mcdowell LM, Stern WL (1989) Initiation of terpenoid synthesis in osmophores of *Gongora truncata* (Orchidaceae): a cytochemical study. Suppl Pap Joint Meet Can Bot Assoc Toronto, 6–10 Aug 1989. Am J Bot 76:31

Curtis JD, Lersten NR (1986) Hydathode anatomy of *Potentilla palustris* (Rosaceae). Nord J Bot 6:739–796

Curtis JD, Lersten NR (1990) Internal secretory structures in *Hypericum* (Clusiaceae): *H. perforatum* L. and *H. balearicum*. New Phytol 114:571–580

Dacey JWH (1980) Internal winds in water lilies: an adaptation for life in anaerobic sediments. Science 210:1017–1019

Dacey JWH (1981) Pressurized ventilation in the yellow waterlily. Ecology 62:1137–1147

Dacey JWH (1987) Knudsen transitional flow and gas pressurization in leaves of Nelumbo. Physiol Plant 85:199–203

Dacey JWH, Klug MJ (1979) Methane efflux from lake sediments through water lilies. Science 203:1253–1255

Dacey JWH, Klug MJ (1982a) Tracey studies of gas circulation in Nuphar: $^{18}O_2$ and $^{14}CO_2$ transport. Physiol Plant 56:361–366

Dacey JWH, Klug MJ (1982b) Ventilation by floating leaves in Nuphar. Am J Bot 69: 999–1003

Dadykin VP, Stepanov LN, Rizhkova VE (1967) Some data about volatile (oxygen-containing) excrets of series of vegetable plants. Kosm Biol Aviakosm Med N 6:48–52

Dalbro S (1955) Leaching of nutrients from apple foliage by rain. In: Proc XIV Int Hortic Congr Wegeningen, vol 1. pp 770–778

Danilova MF, Kashina TK (1987a) Ultrastructure of glandular squamule of *Perilla ocymoides*

(Lamiaceae) in connection with their participance in the synthesis of steroid hormones and gibberellins. Bot Zh (USSR) 72:427–435

Danilova MF, Kashina TK (1987b) Ultrastructure of glandular trichomes of *Perilla ocymoides* (Lamiaceae) in connection with their possible participance in photoperiodical induction of flowering. Bot Zh (USSR) 72:561–568

D'Arcy-Lameta A (1986) Study of soybean and lentil root exudates. II. Identification of some polyphenolic compounds relation with plantlet physiology. Plant Soil 92:113–123

Darvill AG, Albersheim P (1984) Phytoalexins and their elicitors: a defence against microbial infection in plants. Annu Rev Plant Physiol 35:243–275

Datta SC, Ghatterjce AK (1980) Allelopatic potential of *Polygonium orientale* L. in relation to germination and seedling growth of weeds. Flora 169:456–465

Dean JV, Harper JE (1989) Analysis of volatile nitrogen (NO and NO_2) release from plants. In: Linskens HF, Jackson JP (eds) Gases in plant and microbial cells. Modern methods of plant analysis, vol 9. Springer, Berlin Heidelberg New York, pp 240:253

Debata A, Murthy KS (1983) Endogenous ethylene content in rice leaves during senescense. Indian J Plant Physiol 26:425–427

Dell B, McComb AJ (1978a) Plant resins – their formation, secretion and possible functions. In: Woolhouse HW (ed) Advances in botanical research, vol 6. Academic Press, London, pp 277–316

Dell B, McComb AJ (1978b) Biosynthesis of resin terpenes in leaves and glandular hairs of *Newcastelia viscida*. J Exp Bot 29:89–95

Dellacassa E, Soler E, Menendez P, Moyna P (1990) Essential oils from *Lippa alba* (Mill) N.E. Brown and *Aloysia chamaedrifolia* Cham (Verbenaceae) from Uruguay. Flav Fragr J 5:107–108

Dellarco VL (1988) A mutagenicity assessment of acetaldehyde. Mutat Res 195:1–20

Delwiche CC (1970) Carbon monoxide production and utilization by higher plants. Ann NY Acad Sci 174:116–121

Dembitskii AD, Beridze LV, Yurina RA (1982) To the sesquterpene hydrocarbon of *Ocimum gratissimum*. Khim Prir Soedin (USSR) N 3:398–399

Denisova GA (1976) Comparative study of secretory cavities of species *Dictamnus* L. of USSR flora. Rastit Resur 12:24–29

Denisova GA (1989) Terpenoid-containing plant structures, Nauka, Leningrad

Denisova GA, Dranitsina YuA (1962) Localization of coumarin compounds in species of *Archangelica decurrens* Ledb. Dokl AN SSSR 146:954–955

Deus-Neumann B, Zenk MH (1984) A highly selective alkaloid uptake system in vacuoles of higher plants. Planta 162:250–260

Deus-Neumann B, Zenk MH (1986) Accumulation of alkaloids in plant vacuoles does not involve an ion-trop mechanism. Planta 167:44–53

Di Cosmo F, Misawa M (1985) Eliciting secondary metabolism in plant cell cultures. Trends Biotechnol 3:318–322

Dietz VH, Wollenweber E, Favre-Bonvin, J, Smith DM (1981) Two flavonoids from the frond exudate of *Pityrogramma triangularis* var *triangularis*. Phytochemistry 20:1181–1182

Ding Jing, Shen Zhen-de (1985) Cytokinins in root exudate of cotton plants. Acta Phytophysiol Sin 11:249–259

Dinkelaker B, Römpeld V, Marschner H (1989) Citric acid excretion and precipitation of calcium citrate in the rhizosphere of white lupin (*Lupinus albus* L.). Plant Cell Environ 12:285–292

Distelbarth H, Kull U (1985) Physiological investigations of leaf mucilages. II. The mucilage of *Taxus faccata* L. and of *Thuja occidentalis* L. Isr J Bot 34:113–128

Djordjevic MA, Redmond JW, Batley M, Rolfe BG (1987) Clovers secrete specific phenolic compounds which either stimulate or repress nod gene expression in *Rhizobium trifolii*. EMBO J 6:1173–1179

Dmitriev MT (1972) Smell of fresh air. Priroda 3:13–23

Dmitriev MT, Rastyannikov EG, Akimov YuA, Malisheva AG (1988) Chromatography-mass-spectrometric study of volatile excretions of South Crimea. Rastit Resur N 1:81–85

Dobson HEM, Bergström J, Bergström G, Groth I (1987) Pollen and flower volatiles in two *Rosa* species. Phytochemistry 26: 3171–3173

Doireau P (1972) Exhalation d'alcool et d'autres substances volatiles en rapport avec le catabolisme fermentaire, au cours de la germination du haricot. C R Acad Sci D 274: 383–386

Donaldson RP, Soochan P, Zares A (1985) Anaerobic stress in germinating castor bean, ethanol metabolism and effects on subcellular organelles. Plant Physiol 77: 978–983

Downes RW, Tonnet ML (1985) Effect of environmental conditions on growth and rubber production of guayule (*Parthenium argentatum*). Aust J Agric Res 36: 285–294

Drabkin BS (1954) About the mechanism of action of the bird cherry phytoncides. Biokhimiya 19: 385–390

Drew MC, He C-J Morgan PW (1989) Decreased ethylene biosynthesis, and induction of aerenchyma by nitrogen- or phosphate starvation in adventitious roots of *Zea mays*. Plant Physiol 91: 266–271

Drobot'ko VG, Aizenman BE, Shvaiger MO (1958) Antimicrobial substances of higher plants. Kiev, AN Ukr. SSR

Dudley MW, Duaber MT, West CA (1986) Biosynthesis of the macrocyclic diterpene casbene in castor bean (*Ricinus communis* L.) seedlings. Changes in enzyme levels induced by fungal infection and intracellular localization of the pathway. Plant Physiol 81: 335–342

Duffey SS (1986) Plant glandular trichomes: their partial role in defence against insects. In: Juniper B, Southwood R (eds) Insects and the plant surface. Arnold, London, pp 151–172

Duffus CM, Duffus JH (1984) Carbohydrate metabolism in plants. Longman, London

Dumas C, Bowman RB, Gaude T, Gully CM, Heizmann Ph, Roeckel P, Rougier M (1988) Stigma and stigmatic secretion reexamined. Phyton 28: 193–200

Dun NJ, Kiraly M (1983) Capsaicin causes release of a substance P-like peptide in guinea-pig inferior mesenteric ganglia. J Physiol 340: 107–120

Duncan MW, Markey SP, Levy M, Kopin IJ, Norstog K (1991) 2-Amino-3-(methylamino)-propanoic acid (BMAA) cycad leaves: quantification in Australian and Mexican cycads responsible for causing hind limb paralysis in cattle. Neurobiology of Aging 12: 921–925

Duncan MW, Markey SP, Weick Bg, Pearson PG, Ziffer H, Hu YL, Kopin IJ (1992) 2-Amino-3-(methylamino)propanoic acid (BMAA) bioavailability in the primate. Neurobiology of Aging 13: 333–338

Durmishidze SV (1977) Metabolism of some organic compounds polluted atmosphere by plants. Prikl Biokhim Mikrobiol 13: 838–846

Ecker JR, Davis RW (1987) Plant defence genes are regulated by ethylene. Proc Natl Acad Sci USA 84: 5202–5206

Egorov IA, Egofarova RKh (1971) The study of essential oils of blossom pollen of grape. Dokl AN SSSR 199: 1439–1442

Eijk JL (1957) Fytochemisch ondersoak von *Erodium cicutarium*. II. Pharm Weekbl 92: 581–587

Eilert U, Nesbitt LR, Constabel F (1985) Laticifers and latex in fruits of periwinkle, *Catharanthus roseus*. Can J Bot 63: 1540–1546

Eilert, U, Wolters B, Constabel F (1986) Ultrastructure of acridone alkaloid idioblasts in roots and cell cultures of *Ruta graviolens*. Can J Bot 64: 1089–1096

Eisenberg RC (1990) Channels as enzymes. J Membr Biol 115: 1–12

Elgala AM, Amberger A (1986) Root exudate and the ability of corn to utilize insoluble sources of iron. Trans 13th Congr Int Soc Soil Sci, Hamburg, vol 2. pp 279–280

Elliger CA, Zinkel DF, Chan BG, Waiss AC Jr (1976) Diterpene acids as larval growth inhibitors. Experientia 32: 1364–1366

Elstner EF, Konze JP (1976) Effect of pount freezing on ethylene and ethane production by sugar beet leaf discs. Nature 263: 351–352

Emmelin N, Feldberg W (1947) The mechanism of the sting of the common nettle (*Urtica urens*). J Physiol 106: 440–455

Emmelin N, Feldberg W (1949) Distribution of acetylcholine and histamine in nettle plants. New Phytol 48: 143–148

Ermakov AI (1960) Formation and quantitative change of cyanogenous glycosides at the sprouting and ripening of flax. Fiziol Rast (USSR) 7:447–452

Esashi Y, Ooshima Y, Abe M, Kurota A, Satoh S (1986) CO_2-enhanced C_2H_4 production in tissues of imbibed cocklebur seeds. Aust J Plant Physiol 13:417–429

Esau K (1965) Plant anatomy. Wiley, New York

Esau K (1977) Anatomy of seed plants, vol 1–2. Wiley, New York

Eschbach JM, Tupy J, Lacrotte R (1986) Photosynthate allocation and productivity of latex vessels in *Hevea brasiliensis*. Biol Plant 28:321–328

Evans LS, Curry TM, Lewin KT (1981) Responses of leaves of *Phaseolus vulgaris* L. to stimulated acidic rain. New Phytol 88:403–420

Evans NT, Ebert M (1960) Radioactive oxygen in the study of gas transport down the root of *Vicia faba*. J Exp Bot 11:246–257

Evans RC, Tingey DT, Gumperlz ML, Burns WF (1982) Estimates of isoprene and monoterpene emission rates in plants. Bot Gaz 143:304–310

Everaerts C, Bonnard O, Pasteels JM, Roisin Y, König WA (1990) (+)-α-Pinene in the defensive secretion of Nasutiterpenes princeps (Isoptera, Termitidae). Experientia 46:227–230

Fagerstedt KV, Crawford RMM (1986) Changing kinetic properties of barley alcohol dehydrogenase during hypoxic conditions. J Exp Bot 37:857–864

Fahn A (1979) Secretory tissues in plants. Academic Press, London

Fahn A (1987) Secretory tissues and factors influencing their development. Abstr 14th Int Bot Congr, Berlin-Dahlem Botanical Garden, West Berlin. p 25

Fahn A (1988a) Secretory tissues in vascular plants. New Phytol 108:229–257

Fahn A (1988b) Secretory tissues and factors influencing their development. Phyton 28:13–26

Faradey CD, Thomson WW (1986) Functional aspects of the salt glands of the Plumbaginaceae. J Exp Bot 37:1129–1135

Farquhar GD, Mayak S (1984) Physiological responses of cut rose flowers to exposure to low temperature: Changes in membrane permeability and ethylene production. J Exp Bot 35:965–974

Farquhar GD, Wetselaar R, Firth PM (1979) Ammonia volatilization from senescing leaves of maize. Science 203:1257–1258

Feucht W, Treutter D (1990) Flavan-3-ols in trichomes, pistils and phelloderm of some tree species. Ann Bot 65:225–230

Fialkova EYu, Gudkov IN (1989) Radio-modified action of phytohormones – plant growth inhibitors. Abstr 1st All-Union Radiobiol Congr Moscow 21–27 August, vol 2. Department of Scientific and Technical Information, Pushchino (USSR), pp 314–315

Fineran BA (1971) Ultrastructure of vacuolar inclusions in root tips. Protoplasma 72:1–18

Fineran BA (1985) Glandular trichomes in Urticularia: a review of their structure and function. Isr J Bot 34:295–330

Fischer NH, Weidenhamer JD, Bradow JM (1989a) Inhibition and promotion of germination by several sesquiterpenes. J Chem Ecol 15:1785–1793

Fischer NH, Weidenhamer JD, Bradow JM (1989b) Dihydroparthenolide and other sesquiterpene lactones stimulate witchweed germination. Phytochemistry 28:2315–2317

Flath RA, Ohinata K (1982) Volatile components of the orchid *Dendrobium superbum* Rchb F. J Agric Food Chem 30:841–849

Fletcher BS, Bateman MA, Hart NK, Lamberton JA (1975) Identification of a fruit fly attractant in the Australian plant *Zieria smithii*, as o-methyl eugenol. J Econ Entomol 68:815–816

Florin-Christen J, Florin-Christen M, Fiedtke A, Rasmussen L (1990) The role of secreted acid hydrolases in the utilization of complex nutrients by *Tetrohymena*. Microb Ecol 19:311–316

Floyd RA, Ohlrogge AJ (1971) Gel formation on nodal root surfaces of Zea mays. Some observations relevant to understanding of its action at the root-soil interface. Plant Soil 34:595–606

Fluck RA, Jaffe MJ (1976) The acetylcholine system in plants. In: Smith H (ed) Commentaries in plant science. Pergamon Press, Oxford, pp 119–136

Focke U (1985) Untersuchungen zur Aerenchymbildung in den Wurzeln bei 5 Gromineen-Arten in Abhängigkeit vom Standort. Wiss Wilhelm-Pieck-Univ Rostock Naturwiss Reihe 34:76–80

Forrest GI (1980a) Variation in monoterpene composition of the shoot cortical oleoresin within and between trees of *Pinus contorta*. Biochem Syst Ecol 8:337–341

Forrest GI (1980b) Geographical variation in the monoterpenes of Pinus contorta oleoresin. Biochem Syst Ecol 8:343–359

Fountain DW, Couchman K (1984) Volatiles from ripe fruits of karaka (*Corynocarpus laevigatus* J R et G Forst). NZ J Bot 22:341–343

Fowden L (1965) Origin of the amino acids. In: Bonner J, Varner JE (eds) Plant biochemistry. Academic Press, New York, pp 361–390

Franceschi VR (1987) Oxalic acid metabolism and calcium oxalate formation in *Lemna minor* L. Plant Cell Environ 10:397–406

Frank E, Jensen WA (1970) On the formation of pattern of crystal idioblast in *Canavalia ensiformis* DC. IV. The fine structure of the crystal cells. Planta 95:202–217

French JC (1987) Systematic survey of resin canals in roots of Araceae. Bot Gaz 148:360–371

Frey-Wyssling A (1935) Die Stoffausscheidung der höheren Pflanzen. Springer, Berlin Heidelberg New York

Frey-Wyssling A (1973) Comparative organellography of the cytoplasm. Springer, Berlin Heidelberg New York

Frey-Wyssling A, Mühlethaler K (1965) Ultrastructural plant cytology. With an introduction to molecular biology. Elsevier, Amsterdam

Friedman J, Waller GR (1985) Allelopathy and autotoxicity. Trends Biochem Sci 10:47–50

Frischknecht PM, Baumann TW (1985) Stress induced formation of purine alkaloids in plant tissue culture of *Coffea arabica*. Phytochemistry 24:2255–2257

Gaal I (1984) Hlekony metabolitok gazkromatograflas analizis dohany szövettenyeszetben. Bot Kozl 71:199–204

Gafron H, Rubin J (1942) Fermentative and photochemical production of hydrogen in algae. J Gen Physiol 26:219–240

Galatis B, Apostolakos P, Panteris E (1989) Microtubules and lithocyst morphogenesis in *Pilea cadierei*. Can J Bot 67:2788–2804

Galliard T (1978) Lipolytik and lipoxygenase enzymes in plants and their action on wounded tissues. In: Günter K (ed) Biochemistry of wounded plant tissues. De Gruyter, Berlin, pp 155–201

Gärtner R (1970) Die Bewegung des Mesotaenium – Chloroplasten im Starklichtbereich. 1. Zeit- und temperaturabhängige Sekundärprozesse. Z Pflanzenphysiol 63:147–161

Gedalovich E, Fahn A (1985) The development and ultrastructure of gum ducts in *Citrus* plants formed as result of brown-rot gummosis. Protoplasma 127:73–81

Gelmont D, Stein RA, Mead JF (1981) Isoprene – the main hydrocarbon in human breath. Biochem Biophys Res Commun 99:1456–1460

George JB, Marriott J, Palmer JM, Karikari SK (1992) Sensitivity of water stress and ethylene of stored plantain fruits. J Exp Bot 33:1194–1201

Georgiev M, Aleksieva K (1990) The effect of methanol on the growth and biosynthesis of citric acid by *Aspergillus niger*. Biotekhnol Biotekh (Bulgar) N 1:35–38

Gerbaud A (1990) Effect of acetylene on root respiration and acetylene reducing activity in nodulated soya bean. Plant Physiol 93:1226–1229

Gershenzon J, Duffy MA, Karp F, Croteau R (1987) Mechanized techniques for the selective extraction of enzymes from plant epidermal glands. Anal Biochem 163:159–164

Gershenzon J, Maffei M, Croteau R (1989) Biochemical and histochemical localization of monoterpene biosynthesis in the glandular trichomes of spearmint (*Mentha spicata*). Plant Physiol 89:1351–1357

Giannini JL, Holt JS, Briskin DP (1990) The effect of glyceollin on proton leakage in *Phytophthora megasperma* F. *sp glycinea* plasma membrane and red beet tonoplast vesicles. Plant Sci 68:39–45

Gibson RW, Pickett JA (1983) Wild repels aphids by release of an aphid alarm pheromone. Nature 302: 608–609

Gidrol X, Chrestin H, Mounoury G, D'Auzac J (1988) Early activation by ethylene of the tonoplast H^+-pumping ATPase in the latex from *Hevea brasiliensis*. Plant Physiol 86: 899–903

Giesemann A, Biehl B, Lieberei R (1986) Identification of scopoletin as a phytoalexin of the rubber tree *Hevea brasiliensis*. Phytopathol Z 117: 373–376

Glazer AN, Barel AO, Howard JB, Brown DM (1969) Isolation and characterization of fig lysozyme. J Biol Chem 244: 3583–3589

Gleizes M, Pauly G, Carde JP, Marpeau A, Bernard-Dagan C (1983) Monoterpene hydrocarbon biosynthesis by isolated leucoplasts of *Citrofortunella mitis*. Planta 159: 373–381

Glinka Z, Reinhold L (1962) Rapid changes in permeability of cell membranes to water brought about by carbon dioxide and oxygen. Plant Physiol 37: 481–486

Ghosh SS, Purkayastha SK (1962) Anatomical studies of wood and bark of *Acacia senegal* Willd trees with special reference to gum exudation. Indian Forester 88: 92–99

Goffreda JC, Mutschler MA, Ale DA, Tingey WM, Steffens JC (1989) Aphid deterrence by glucose esters in glandular trichome exudate of the wild tomato, *Lycopersicon pennellii*. J Chem Ecol 15: 2135–2147

Gogala N (1988) Growth substances in root exudate of *Pinus sylvestris* – their influence on mycorrhizal fungi. Abstr 6th Congr FESPP, Split Yugoslavia, 4–10 Sept. p 719

Goldstein AH, Danon A, Baertlein DA, McDaniel RG (1988) Phosphate starvation inducible metabolism in *Lycopersicon esculentum*. 111. Characterization of the phosphate starvation by inducible-excreted phosphatase. Plant Physiol 87: 716–720

Golik KN (1965) Gas exchange and transpiration of upper and lower sides of leaves in some species of putamen and fruit plants. In: Vlasyuk PA (ed) Photosynthesis and productivity of plants. Naukova Dumka, Kiev, pp 146–152

Golubinskii IN (1974) Biology of pollen sprouting. Naukova Dumka, Kiev

Gommers FJ, Bakker J (1988) Mode of action of α-terthienyl and related compounds may explain the suppressant effects of *Tagetes* species on populations of free-living endoparasitic plant nematodes. In: Lam J, Breteler H, Arnason T, Hansen L (eds) Chemistry and biology of naturally occurring acetylenes and related compounds (Noarc). Bioactive molecules, vol 7. Elsevier, Amsterdam, pp 61–69

Good BJ, Patrick WH Jr (1987) Gas composition and respiration of water oak (*Quercus nigra* L.) and green ash (*Fraxinus pennsylvanica* Marsh). Plant Soil 97: 419–427

Goodwin TW, Mercer EI (1983) Introduction to plant biochemistry, vol 1–2. Pergamon Press, Oxford

Gorbavtsov VA (1970) To a question about role of the surface events in gas-exchange of plant leaves with environment. Fiziol Rast (USSR) 17: 663–672

Gorham J (1987) Photosynthesis, transpiration and salt fluxes through leaves of *Leptochloa fusca* L. Kunth. Plant Cell Environ 10: 191–196

Gottsberger G, Arnold T, Lenskens HF (1989) Are amino acid and sugar concentration correlated in floral nectar? Proc Ned Akad Wet C 92: 461–464

Goudey JS, Tittle FL, Spencer MS (1989) A role for ethylene in the metabolism of cyanide by higher plants. Plant Physiol 89: 1306–1310

Govindappa T, Govardham L, Jyothy PS, Veerabhadrappa PS (1987) Purification and characterization of acetylcholinesterase isozymes from the latex of *Synadenium grantii* Hook, "f". Indian J Biochem Biophys 24: 209–217

Govindjee, van Rensen JJ (1978) Bicarbonate effects on the electron flow in isolated broken chloroplasts. Biochim Biophys Acta 505: 183–213

Gracza L (1985) Biochemisch-pharmakologische Untersuchung pflanzlicher Arzneistoffe. II. Hemmung der Acetylcholinesterase durch Monoterpenderivate in vitro. Z Naturforsch 40c: 151–153

Gray JC (1987) Control of isoprenoid biosynthesis in higher plants. In: Callow JA (ed) Advances in botanical research, vol 14. Academic Press, London, pp 25–91

Greca M della, Monaco P, Previtera L, Aliotta G, Pinto G, Pollio A (1989) Allelochemical activity of phenylpropanes from *Acorus gramineus*. Phytochemistry 28:2319–2321
Greenaway W, Whatley FR (1991) Analysis of phenolics of bud exudate of *Populus ciliata* by GC-MS. Phytochemistry 30:1887–1889
Greenaway W, Scaysbrook T, Whatley FRS (1987) The analysis of bud exudate of *Populus euramericana* and of propolis by gas chromatography – mass spectrometry. Proc R Soc Lond B 232:249–272
Greenaway W, Wollenweber E, Scaysbrook T, Whatley FR (1988a) Novel isoferulate esters identified by gas chromatography – spectrometry in bud exudate of *Populus nigra*. J Chromatogr 448:284–290
Greenaway W, Wollenweber E, Scaysbrook T, Whatley FR (1988b) Esters of caffeic acid with aliphatic alcohols in bud exudate of *Populus nigra*. Z Naturforsch 43c:795–798
Greenaway W, English S, Whatley FR (1990a) Phenolic composition of bud exudates of *Populus deltoides*. Z Naturforsch 45c:587–593
Greenaway W, English S, Whatley FR (1990b) Variation in bud exudate composition of *Populus nigra* assessed by gas chromatography-mass spectrometry. Z Naturforsch 45c:931–936
Greenaway W, Davidson CG, Scaysbrook T, May J, Whatley FR (1990c) Hybrid origin of *Populus* x *Jackii* confirmed by gas chromatography-mass spectrometry analysis of its bud exudate. Z Naturforsch 45c:594–598
Greenaway W, Gümüsdere I, Whatley FR (1991) Analysis of phenolic of bud exudates of *Populus euphratica* by GC-MS. Phytochemistry 30:1883–1885.
Gregory P, Ave DA, Bouthyette PY, Tingey WM (1986) Insect-defensive chemistry of potato glandular trichomes. In: Juniper B, Southwood R (eds) Insects and the plant surface. Arnold, London, pp 173–183
Griffing LR, Mersey BG, Fowke LC (1986) Cell-fractionation analysis of glucan synthase I and II distribution and polysaccharides secretion in soybean protoplasts. Evidence for the involvement of coated vesicles in wall biogenesis. Planta 167:175–182
Grineva G (1975) Regulation of metabolism in plants at oxygen deficit. Nauka, Moscow
Grisebach H, Zilg H (1968) Über die Bedeutung der Isoflavonene bei der Isoflavonobiosynthese. Z Naturforsch 23b:494–504
Grodzinski B, Woodrow L (1989) Ethylene and carbon dioxide exchange in leaves whole plants: In: Clijsters H, De Proft M, Marcell R, von Poucke M (eds) Biochemical and physiological aspects of ethylene production in lower and higher plants. Kluwer Academic, Dordrecht, pp 271:278
Grodzinskii AM (1965) Allelopathy in plants life and their communities. Naukova Dumka, Kiev
Grodzinskii AM (1981) About new conception of allelopathy. In: Grodzinskii AM (ed) Chemical interactions of plants. Naukova Dumka, Kiev, pp 3–18
Grodzinskii AM (1991) Allelopathy of plants and soil exhaustion. Naukova Dumka, Kiev
Grodzinskii DM (1985) Biological recognizing in plant physiological processes. Fiziol Biokhim Kul't Rast 17:211–219
Groeneveld HW, Furr M, Mahlberg PG (1987) Sites of in-vitro triterpene synthesis in *Euphorbia latex*. Acta Bot Neerl 36:71–80
Grosse W (1982) Function of serotonin in seeds of walnuts. Phytochemistry 21:819–822
Grosse W, Mevi-Schütz J (1987) A beneficial gas transport system in *Nymphoides peltata*. Am J Bot 74:947–952
Grosse W, Schröder P (1984) Oxygen supply of roots by gas transport in alder trees. Z Naturforsch 39c:1186–1188
Grosse W, Schröder P (1986) Pflanzenleben unter anaeroben Umweltbedingungen, die physikalischen Grundlagen und anatomischen Voraussetzungen. Ber Dtsch Bot Ges 99:367–381
Grosse W, Sica S, Lattermann S (1990) Oxygen supply of roots by thermo-osmotic gas transport in *Alnus glutinosa* and other wetland trees. In: Werner D, Müller P (eds) Fast-growing trees and nitrogen-fixing trees. Int Conf, Marburg, 8–12 October 1989. Fischer, Stuttgart, pp 246–249
Grudzinskaya A (1980) Family Urticaceae. In: Takhtadzhyan AL (ed) Plant life, vol 5, part 1. Prosveshchenie, Moscow, pp 282–289
Grümmer G (1955) Die gegenseitige Beeinflussung höherer Pflanzen – Allelopathie. Fischer, Jena

Gubareva VA (1962) Interconnection of water rotation with its chemical content, root, leaf excrets and oak growth. Soobsh Labor Lesov AN SSSR N 7:48–52
Gubler F, Jacobsen JV, Ashford AE (1986) Involvement of the Golgi apparatus in the secretion of α-amylase from gibberellin-treated barley aleurone cells. Planta 168:447–452
Guinel FC, McCully ME (1986) Some water-related physical properties of maize root-cap mucilage. Plant Cell Environ 9:657–666
Gunderson CA, Tayler GE Jr (1988) Kinetics of inhibition of foliar gas exchange by exogenous ethylene: an ultrasensitive response. New Phytol 110:517–524
Gunningham RT, Chambers DL, Forbes AG (1975) Oriental fruit fly: thickened formulations of methyl eugenol in spot applications for male annihilation. J Ecol Entomol 68:861–863
Gur'ev AN (1981) Insecticidic characteristics of some aquatic and moisture-loving plants. In: Aisenman BE (ed) Phytoncides: role in biogenocenosis, significance for medicine. Naukova Dumka, Kiev, pp 310–312
Gurielidze KT, Vasil'eva IS, Paseshnichenko VA (1988) The determination of furastanol steroid glucosides and their finding in plant tissues using Erlich reagent. In: Kretovich VL, Scholtz KF (eds) New methods of practical biochemistry. Nauka, Moscow, pp 146–150
Guy M, Kende H (1984) Ethylene formation in *Pisum sativum* and *Vicia faba* protoplasts. Planta 160:276–280
Guzev VS, Golishin PN, Ivanov PI, Zvyagintzev DG (1990) Role of plant excretions in their defence against microbial damages. Izv AN SSSR Ser Biol N1:91–96
Haard NF (1983) Stress metabolites. In: Lieberman M (ed) Post-harvest physiology and crop preservation. Plenum Press, New York, pp 299–314
Hahn MG, Bucheli P, Cervone F, Doares SH, O'Neill RA, Darvill A, Albersheim P (1989) Roles of cell wall constituents in plant-pathogen interactions. In: Kosuge T, Nester EW (eds) Plant-microbe interactions, vol 3. McGraw Hill, New York, pp 131–181
Hahn MG, Darvill A, Albersheim P, Bergman C, Cheong JJ, Keller A, Veng-Meng Lo (1992) Preparation and characterization of oligosaccharide elicitors of phytoalexin accumulation. In: Bowles DJ, Gurr S, McPherson M (eds) Molecular plant pathology: a practical approach, vol 2. pp 1–44. IRL Press, Oxford Washington
Hall DH (1967) The ultrastructure of wax deposits of plant leaf surfaces. II. Cuticular pores and wax formation. J Ultrastruct Res 17:34–44
Haller Th, Stolp H (1985) Qunatitative estimation of root exudation of maize plants. Plant Soil 86:207–216
Hamilton-Kemp TR, Andersen RA (1984) Volatile compounds from *Triticum aestivum*. Phytochemistry 23:1176–1177
Hamilton-Kemp TR, Rodriguez JG, Archbold DD, Andersen RA, Loughrin JH, Patterson CG, Lowry SR (1989) Strawberry resistance to *Tetranychus urticae* Koch: effects of flower fruit and foliage removal – comparison of airvs nitrogen-entrained volatile compounds. J Chem Ecol 15:1465–1473
Hanlidou E, Kokkini S, Bosabalidis AM, Bessiere JM (1991) Glandular trichomes and essential oil constituents of *Calamintha menthifolia* (Lamiaceae). Plant Syst Evol 177:17–26
Harborne JB (1982) Introduction to ecological biochemistry, 2nd edn. Academic Press, London
Harborne JB (1988) Introduction to ecological biochemistry, 3rd edn. Academic Press, London
Harper JE (1981) Evolution of nitrogen oxide(s) during in vivo nitrate reductase assay of soybean leaves. Plant Physiol 68:1488–1493
Harr J, Guggenheim R, Boller T (1984) High pH-values and secretion of ions on leaf surfaces: a characteristic of the phylloplane of Malvaceae. Experientia 40:935–937
Harris PJ, Hartley RD (1976) Detection of bound ferulic acid in cell walls of the Gramineae by ultraviolet fluorescence microscopy. Nature 259:508–510
Hartley RD, Buchan H (1979) High-performance liquid chromatography of phenolic acids and aldehydes derived from plants or from the decomposition of organic matter in soil. J Chromatogr 180:139–143
Hartmann E, Gupta R (1989) Acetylcholine as a signaling system in plants. In: Boss WF, Morve DI (eds) Second messengers in plant growth and development. Allan Liss, New York, pp 257–287

Hartmann T (1985) Prinzipien des pflanzlichen Sekundärstoffwechsels. Plant Syst Evol 150:15–34

Haslam E (1985) Metabolites and metabolism: a commentary on secondary metabolism. Clarendon Press, Oxford

Hatanaka A, Sekiya J, Kajiwara T (1978a) Distribution of an enzyme system producing cis-3-hexenal and n-hexanal from linolenic and linoleic acid in some plants. Phytochemistry 17:869–872

Hatanaka A, Kajiwara T, Sfkiya J, Koda F (1978b) Specificity of enzyme system producing C_6-aldehyde in *Thea* and *Farfugum chloroplasts.* Phytochemistry 17:548–549

Hatanaka A, Kajiwara T, Matsui K, Matsunaga T (1989) Non-enzymatic isomerization of 12-hydroxy-(3)-dodecenal to the (2E)-isomer after enzymatic cleavage of 13-hydroperoxylinoleyl alcohol in tea chloroplasts. Z Naturforsch 44c:161–164

Hauck C, Schildknecht H (1990) Separation of enantiomers of the germination stimulant strigol on cellulose triacetate and determination of their biological activity. J Plant Physiol 136: 126–128

Haupt I (1976) Separation of the sites of synthesis and accumulation of 3,4-dihydroxyphenylalanine in *Euphorbia lathyris* L. In: Luckner M, Mothes K, Nover D (eds) Secondary metabolism and coevolution. Nova Acta Leopoldina Suppl, N 7. Dtsch Akad Naturforsch Leopoldina, Hale, pp 129–132

Hauser MT, Wink M (1990) Uptake of alkaloids by latex vesicles and isolated mesophyll vacuoles of *Chelidonium majus* (Papaveraceae). Z Naturforsch 45c:949–957

Havir EA, McHale NA (1989) Regulation of catalase activity in leaves of *Nicotiana sylvestris* by high CO_2. Plant Physiol 89:952–957

Hayashi N, Komae H (1980) Chemistry and distribution of sesquiterpene furans in Lauraceae. Biochem Syst Ecol 8:381–383

Hayden WJ, Clough J (1990) Methyl salicylate secretory cells in roots of *Viola arvensis* and *V. rafinesquii* (Violaceae). Castanea 55:65–70

Hegnauer R (1990) Chemotaxonomie der Pflanzen Bd 9. Birkhäuser, Basel

Heilbronner E (1959) Asulens. In: Ginsburg D (ed) Non-benzenoid aromatic compounds. Intersci, New York, pp 171–276

Heinrich G (1973) Die Feinstruktur der Trichom-Hydathoden von Monarda fistulosa. Protoplasma 77:271–278

Heinrich G (1989) Analysis of cations in nectars by means of a laser microprobe mass analyser (LAMMA). Beitr Biol Pflanz 64:293–308

Heinrich G, Schultze W (1985) Composition and site of biosynthesis of the essential oil in fruits of *Phellodendron amurense* Rupr. (Rutaceae). Isr J Bot 34:205–217

Heinrich G, Schultze W, Wegner R (1980) Zur Kompartmentierung der Synthese von Mono- und Sesquiterpenen des ätherichen Öls bei *Poncirus trifoliata.* Protoplasma 103:115–129

Heinrich G, Schultze W, Schröder W (1986) LAMMA-Ionenspectren der Milchsäfte höherer Pflanzen. Biochem Physiol Pflanz 181:227–236

Heisey RM, Delwiche CC (1984) Phytotoxin volatiles from *Trichostema lanceolatum* (Labiatae). Am J Bot 71:821–828

Hellebust JA (1974) Extracellular products. Botanical Monograf, vol 10. Algal physiology and biochemistry. pp 838–863

Helsper J, Ketel DH, Hulst AC, Breteler H (1988) Secretion of thiophenes by differentiated cell cultures of *Tagetes* species. In: Pais MSS, Mavituna F, Novais JM (eds) Plant cell biotechnology. NATO ASI Ser, vol H18. Springer, Berlin Heidelberg New York, pp 421–424

Henkel PA (1976) The observation of guttation of lady's mantle *Alchemilla vulgaris* L. in natural conditions. Fiziol Rast (USSR) 23:325–330

Herrera J (1985) Nectar secretion patterns in southern Spanish mediterranean scrublands. Isr J Bot 34:47–58

Heslop-Harrison J, Heslop-Harrison Y (1985) The secretory system of the stigma of *Oenothera organensis* Munz: some developmental and quantitative features. Isr J Bot 34:187–204

Hewitt CN, Monson RK, Fall R (1990) Isoprene emission from the grass *Arundo donax* L. are not linked to photorespiration. Plant Sci 66:139–144

Hiltz P (1951) Contribution àl'étude des lithocystes et des cystolithes de *Ficus elastica.* Dipl. Et Sup Paris, librairie gen del l'Enseignement, p 27
Hodson MJ (1986) Silicon deposition in the roots, culm and leaf of *Phalaris canariensis* L. Ann Bot 58:167–177
Hogan ME, Swift IE, Done J (1983) Urease assay and ammonia release from leaf tissues. Phytochemistry 22:663–667
Holloway PJ (1982a) Structure and histochemistry of plant cuticular membranes: an overview. In: Cutler DF, Alvin KL, Price CE (eds) The plant cuticule. Linnean Soc Symp Ser N10. Academic Press, London, pp 1–32
Holloway PJ (1982b) The chemical composition of plant cutins. In: Cutler DF, Alvin KL, Price CE (eds) The plant cuticule. Linnean Soc Symp Ser N 10. Academic Press, London, pp 45–85
Hong YCA Haung LC, Reineccius GA, Harlander SK, Labuza TP (1990) Production of aroma compounds from strawberry cell suspension cultures by addition of precursors. Plant Cell Tissue Organ Cult 21:245–251
Hook DD, Brown CL (1972) Permeability of the cambium to air in trees adapted to wet habitats. Bot Gaz 133:304–310
Hook DD, Brown CL, Wetmore RH (1972) Aeration in trees. Bot Gaz 133:443–454
Horiguchi T (1987) Protein metabolism in rice (Oryza sativa) seedlings. IV. Peptides in the guttation liquid. Mem Fac Agric Kagoshima Univ 23:63–69
Horlacher R, Poskuta J (1986) Der Einfluß der Wurzelanaerobiose auf Äthanol-Lactat- und Glukosegehalte sowie auf die Aktivitäten von ADH und LDH in Weizenkeimpflanzen. Biol Plant 28:130–136
Horst WJ, Wagner A, Marscher H (1982) Mucilage protects root meristems from aluminium injury. Z Pflanzenphysiol 105:435–444
Horton RF (1985) Carbon dioxide flux and ethylene production in leaves. In: Roberts JA, Tucker GA (eds) Ethylene and plant development. Butterworths, London, pp 37–46
Horvat BJ, Chapman GW (1990) Comparison of volatile compounds from peach fruit and leaves (cv. Monroo) during maturation. J Agric Food Chem 38:1442–1444
Hosokava D, Ohashi Y (1988) Immunochemical localization of pathogenesis-related proteins secreted into the intercellular spaces of salicylate-treated tobacco leaves. Plant Cell Physiol 29:1035–1040
Houslay MD, Stanley KK (1982) Dynamics of biological membranes. Wiley, Chichester
Howard JB, Glazer AN (1969) *Papaya* lysozyme. Terminal sequences and enzymic properties. J Biol Chem 244:1399–1409
Hradetzky D, Wollenweber E (1987) Flavonoids from the leaf resin of snakeweed, *Gutierrezia sarothrae.* Z Naturforsch 42c:71–76
Hrazdina G, Wagner GJ (1985) Compartmentation of plant phenolic compounds: Sites of synthesis and accumulation. In: van Sumere CF, Lea PJ (eds) Ann Proc Phytochem Soc Europe, vol 25. Clarendon Press, Oxford, pp 119–133
Hudson JB, Towers GHN (1988) Antiviral properties of acetylenes and thiophenes. In: Lam J, Breteler H, Arnason T, Hansen L (eds) Chemistry and biology of naturally occurring acetylenes and thiophenes. Elsevier, Amsterdam, pp 315–339
Huesing J, Jones D, Deverna J, Myer J, Collins G, Severson R, Sisson V (1989) Biochemical investigations of antibiosis material in leaf exudate of wild *Nicotiana* species and interspecific hybrids. J Chem Ecol 15:1203–1217
Hutchinson TC, Adams CM (1987) Comparative abilities of leaf surfaces to neutralize acid rain drops. 1. The influence of calcium nutrition and charcoal-filtered air. New Phytol 106:169–183
Hystee RB, Turcon G (1973) Rapid release of peroxidase by peanut cells in suspension culture. Can J Bot 51:1169–1175
Iinuma M, Hamada K, Mizuna M (1986) Complex flavonoids from *Pityrogramma* frond exudates: synthesis of two flavones with C–C-linked dihydrocinamoyl substituents. Z Naturforsch 41c:681–683
Imaseki H (1970) Induction of peroxidase activity by ethylene in sweet potato. Plant Physiol 46:172–174

Inamdar JA, Subramanian RB, Mohan JSS (1986) Studies on the resin glands of *Azadirachta indica* A. Juss (Meliaceae). Ann Bot 58:425–429

Ishikawa M, Kaneko A, Kashiwa T, Watanabe H (1987) Participation of β-myrcene in the susceptibility and/or resistance of pine trees to the pine wood nematode, *Bursaphelenchus xylophilus*. Agric Biol Chem 51:3187–3197

Ish-Shalom-Gordon N, Dubinsky Z (1990) Possible modes of salt secretion in *Avicennia marina* in the Sinai. Plant Cell Physiol 31:27–32

Isidorov VA, Zenkevich JBB (1985) Volatile organic compounds in the atmosphere of forests. Atmos Environ 19:1–8

Ivanchenko VA, Grodzinskii AM, Cherevchenko TM, Lebeda AF, Makarchuk NM, Snezhko VV (1989) Phytoergonomics. Naukova Dumka, Kiev

Ivanov VP (1973) Plant excretions and their significance in the life of phytocenosis. Nauka, Moscow

Iwanami Y (1981) Inhibiting effects of volatile constituents of plants on pollen growth. Experientia 37:1280–1281

Iwanami Y, Hyodo H (1983) Effects of volatiles evolved from excised plant tissues and lemon oil on the growth of mung bean seedlings. Plant Cell Physiol 24:473–477

Iyengar S, Arnason JT, Philogene BJR, Morand P, Westiuk NH, Timmins G (1987) Toxicokinetics of the phytotoxic allelochemical α-terthienyl in three herbivorous Lepidoptera. Pestic Biochem Physiol 29:1–9

Izawa S, Kraayenhof RK, Ruuge EK, Devault D (1973) The site of KCN inhibition in the photosynthetic electron transport pathway. Biochim Biophys Acta 314:328–339

Jackson MB (1989) Regulation of aerenchyma formation in roots and shoots by oxygen and ethylene. In: Osborne DJ, Jackson MB (eds) Cell separation in plants. NATO Ser H: Cell Biol, vol 35. Springer, Berlin Heidelberg New York, pp 263–274

Jackson MB, Campbell DJ (1975) Movement of ethylene from roots to shoots, a factor in the responses of tomato plants to waterlogged soil conditions. New Phytol 74:397–406

Jackson MB, Osborn DJ (1970) Ethylene, the natural regulator of leaf abscission. Nature 225:1019–1022

Jacob DJ, Wofsy SC (1988) Photochemistry of biogenic emissions over the Amazon forest. J Geophys Res 93:1477–1486

Jacobsen JV, Yamaguchi M, Howard FD, Bernard RA (1986) Product inhibition of the cysteine sulfoxide lyase of *Tulbaghia violacea*. Arch Biochem Biophys 127:252–258

Jauregui MA, Reisenauer HM (1982) Dissolution of oxides of manganese and iron by root exudate components. Soil Sci Am J 46:314–317

Jeffree CE (1986) The cuticle, epicuticular waxes and trichomes of plants, with reference to their structure, functions and evolution. In: Juniper B, Southwood R (eds) Insects and the plant surface. Arnold, London, pp 23–64

Jeffree CE, Dale JE, Fry SC (1986) The genesis on intercellular spaces in developing leaves of *Phaseolus vulgaris* L. Protoplasma 132:90–98

Jensen CR, Letey J, Stolzy LW (1964) Labelled oxygen: transport through growing corn roots. Science 144:550–552

Jensen EF (1965) Ethylene and polyacetylenes. In: Bonner J, Varner JE (eds) Plant biochemistry. Academic Press, New York, pp 641–664

Jensen TE, Welbourne F (1962) The cytological effects of growth inhibitors of excised roots of *Vicia faba* and *Pisum sativum*. Proc S Dak Acad Sci 41:131–136

Jente R, Richter E, Bosold F, Olatunji CA (1988) Experiments on biosynthesis and metabolism of acetylenes and thiophenes. In: Lam J, Breteler H, Arnason T, Hansen L (eds) Chemistry and biology of naturally occurring acetylenes and related compounds (NOARC). Elsevier, Amsterdam, pp 187–199

Joel DM (1986) Glandular structures in carnivorous plants: Their role in mutual and unilateral exploitation of insects. In: Juniper B, Southwood R (eds) Insects and the plant surface. Arnold, London, pp 219–234

Joel DM, Juniper BE (1982) Cuticular gaps in carnivorous glands. In: Cutler DF, Alvin KL, Price CE (eds) The plant cuticle. Linnean Soc Symp Ser N10. Academic Press, London, pp 121–130

Joel DM, Heide-Jorgensen HS (1985) Ultrastructure and development of the pitcher epithelium of *Sarracenia*. Isr J Bot 34:331–349.

Jones RL (1981) Cyanide and coevolution. In: Vennesland B, Conn EE, Knowles CJ, Westly J, Wissing F (eds) Cynanide in biology. Academic Press, London, New York, pp 509–516

Johnen BG, Sauerbeck DR (1977) A tracer technique for measuring growth, mass and breakdown of plant roots during vegetation. Soil organisms as components of ecosystem. In: Lohm V, Persson T (eds) Proc IV Int Soil Zoo Coll Ecol Bull, Stockholm, vol 25. p 366

Jones RL (1985) Protein synthesis and secretion by the barley aleurone: a perspective. Isr J Bot 34:377–395

Jones CA, Rasmussen RA (1975) Production of isoprene by leaf tissue. Plant physiol 55:982–987

Jones RL, Robinson DG (1989) Protein secretion in plants. New Phytol 111:567–597

Jones RS, Stutte CA (1986) Acetylcholine and red-light influence on ethylene evolution from soybean leaf tissues. Ann Bot 57:897–900

Jorgensen LB, Behnke HD, Mabry TJ (1977) Protein-accumulating cells and dilated cisternae of the endoplasmic reticulum in three glucisinolate-containing genera: Armoraceae, Capparis, Drypetes. Planta 137:215–224

Joseleau JP, Ullmann G (1990) Biochemical evidence for the site of formation of gum arabic in *Acacia senegal*. Phytochemistry 29:3401–3405

Junge CE (1963) Air chemistry and radioactivity. Academic Press, New York

Juniper BE (1986) The path to plant carnivory. In: Juniper BE, Southwood R (eds) Insects and plant surface. Arnold, London, pp 195–218

Juniper BE, Robins RE, Joel DM (1989) Carnivorous plants. Academic Press, London, New York

Justin SHFW, Armstrong W (1987) The anatomical characteristics of roots and plant response to soil flooding. New Phytol 106:465–495

Justin SHFW, Armstrong W (1991) Evidence for the involvement of ethene in aerenchyma formation in adventitious roots of rice (*Oryza sativa* L.). New Phytol 118:49–62

Jutner F (1988) Changes of monoterpene concentration in needles of pollution-injured *Picea abies* exhibiting montane yellowing. Physiol Plant 72:48–56

Kakes P (1990) Properties and functions of the cyanogenic system in higher plants. Euphytica 48:25–43

Kallarackal J, Garlick PR, Milburn JA (1986) Characterization of the structural inclusions in the latex of banana (*Musa* sp.). Can J Bot 64:2591–2601

Kameoka H, Iida H, Hashimoto S, Miyazawa M (1984) Sulphides and furanones from steam volatile oils of *Allium fistulosum* and *Allium chinense*. Phytochemistry 23:155–158

Kanchan SD, Javachandra (1980) Allelopathic effects of Parthenium hysterophorum L. II. Leaching of inhibitors from aerial vegetative parts. Plant Soil 55:61–66

Kandasamy MK (1986) The stigmatic exudate of *Cyamopsis tetragonoloba* Taub. Biochem Physiol Pflanz 181:571–575

Kandra L, Wagner GL (1988) Studies of the site and mode of biosynthesis of tobacco trichome exudate components. Arch Biochem Biophys 265:425–432

Karp F, Harris JL, Croteau R (1987) Metabolism of monoterpenes: Demonstration of the hydroxylation of (+)-sabinene to (+)-cis-sabinol by an enzyme preparation from sage (*Salvia officinalis*) leaves. Arch Biochem Biophys 256:179–193

Kartashova NN (1965) The structure and function of the flower nectaries of dicotyledons. Tomskii University, Tomsk (USSR)

Kartmazova LS, Lyapunova PN (1976) Anatomical structure of generative organs of herbaceous periwinkle and localization of alkaloids in their tissues. Rastit Resur 12:59–67

Kasperska A, Kubacka-Zebalska M (1989) Stress ethylene metabolism as related to degree of tissue injury. In: Clijsters H, De Proft M, Marcelle R, van Poucke M (eds) Biochemical and physiological aspects of ethylene production in lower and higher plants. Kluwer, Dordrecht, pp 211–218

Kaverzneva LN (1986) Role of exometabolites of pine roots in biogeocenosis. In: Common problems of biogenocenology. Abstr 2nd All-Union Meet, 11–13 November 1986, vol 1. Nauka, Moscow, pp 242–243

Kawano Y, Mitchell WC, Matsumoto H (1968) Identification of the male oriental fruit fly attractant in the golden shower blossom. J Econ Entomol 61:986–988

Kawasaki S (1991) Extensin secreted into the culture medium of tobacco cells. 11. Extensin subfamilies of similar composition. Plant Cell Physiol 32:721–728

Kawase M (1981) Effect of ethylene on aerenchyma development. Am J Bot 68:651–658

Kay Y, Pallas JE (1980) Inhibition of photosynthesis by ethylene. Nature 285:51–52

Kazarov AA, Plieva LS (1989) Evolution of ions into atmosphere by plants of some zones of the Northern Caucasus. Fiziol Rast 36:948–954

Keen NT, Lyne RL, Hymovitz T (1986) Phytoalexin production as a chemosystematic parameter within the genus *Glycine*. Biochem Syst Ecol 14:481–486

Kefeli VI, Turetskaya RKh (1964) About mechanism of action of natural plant growth inhibitors. Usp Sovrem Biol 57:99–114

Kelsey RG, Shafizadeh F (1980) Glandular trichomes and sesquiterpene lactones of *Artemisia nova* (Asteraceae). Biochem Syst Ecol 8:371–377

Kelsey RG, Reynolds CW, Rodriguez E (1984) The chemistry of biologically active constituents secreted and stored in plant glandular trichomes. In: Rodriguez E, Healey PL, Mehtha I (eds) Biology and chemistry of plant trichomes. Plenum Press, New York, pp 187–241

Kemp MS, Burden RS (1986) Phytoalexins and stress metabolites in the sapwood of trees. Phytochemistry 25:1261–1269

Kende H (1989) Enzymes of ethylene biosynthesis. Plant Physiol 91:1–14

Kennard EH (1938) Kinetic theory of gases. McGraw-Hill, New York

Kenyon WH, Black CC (1986) Electrophoretic analysis of protoplast, vacuole, and tonoplast vesicle proteins in Crassulacean acid metabolism plants. Plant Physiol 82:916–924

Kevan PG, Eisikowitch D, Rathwell B (1989) The role of nectar in the germination of pollen in *Asclepias syriaca* L. Bot Gaz 150:266–270

Khailov KM (1971) Ecological metabolism in the sea. Naukova Dumka, Kiev

Khan AA, Akbar M, Seshu DV (1987) Ethylene as an indicator of salt tolerance in rice. Crop Sci 27:1242–1247

Khavari-Nejad RA (1986) Carbon dioxide enrichment preconditioning effects on chlorophyll contents and photosynthetic efficiency in tomato plants. Photosynthetica 20:315–317

Kijne JW, Diaz CL, Lugtenberg BJJ (1990) Role of lectin in the pea-*Rhizobium* symbiosis. In: Lugtenberg BJJ (ed) Signal molecules in plants and plant microbe interactions. Springer, Berlin Heidelberg New York, pp 351–358

Kimmerer TW, Kozlowski TT (1982) Ethylene, ethane, acetaldehyde and ethanol production by plants under stress. Plant Physiol 69:840–847

Kimmerer TW, MacDonald RC (1987) Acetaldehyde and ethanol biosynthesis in leaves of plants. Plant Physiol 84:1204–1209

Kimmerer TW, Stringer MA (1988) Alcohol dehydrogenase and ethanol in the stems of trees. Evidence for anaerobic metabolism in the vascular cambium. Plant Physiol 87:693–697

King RR, Pelletier Y, Singh RP, Calhoun LA (1986) 3,4-Di-o-isobutyl-6-o-capryl-sucrose: the major component of a novel sucrose ester complex from the type-B-glandular trichomes of *Solanum berthaultii* Hawkes (PL 473340). J Chem Soc Chem Commun 14:1078 –1079

King RR, Singh RP, Boucher A (1987) Variation in surcrose esters from the type B glandular trichomes of certain wild potato species. Am Potato J 64:529–534

Kirchoff BK, Kennedy H (1985) Foliar non-structural nectaries in the Marantaceae. Can J Bot 63:1785–1788

Kirillova TV (1964) The release of water-soluble compounds of plant tissues. Usp Sovrem Biol 57:463–476

Kisser JG (1958) Die Ausscheidung von ätherischen Ölen und Harzen. Handb Pflanzenphysiol 10:91–131

Klein RR, Miller DA (1980) Allelopathy and its role in agriculture. Commun Soil Sci Plant Anal 11:43–56

Klepper LA (1979) Nitric oxide (NO) and nitrogen dioxide (NO_2) emissions from herbicide-treated soybean plants. Atmos Environ 13:537–542

Klinger H, Frosch S, Wagner E (1991) In vitro effects of monoterpenes on chloroplast membranes. Photosynth Res 28: 109–118

Klochenko PD, Golovnya RV, Terenina MB, Zhuravleva IL, Sakevich AI (1990) Dynamics of changes in the volatile amines content during the growth of some species of green and cyanellae algae. Izv AN SSSR Ser Biol N4: 503–510

Kloss M, Iwannek KH, Fedrik I, Niemann EG (1984) Organic acides in the root exudates of *Diplachne fusca* (Linn) Beauv. Environ Exp Bot 24: 179–188

Knobloch K, Weigand H, Weis N, Schwarm HM, Vigenschow H (1986) Action of terpenoids on energy metabolism. In: Brunke EJ (ed) Progress in essential oil research. de Gruyter, Berlin, pp 429–445

Knobloch K, Iberl B, Pauli A, Weigand H, Weis N (1987) Mode of action of essential oils. An experimental approach to antimicrobial activity. Abstr 14th Int Bot Congr Berlin-Dahlem Botanical Garden, Berlin p 54

Knox RB (1984) Pollen-pistil interactions. In: Linskens HF, Heslop-Harrison J (eds) Cellular interaction. Encycl Plant Physiol, vol 17. Springer, Berlin Heidelberg New York, pp 508–608

Knox RB, Kendrick J, Bernhardt P, Marginson R, Beresford G, Baker I, Baker HG (1985) Extrafloral nectaries as adaptations for bird pollination in Acacia terminalis. Am J Bot 72: 1185–1196

Knox RB, Marginson R, Kendrick J, Beattie AJ (1986) The role of extrafloral nectaries in *Acacia*. In: Juniper B, Southwood R (eds) Insects and the plant surface. Arnold, London, pp 295–307

Koenze JR, Elstner EF (1976) Pyridoxalphosphate-dependent ethylene production from methionine by isolated chloroplasts. FEBS Lett 66: 8–11

Kojima M, Poulton IE, Thayer SS, Conn EE (1979) Tissue distributions of dhurrin and of enzymes involved in its metabolism in leaves of *Sorghum bicolor*. Plant Physiol 63: 1022–1028

Kolesnichenko MV (1976) Biochemical interactions of woody plants. Lesnaya Promyshlennost', Moscow

Koller AL, Rost TL (1988) Leaf anatomy in *Sansevieria* (Agavaceae). Am J Bot 75: 615–633

Kolpikov DI (1969) About study of phenomena of out-stomatal secretion and out-vascular microcirculation of liquids and gases in xeromorphic and other green terrestrial plants. In: Putintsev NI (ed) The study of flora and vegetation of the North Caucasus. Stavropol University, Stavropol (USSR), pp 61–69

Komai K, Tang Ch (1989) Chemical constituents and inhibitory activities of essential oils from *Cyperus brevifolins* and *C. kyllingia*. J Chem Ecol 15: 2171–2176

Komissarenko NF, Makarevich IF, Kolesnikov LG (1968) Polyphenolic compounds of smoke tree (*Cotinus coggygria* Scop). In: Kursanov AL, Zaprometov MN (eds) Phenolic compounds and their Biological functions. Nauka, Moscow, pp 78–79

Königs R, Gülz PG (1974) Die Monoterpene im ätherischen Öl der Blätter von *Cistus ladaniferus*. Z Pflanzenphysiol 72: 237–248

Kossakovskaya IV, Moroz PA, Chernyad'ev II, Doman NG (1983) Influence of allelophatic factor on the carboxylase activity of ribulose bisphosphate carboxylase of the apple-tree seedlings leaves. In: Abstr Int Symp Regulation of Metabolism of Primary and Secondary Products of Photosynthesis. Department of Scientific and Technical Information, Pushchino (USSR), pp 6–7

Kouji H, Masuda T, Matsunaka S (1990) Mechanism of herbicidal action and soybean selectivity of AKH-7088, a novel diphenyl ether herbicide. Pestic Biochem Physiol 37: 219–226

Kovalenko SG (1972) To study of content and characteristics of volatile and water-soluble excretions of some woody-scrubby plants. In: Grodzinski AM (ed) Base of chemical interactions of plants in phytocenosis. Trans 3rd All-Union Meet Naukova Dumka, Kiev, pp 72–73

Kozel PC, Tukey HB Jr (1968) Loss of gibberellins by leaching from stems and foliage of *Chrysanthemum morifolium* Princess Anne. Am J Bot 55: 1184–1189

Krall AR, Tolbert NE (1957) A comparison of the light dependent metabolism of carbon monooxide by barley leaves with that of formaldehyde, formate and CO_2. Plant Physiol 32: 321–326

Kramer PJ (1956) Physical and physiological aspects of water absorption. In: Buhland W (ed) Encyclopedia off plant physiology, vol 3. Springer, Berlin Heidelberg New York, pp 124–129

Kramer PJ, Jackson WT (1954) Causes of injury to flooded tobacco plants. Plant Physiol 29:241–245

Krasil'nikov NA (1958) Microorganisms of soil and higher plants. AN SSSR, Moscow

Kravchenko LV, Makarova NM, Berestetskii OA (1987) Content of free amino acids in root excretions of various pea cultivars. Fiziol Biokhim Kul't Rast 19:181–186

Kreis TE (1990) Role of microtubules in the organization of the Golgi apparatus. Cell Motil Cytoskel 15:67–70

Kretovich VL (1971) Base of plant biochemistry. Vysshaya Shkola, Moscow

Krisai I, Mrazek E (1986) Calcium oxalate crystals in *Geastrum*. Plant Syst Evol 154:325–341

Kristen U, Lockhausen J, Kandasamy MK (1988) Structural aspects of protein secretion in higher plant cells. Phyton 28:183–191

Kronested EC, Robards AW (1987) Sugar secretion from the nectary of *Strelitzia*: a structural and physiological study. Protoplasma 137:168–182

Kronested-Robards EC, Greger M, Robards AW (1989) The nectar of the *Strelitzia reginae* flower. Physiol Plant 77:341–346

Kuc J, Lisker N (1978) Terpenoides and their role in wounded and infected plant storage tissue. In: Günder K (ed) Biochemistry of wounded plant tissues. De Gruyter, Berlin, pp 203–242

Kucera LJ (1985) Zur Morphologie der Interzellularen in den Markstrahlen. Stand der Kenntnisse uber die Interzellularen. Vierteljahrsschr Naturforsch Ges Zuer 130:35–74

Kudryavtseva VM, Volinets AP (1981) Influence of flavonoid compounds on the sprouting of the tulip (*Tulipa* L.) Pollen. Vestn AN Beloruss SSR Ser Biol N 1:10–15

Kullenberg B, Bergström G (1975) Chemical communication between living organisms. Endeavour 34:59–69

Kurosami K (1961) Electron microscopic analysis of the secretion mechanism. Int Rev Cytol 11:1–117

Kursanov AL (1976) Transport of assimilates in plants. Nauka, Moscow

Kursanov AL, Kryukova NN, Vartapetyan BB (1952) Intraplant movement of carbonic acid absorbed by roots. Dokl AN SSSR 85:913–916

Kutchan TM, Rush M, Coscia CJ (1986) Subcellular localization of alkaloids and dopamine in different vacuolar compartment of *Papaver bracteatum*. Plant Physiol 81:161–166

Ladd TL Jr, Buriff CR, Beroza M, McGovern TP (1975) Japanese beetles attractancy of mixtures of lures containing phenethylpropionate and eugenol. J Econ Entomol 68:819–820

Lakhno EE, Kozlova NV (1967) About chemical content of volatile phytoncides from woody plants. In: Aisenman BE (ed) Phytoncides, their biological role and significance for medicine and national economy. Naukova Dumka, Kiev, pp 135–137

Lakhno ES, Vikhrestyuk TA (1979) Mass-spectrometric study of chemical composition of volatile plant excretions. Ukr Bot J 36:125–129

Lander V, Schreiner P (1990) Acorenone and γ-asarone: Indicators of the origin of calamus oils (*Acorus calamus* L.). Flav Fragr J 5:75–79

Lang DR, Racker E (1974) Effect of quercetin and F_1 inhibitor on mitochondrial ATPase and energy-linked reactions in submitochondrial particles. Biochim Biophys Acta 333:180–186

Langdon SC (1917) Carbon monooxide occurrence free in kelp (*Nereocystis lutkeana*). J Am Chem Soc 39:149–156

Langenheim JH, Foster CE, McGinley RB (1980) Inhibitory effects of different quantitative composition of *Hymenaea* leaf resins on a generalist herbivore *Spodoptere exigua*. Biochem Syst Ecol 8:385–396

Langer G, Bohatka S, Bercecz I, Schlenk B, Bornemisza-Pausperth P, Kiss K, Buzas I, Partay G, Mozsik L, Sagi F (1984) Mass-spectrometric determination of gases in plants. Vacuum 34:757–758

Larigauderie A, Roy J, Berger A (1986) Long-term effect of high CO_2 concentration on photosynthesis of water hyacinth [*Eichornia orassipes* (Mart) Solms]. J Exp Bot 37:1303–1312

Larigauderie C, Latche A, Pech JC, Triantaphylides C (1990) Short-term effects of γ-irradiation on 1-aminocyclopropane-1-carboxylic acid metabolism in early climateric cherry tomatoes. Comparison with wounding. Plant Physiol 92:577–581

Laties GG (1982) The cyanide-resistant, alternative path in higher plant respiration. Annu Rev Plant Physiol 33:519–555

Lauter DJ, Munns DN (1986) Water loss via the glandular trichomes of chickpea (Cicer arietinum L.). J Exp Bot 37:640–649

Lee E, Speirs J, McGlasson WB, Brady CJ (1987) RNA changes in tomato fruit pericarp in response to propylene, wounding or ripening. J Plant Physiol 129:287–299

Leelavathy KM (1970) Amino acids and sugars of the exudates from germinating seeds of rice. Proc Ind Acad Sci 72:81–90

Lefebvre R (1978) Actions comparés de l'ethylene et del'ethane sur le bourgeonnement. CR Acad Sci Paris 287D:25–27

Lersten NR, Curtis JD (1974) Colleter anatomy in red mangrove Rhizophora mangle (Rhizophoraceae). Can J Bot 52:2277–2278

Lersten NR, Curtis JD (1988) Secretory reservoirs (ducts) of two kind in giant ragweed (*Ambrosia trifida*, Asteraceae). Am J Bot 75:1313–1323

Lersten NR, Curtis JD (1989) Polyacetylene reservoir (duct) development in *Ambrosia trifida* (Asteraceae) staminate flowers. Am J Bot 76:1000–1005

Lersten NR, Curtis JD (1991) Laminar hydathodes in Urticaceae: Survey of tribes and anatomical observations on *Pilea pumila* and *Urtica dioica*. Pl Syst Evol 176:179–203

Levin DA (1973) The role of trichomes in plant defence. Quart Rev Biol 48:3–15

Levitt J, Lovett JV (1985) Alkaloids, antagonisms and allelopathy. Biol Agr and Hort 2:289–301

Leuning R (1983) Transport of gases into leaves. Plant Cell Environ 6:181–194

Lewis JA, Moore CJ, Fletcher MT, Drew RA, Kitching W (1988) Volatile compounds from the flowers of *Spathiphyllum cannaefollium*. Phytochemistry 27:2755–2757

Lieberei R (1986) Cyanogenesis of *Hevea brasiliensis* during infection with *Microcyclus ulei*. Phytopathol Ztschr 115:134–146

Lieberei R, Selmar D, Biehl B (1985) Metabolization of cyanogenic glycosides in *Hevea brasiliensis*. Plant Syst Evol 150:49–63

Ligero F, Lluch C, Olivares J (1987) Evolution of ethylene from roots and nodulation rate of alfalfa (*Medicago sativa* L) plant inoculated with Rhizobium meliloti as affected by the presence of nitrate. J Plant Physiol 129:461–467

Lin SYH, Trumble JT, Kumamoto J (1987) Activity of volatile compounds in glandular trichomes of *Lycopersicon* species against two insect herbivores. J Chem Ecol 13:837–850

Lincoln DE (1990) Leaf resin flavonoides of *Diplacus aurantiacus*. Biochem Syst Ecol 8: 397–400

Lincoln DE, Walla MD (1986) Flavonoids from *Diplacus aurantiacus* leaf resin. Biochem Syst Ecol 14:195–198

Lipton DS, Blanchar RW, Blevins DJ (1987) Citrate, malate, succinate concentration in exudates from P-sufficient and P-stressed *Medicago sativa* L. seedlings. Plant Physiol 85:315–317

Loewis MW, Delwiche CC (1963) Carbon monooxide production by algae. Plant Physiol 38:371–374

Lookado SE, Pollard AJ (1991) Chemical contents of stinging trichomes of *Cnidoscolus texanus*. J Chem Ecol 17: 1909–1916

Lorber P, Muller WH (1976) Volatile growth inhibitors produced by *Saliva leucophylla*: effect on seedling root tip ultrastructure. Am J Bot 63:196–200

Lovett JV (1987) Plant defence and allelopathy. Abstr 14th Int Bot Congr Berlin-Dahlem Botanical Garden, Berlin. p 414

Lovett JV, Potts WS (1987) Primary effects of allelochemicals of *Datura Stramonium* L. Plant Soil 98:137:144

Lovett JV, Levitt J, Duffield AM, Smith NG (1981) Allelopathic potential of *Datura stramonium* L. (thorn-apple). Weed Res 21:165–170

Luca de V, Cutler AJ (1987) Subcellular localization of enzymes involved in indole alkaloid biosynthesis in *Catharanthus roseus*. Plant Physiol 85:1099–1102

Luchko AS, Porutskii GV, Yavorskii AG (1964) Gaseous excretions and the amino acid composition of green pea. Fiziol Rast 11:53–58

Luckner M (1977) Secondary metabolism in plants and animals (Der Sekundärstoffwechsel in Pflanze und Tier). Chapman and Hall, London

Luckner M (1990) Secondary metabolism in microorganisms, plants and animals, 3rd edn. Springer-VEB Gustav Fisher, Berlin-Jena

Luckner M, Dietrich B, Lerbs W (1980) Cellular compartmentation and channelling of secondary metabolism in microorganisms and higher plants. In: Reinolds L, Harborne JB, Swain T (eds) Progress in phytochemistry, vol 6. Pergamon Press, Oxford, pp 103–142

Lund ED, Kirkland CL, Shaw PE (1981) Methanol, ethanol and acetaldehyde contents of citrus products. J Agric Food Chem 29: 361–366

Lundegardh H, Stenlid G (1944) On the exudation of nucleotides and flavonone from living roots. Ark Bot 31 A: 1–27

Lundgren L, Norelius G, Stenhagen G (1985) Leaf volatiles from some wild tomato species. Nord J Bot 5: 315–320

Lüttge U, Higinbotham N (1979) Transport in plants. Springer, Berlin Heidelberg New York

Lynn KR (1989) Four lysozymes from latex of *Asclepias syriaca.* Phytochemistry 28: 1345–1348

Lynn KR, Clevette-Radford NA (1986a) Lectins from lattices of *Euphorbia* and *Elaeophorbia* species. Phytochemistry 25: 1553–1557

Lynn KR, Clevette-Radford NA (1986b) Hevains: serine-centered proteases from the latex of *Hevea brasiliensis.* Phytochemistry 25: 2279–2282

Lynn KR, Clevette-Radford NA (1988) Proteases of Euphorbiaceae. Phytochemistry 27: 45–50

Lycon CJ (1970) Ethylene inhibition of auxin transport by gravity in leaves. Plant Physiol 45: 644–646

Lyr H, Banasiak L, Casperson G (1987) Alkenale-eine Fungizide Wirkstoffgruppe. Biol Rundsch 25: 239–249

Lyubimov VYu, Nazarova GN, Muzafarov EN (1986) Influence of flavonoids on photosynthetic CO_2-assimilation by isolated spinach chloroplasts. In: Romanova AK, Muzafarov EN (eds) Characteristics of flavonoids and their functions in plant cell metabolism. Department of Scientific and Technical Information, Pushchino (USSR), pp 104–115

MacDougal DT, Working EB (1933) The pneumatic system of plants, especially trees. Carnegie Inst Wash Publ N 441: 1–87

MacFarlane WV (1963) The stinging properties of *Laportea.* Econ Bot 17: 303–311

MacLeod AJ, MacLeod G, Subramanian G (1988) Volatile aroma constituents of celery. Phytochemistry 27: 373–375

Macnab RM (1979) Chemotaxis in bacteria. In: Haupt W, Fienleib ME (eds) Encyclopedia of plant physiology. New Ser. Physiology of movement, vol 7. Springer, Berlin Heidelberg New York, pp 310–334

Macrae S, Gilliland MG, Staden JV (1986) Rubber production of rubber-producing potential. Plant Physiol 8: 1027–1032

Maestrin SY, Bonn DJ (1981) Allelopathic relationships between phytoplancton species. Can Bull Fish Aquat Sci 210: 323–346

Maffei M, Chialva F, Sacco T (1989) Glandular trichomes and essential oils in developing peppermint leaves. I. Variation of peltate trichome number and terpene distribution within leaves. New Phytol 111: 707–716

Maheshwary SC, Gupta R, Gharyal PK (1982) Cholinesterase in plants. In: Sen SP (ed) Recent developments in plant sciences: SSSMMM Sircar memorial volume. Today and Tomorrows, New Delhi, pp 145–160

Mahlberg PG, Hammond CT, Turner JC, Hemphill JK (1984) Structure, development and composition of glandular trichomes of *Cannabis sativa* L. In: Rodriguez E, Healey PL, Mehtha I (eds) Biology and chemistry of plant trichomes. Plenum Press, New York, pp 23–51.

Mailänder HC (1974) Über den Einfluß ätherischer Öle auf Atmungsvorgänge in Blättern. Beitr Biol Pflanz 50: 163–176

Maitra RK, Sen SP (1989) Production of ethylene and ethane by callus tissues of *Daucus carota* L. in presence of 2,4-disubstituted phenols. Plant Sci 62: 33–35

Major RT (1967) The ginkgo. The most ancient living tree. Science 157:1270–1273

Major RT, Thomas M (1972) Formation of 2-hexenal from linolenic acid by macerated *Ginkgo* leaves. Phytochemistry 11:611–617

Major RT, Marchini P, Sproston T (1960) Isolation from *Ginkgo biloba* L. of an inhibitor of fungus growth. J Biol Chem 235:3298–3299

Major RT, Marchini P, Boulton AJ (1963) Observation on the production of α-hexenal by leaves of certain plants. J Biol Chem 238: 1813–1816

Major RT, Collins OD, Marchini P, Schnabel HW (1972) Formation of 2-hexenal by leaves. Phytochemistry 11:607–610

Makarchyuk NM, Leshchinskaya YaC, Akimov YuA, Lebeda AF, Chekman IC, Golota LG, Andrushchyuk AA, Daletskaya LP (1990) Phytoncides in medicine. Naukova Dumka, Kiev

Makovec P, Sindelar L (1984) The effect of phenolic compounds on the activity of respiratory chain enzymes and on respiration and phosphorylation activities of potato tuber mitochondria. Biol Plant 26:415–422

Maksymovych R, Ledbetter MC (1987) Fine structure of epithelial canal cells in petioles of *Xanthium pensylvanicum*. Am J Bot 74:65–73

Mal'tsev SV, Anan'ev GM, Krasnovskii AA (1986) The effect of the hydrogen photoformation by chloroplasts of higher plants. Biofizika 31:529–530

Mal'yan AN, Muzafarov EN, Akulova EA (1977) Interaction of flavonols (aglycone, glycosidated and acyl derivatives) and photophosphorylation coupling factor CF_1. In: Akulova EA, Muzafarov EN (eds) Regulation of energy exchange of chloroplasts and mitochondria by endogenous phenolic inhibitors. Department of Scientific and Technical Information, Pushchino (USSR), pp 41–55

Mansfield JW, Bailey JA (1982) phytoalexins: Current problems and future prospects. In: Bailey JA, Mansfield JVW (eds) phytoalexins. Blackie, Glasgow, pp 319–323

Marginson R, Sedgley M, Douglas TJ, Knox RB (1985) Structure and secretion of the extrafloral nectaries of Australian acacias. Isr J Bot 34:91–102

Margna UV (1980) About site of the flavonoid biosynthesis in whole system of plant metabolism. Zh Common Biol (USSR) 41:68–79

Mariani P, Rascio N, Baldan B, Paiero P, Urso T (1988) Epidermal mucilage cells in leaves of *Salix* species. Flora 181: 137–145

Mariani P, Cappelletti EM, Campoccia D, Baldon B (1989) Oil cell ultrastructure and development in *Liriodendron tulipifera* L. Bot Gaz 150:391–396

Martin B, Malmberg G, Gref R, Oquist G (1984) Identification of the photosynthetic CO_2 fixation inhibitors in isolated pine chloroplast as resin acids. Physiol Plant 62:110–114

Martin MN (1991) The latex of *Hevea brasiliensis* contains high levels of both chitinases and chitinases/lysozymes. Plant Physiol 95:469–476

Masquelier J, Vitte G (1967) Sur la production aerobic d'alcogol et de divers corps volatils par les vegetaux. C R Acad Sci Paris 265D:1924–1927

Matern U (1987) Die Isomerenfalle für Sekundärmetabolite, eine Alternative zum Lonenfallen-Modell. Biol Unserer Zeit 17:148–152

Matile Ph (1976) Localization of alkaloids and mechanism of their accumulation in vacuoles of *Chelidonium majus* laticifers. In: Secondary metabolism and coevolution. Nova Acta Leopoldina Suppl N 7. Dtsch Akad Naturforsch Leopoldina, Halle, pp 139–156, Bazth, Leipzig

Matile Ph (1978) Biochemistry and function of vacuoles. Annu Rev Plant Physiol 29:193–213

Matile Ph (1984) Das toxische Kompartiment der Pflanzenzelle. Naturwissenschaften 71:18–24

Matile P (1987) The sap of plant cells. New Phytol 105:1–26

Matile P, Altenburger R (1988) Rhythms of fragrance emission in flowers. Planta 174:242–247

Matile P, Jans B, Rickenbacher R (1970) Vacuoles of *Chelidonium latex*: lysosomal property and accumulation of alkaloids. Biochem Physiol Pflanz 161:447–458

Mattoo AK, Baker JE, Chalutz E, Lieberman M (1977) Effect of temperature on the ethylene-synthezing systems in apple, tomato and *Penicillium digitatum*. Plant Cell Physiol 18:715–719

Matsuoka H (1987) Response of plant leaves to odors. Frag J 15:37–42

Matsuoka H (1988) Chemical communication between plants. Koryo 158:73–81
Matveev NM (1985) Role of plant excretions in formation of forest communities in the steppe zone. PhD Thesis, Tartu University, Tartu, p 47
Maxwell CA, Hartwig UA, Joseph CM, Phillips DA (1989) A chalcone and two related flavonoids released from alfalfa roots induce no genes of *Rhizobium meliloti*. Plant Physiol 91: 842–847
Mayne RG, Kende H (1986) Ethylene biosynthesis in isolated vacuoles of *Vicia faba* L. – requirement for membrane integrity. Planta 167:159–165
McDonald RE, Kushand MM (1986) Accumulation of putrescine during chilling injury of fruits. Plant Physiol 82:324–326
McDougall BM, Rovira AD (1970) Sites of exudation of ^{14}C-labelled compounds from wheat roots. New Phytol 69:999–1003
McKersie BD, McDermont BM, Hunt LA, Poysa V (1982) Changes in carbohydrate levels during ice encasement and flooding of winter cereals. Can J Bot 60:1822–1826
McQueen JK (1987) Classical transmitters and neuromodulators. In: Flückiger E, Müller EE, Thorner MO (eds) Basic and clinical aspects of neurosciences, vol 2. Transmitter molecules in the brain. Springer, Berlin Heidelberg New York, pp 7–16
Mecklenburg RA, Tukey HB Jr (1964) Influence of foliar leaching on root uptake and translocation of calcium -45 to the stems and foliage of *Phaseolus vulgaris*. Plant Physiol 39:533–536
Mecklenburg RA, Tukey HB Jr, Morgan JV (1966) A mechanism for the leaching of calcium from foliage. Plant Physiol 41:610–613
Menetrez ML, Spurr HW Jr, Danehower DA, Lawson DR (1990) Influence of tobacco leaf surface chemicals on germination of *Peronospora tabacina* Adam sporangia. J Chem Ecol 16:1565–1576
Mekurina ON, Ronitskaya NV, Kozlova IV (1990) The study of the composition of oblepikha root exudates at the interaction with the actinomycetes of genus *Frankia*. In: Matveev NM (ed) Problems of forest biogeocenology, ecology and nature preservation in the steppe zone. Kuibyshev State Univ, Kuibyshev, pp 100–103
Melan MA, Whittier DP (1989) Characterization of mucilage on the proximal cells of young gametophytes of *Botrychium dissectum* (Ophioglossaceae). Am J Bot 76:1006–1014
Melburne JA, Kallarackal J, Baker DA (1990) Water relations of the banana. 1. Predicting the water relations of the field-grown banana using exuding latex. Aust J Plant Physiol 17:57–68
Melon JE, West CA (1979) Diterpene biosynthesis in maize seedlings in responses to fungal infection. Plant Physiol 64:406–410
Mench M, Morei JL, Guckeri A (1987) Metal binding properties of high molecular weight soluble exudates from maize (*Zea mays* L.) roots. Biol Fertil Soils 3:165–169
Mende P, Wink M (1987) Uptake of the quinolizidine alkaloid lupanine by protoplasts and isolated vacuoles of suspension-cultured *Lupinus polyphyllus* cells. Diffusion or carrier-mediated transport? J Plant Physiol 129:229–242
Menon ARS, Babu AM (1989) Structure and development of ethephon in the stem of *Sterculia urens* Roxb. Phyton 29:41–47
Mer CL (1961) Carbon dioxide and ethanol as factors controlling the growth of etiolated oat seedlings. Nature 191:260–261
Meskhi AB, Dzhokhadze DI (1973) Some characteristics of the simple phenols action on life activity of different systems. Fiziol Rast 20:1253–1256
Metcalfe CR (1967) Distribution of latex of the plant kingdom. Econ Bot 21:115–125
Metcalfe CR, Chalk L (1950) Anatomy of the dicotyledons, vol 2. Clarendon Press, Oxford
Metlitskii LV (1987) Immunologic control in plant life. 15-Timiryazev Lecture. Nauka, Moscow
Metzler DE (1977) Biochemistry. The chemical reactions of living cells. Academic Press, New York
Miernyk JA (1987) Extracellular secretion of acid hydrolases by maize endosperme cells grown in liquid medium. J Plant physiol 129:19–32
Mihaliak CA, Lincoln DE (1989) Changes in leaf mono- and sesquiterpene metabolism with nitrate availability and leaf age in *Heterotheca subaxillaris*. J Chem Ecol 15:1579–1588
Mikhelson MJ, Zeimal' EV (1973) Acetylcholine. An approach to the molecular mechanism of action. Pergamon Press, Oxford

Miki NK, Clarke KJ, McCully ME (1980) A histological and histochemical comparison of the mucilages on the root tips of several grasses. Can J Bot 58:2581–2593

Miller JM, Conn EE (1980) Metabolism of hydrogen cyanide by higher plants. Plant Physiol 65:1199–1202

Miller L, Moore R (1990) Defective secretion of mucilage is the cellular basis for agravitation in primary roots of *Zea mays* cv. ageotropic. Ann Bot 66: 169–178

Miller RH (1985) The prevalence of pores and canals in leaf cuticular membranes. Ann Bot 55:459–471

Mina VN (1965) Leaching of some substances from woody plants by atmosphere precipitation and its importance in biological rotation. Pochvovedenie N 6:7–17

Miroslavov EA (1974) Structure and function of leaf epidermis in angiosperms. Nauka, Leningrad

Mokhnachev IG, Kuz'min MP (1966) Volatile substances of food products. Pishchevaya Promyshlennost', Moscow

Molisch H (1937) Der Einfluß einer Pflanze auf die andere – Allelopathie. Fisher, Jena

Molotok GP, Britikov EA, Sinyukhin AM (1968) The electropysiological and functional activity of nectary glands in the lime tree subjected to mechanical stimulation. Dokl AN SSSR 181:750–753

Monson RK, Fall R (1989) Isoprene emission from aspen leaves. Influence of environment and relation to photosynthesis and photorespiration. Plant Physiol 90:267–274

Monson RK, Hills AJ, Zimmerman PR, Fall RR (1991) Studies of the relationship between isoprene emission rate and CO_2 or photon-flux density using a real-time isoprene analyser. Plant Cell Environ 14:517–523

Moody SF, Clarke AE, Bacic A (1988) Structural analysis of secreted slime from wheat and cowpea roots. Phytochemistry 27:2857–2861

Moore R, McCleven CE (1985) The involvement of glucose-6-phosphatase in mucilage secretion by root cap cells of *Zea mays*. Ann Bot 56:139–142

Morand P (1990) Comparative metabolism of the phytotoxic allelochemical α-terthienyl in three species of lepidopterans. Pestic Biochem Physiol 37:154–164

Moreland DE, Novitzky WP (1987a) Interference by luteolin, quercetin- and taxifolin with choloroplast-mediated electron transport and photophosphorylation. Plant Soil 98:145–159

Moreland DE, Novitzky WP (1987b) Effects of phenolic acids coumarins and flavonoids in isolated chloroplasts and mitochondria. In: Waller GR (ed) Allelochemicals role in agriculture and forestry. ACS Symp Ser, vol 330. Am Chem Soc, Washington pp 247–261

Moreno OAV, Vazquez-Duhalt R, Nolasco H (1990) Extracellular accumulation of high specific-activity peroxidase. Plant Cell Rep 9:147–150

Morre DJ, Jones DD, Mollenhauer HH (1967) Golgi-apparatus-mediated polysaccharide secretion by outer root cap cells of *Zea mays*. 1. Kinetics and secretory pathway. Planta 74:286–301

Morre DJ, Sellden G, Ojanpera K, Sandelius AS, Egger AM, Morre DM, Chalko CM, Chalko RA (1990) Peroxisome proliferation in Norway spruce induced by ozone. Protoplasma 155:58–65

Morrison JC, Greve LC, Labavitch JM (1987a) The role of cell wall-degrading enzymes in the formation of gum ducts in almond fruit. J Soc Hortic Sci 112:367–372

Morrison JC, Labavitch JM, Greve LC (1987b) The role of ethylene in initiating gum duct formation in almond fruit. J Am Soc Hortic Sci 112:364–367

Motoda S (1979) Formation of aldehydes from amino acids by polyphenol oxidase. J Ferment Technol 57:395–399

Mueller LE, Carr PH (1954) The submicroscopic of plant surfaces. Am J Bot 41:593–600

Muller CH (1965) Inhibitory terpenes volatilized from *Salvia* shrubs. Bull Torrey Bot Club 92:38–45

Muller CH, Chou CH (1972) Phytotoxins: an ecological phase of phytochemistry. In: Harborne JB (ed) Phytochemical ecology. Academic Press, London, pp 201–216

Muller WH, Muller CH (1964) Volatile growth inhibitors produced by *Salvia* species. Bull Torrey Bot Club 91:327–330

Muller WH, Lorber P, Haley B, Johnson K (1969) Volatile growth inhibitors produced by *Salvia leucophylla*: effect on oxygen uptake by mitochondrial suspensions. Bull Torrey Bot Club 96:89–96

Mulvaney CS (1984) Evolution of nitrogen compounds from soybean. PhD Thesis, University of Illinois, Champain-Urbana

Müntz K (1984) Stickstoffmetabolismus der Pflanzen. Fischer, Jena

Murphy J (1985) Acetone production during the germination of fatty seeds. Physiol Plant 63:231–234

Muzafarov EN, Zaletskya OYu (1977) Mechanism of action of flavonols on electron transfer and photophosphorylation in connection with their structural complication. In: Akulova EA, Muzafarov EN (eds) Regulation of energy exchange of chloroplasts and mitochondria by endogenous phenolic inhibitors. Department of Scientific and Technical Information, Pushchino (USSR), pp 7–27

Muzafarov EN, Khristin MS, Strubitskii IV (1983) Redox-conversions of phenolic substances in chloroplasts. Biofizika 28:621–624

Nabeta K, Aoki Y, Sugisawa H, Miyaka M (1988) Biosynthetic study of volatile terpenes by tissue culture. Front flavour. Proc Flavour Conf, Porto Karras, Chalkidiki, 1–3 July 1987. Elsevier. Amsterdam, pp 603–621

Nabeta K, Aoki Y, Sugisawa H, Miyaka M (1989) Biosynthetic study of volatile terpenes by tissue culture . In: Charalambous G, Doxastakis G (eds) Frontiers of flavour Conf, Porto Karras, Chalkidiki, 1–3 July 1987. Elsevier. Amsterdam, pp 603–621

Nahrstedt A (1985) Cyanogenic compounds as protecting agents for organisms. Plant Syst Evol 150:35–47

Najera MT (1974) Estructur as secretor as de las especies Argentinas del genero *Bromelia* (Bromeliaceae), Boletin de la Sociedad Argentina de Botanica 15, pp. 384–392

Nasrallah JB, Kao TH, Goldberg ML, Nasrallah ME (1985) A DNA Clone encording an S-locus-specific glycoprotein from *Brassica oleracea* L. Nature 318:263–267

Neave IA, Dawson JO (1989) Juglone reduces growth, nitrogenase activity, and root respiration of actinorhizal black alder seedlings. J Chem Ecol 15:1823–1836

Nelson RS, Ryan SA (1983) Soybean mutants lacking constitutive nitrate reductase activity. 1. Selection and initial plant characterization. Plant Physiol 72:503–509

Nemeryuk G (1970) Migration of salts in atmosphere at transpiration. Fiziol Rast 17:673–679

Nemethy EK, Calvin M (1983) Terpenoid biosynthesis in *Euphorbia* latex. In: Thomson WW, Mudd JB, Gibbs M (eds) Biosynthesis and function of plant lipids. Waverly Press, Baltimore, pp 216–228

Neumann D (1985) Storage of alkaloids. In: Mothes K, Schütte HR, Luckner M (eds) Biochemistry of alkaloids. VEB Deutsch Verl Wiss, Berlin, pp 49–55

Neumann D, Müller E (1972) Beiträge zur Physiologie der Alkaloide. III. *Chelidonium majus* L und *Sanguinaria canadensis* L: Ultrastruktur der Alkaloidbehalter, Alkaloidaufnahme und Verteilung. Biochem Physiol Pflanz 163:375–391

Nichol P, Hall JL (1988) Characteristics of nectar secretion by the extrafloral nectaries of *Ricinus communis*. J Exp Bot 39:573–586

Nichols R (1966) Ethylene production during senescence of flowers. J Hortic Sci 41:279–290

Nicolova R, Izvorska N, Bakardjeva N (1990) Characteristics and secretion of superoxide dismutase from tobacco callus tissue. Comp Rend Acad Bulgar Sci 43:65–68

Nijus D, Kelley PM, Harnadek GJ (1986) Bioenergetics of secretory vesicles. Biochim Biophys Acta 853: 237–265

Nikitin AA, Pankova IA (1982) Anatomical atlas of useful and some toxic plants. Nauka, Leningrad.

Nobel PS (1970) Plant Cell Physiology. A physico-chemical approach. Freeman, San Francisco

Novak K, Stanek M (1986) Phytoalexins in plant root exudates. Sb ref 10 Ceskol Konf Ochri Rostlins, Brno, 2–5 zari 1986. Academia, Praha, pp 81–82

Novikov NA, Solov'ev (VA) (1973) Daily changes of the gaseous mixture composition by some woody-injured fungi. Vestn AN Beloruss SSR Ser Biol N 5:52–55

Novitskaya YuE (1966) About excretion of chemical substances by leaves of woody plants. In: Grodzinskii AM (ed) Physiologo-biochemical base of interaction between plants in phytocenosis. Naukova Dumka, Kiev, pp 234–238

Nunomura N, Sasaki M, Asao Y, Lokotsuka T (1976) Identification of volatile components of Shoyu (soy sauce) by gas-chromatography-mass spectrometry. Agric Biol Chem 40:485–490

Okazaki M, Setoguchi H, Aoki H, Suga S (1986) Application of soft X-ray microradiography to observation of cystoliths in the leaves of various higher plants. Bot Mag Tokyo 99:281–287

Okoli BE, McEuen AR (1986) Calcium-containing crystals in *Telfairia* Hooker (Cucurbitaceae). New Phytol 102:199–207

Olifer VA (1972) The washing away of nutrient elements by rain from field cultures. Agrokhimiya N 8: 70–72

Olson AO, Spencer M (1968) Some investigations on the effect of ethylene on mitochondrial system. In: Wighman F, Setterfield G (eds) Biochemistry of physiology of plant growth substances. Runge Press, Ottawa, pp 1243–1254

Omori S, Yoshida E, Taneda K (1983) Ubate gairacu horacubu hococu. Fac Agric Jawate Univ 16:169–177

Oparina LA, Ruzieva RKh (1986) Interaction of $NADP^+$ - malic enzyme from maize leaves with endogenous phenolic compounds in vitro. In: Romanova AK, Muzafarov EN (eds) Characteristics of flavonoids and their functions in plant cell metabolism. Department of Scientific and Technical Information, Pushchino (USSR), pp 116–121

Ormerod JG, Nesbakken G, Beale SI (1990) Specific inhibition of antenna bacteriochlorophyll synthesis in *Chlorobium Vibrioforme* by anesthetic gases. J Bacteriol 172:1352–1360

Oross JW, Leonard RT, Thomson WW (1985) Flux rate and secretion model for salt glands of grasses. Isr J Bot 34:69–77

Ortega CR, Anaya AL, Gavilanes-Ruiz M, Nieto SS, Estrada MJ (1990) Effect of diacetyl piquerol on H^+-ATPase activity of microsomes from *Ipomoea purpurea*. J. Chem Ecol 16:2253–2261

Osmond CB, Lüttge U, West KR, Pallaghy CK, Shacker-Hill B (1969) Ion absorption in *Atriplex* leaf tissue. II. Secretion of ions to epidermal bladders. Aust J Biol Sci 22:797–814

Oster U, Spraul M, Rudiger W (1990) Natural inhibitors of germination and growth. V. Possible allelopathic effects of compounds from *Thuja occidentalis*. Z. Naturforsch 45c:835–844

Outer RWD, van Veenendaal WLH (1986) Distribution and development of secretory ducts in *Trichoscypha* (Anacardiaceae). Acta Bot Neerl 35:425–435

Ozarowski A (1956) Asarum europaeum L im Lichte der phytochemischen, toxikologischen, pharmakodynamischen und klinischen Forschungen. Pharmazie 11:63–69

Ozeretskovskaya OL, Chalova LI (1985) Biogenic inductors of the phytoalexin formation in plants. Mikol Fitopatol 19:260–267

Pahlke KM (1989) Nektar bei *Vaccinium myrtillus* L: Zuckerproduktionsrate als flächenbezogenes Maß. Beitr Biol Pflanz 64: 329–334

Pallaghy CK (1971) Stomatal movement and potassium transport in epidermal strips of *Zea mays*. The effect of CO_2. Planta 101:287–295

Pallas JE, Kays SJ (1982) Inhibition of photosynthesis by ethylene – a stomatal effect. Plant Physiol 70:598–601

Palma-Fleming HA, Kepner RE (1983) Volatile components of California live oak, *Quercus agrifolia*. Phytochemistry 22:1503–1505

Panenko ID, Pilipenko AD (1981) Influence of overhead irrigation technology on the yield of some cultures in connection with leaching of substances from plants. In: Grodzinsky AM (ed) Allelopathy in natural and artificial phytocenosis. Naukova Dumka, Kiev, pp 47–51

Parijus J, Broekaert WF, Goldstein IJ, Peumans WJ (1991) Hevein: an antifungal protein from the rubber tree (*Hevea brasiliensis*) latex. Planta 183: 258–264

Parker GG (1983) Through-fall and stem-flow in the forest nutrient cycle. Adv Ecol Res 13:57–113

Parr AJ, Robins RJ, Rhodes MJC (1986) Product release from plant cells grown in culture. In: Morris P, Scragg AN, Stafford A, Fowler MW (eds) Secondary metabolism in plant cell cultures. Cambridge University Press, Cambridge, pp 173–177

Passannanti S, Paternostro M, Piozzi F, Faarax MX (1985) Terpenoids from latex of *Euphorbia ruspolii*. Fitoterapia 56:47–50

Pate JS, People MB, Storer PJ, Atkins CA (1985) The extrafloral nectaries of cowpea (*Vigna unguiculata* L.) walp. II. Nectar composition origin of nectar solutes and nectary functioning. Planta 166:28–38

Paterson-Jones JC, Gilliland MG, van Staden J (1990) The biosynthesis of natural rubber. J Plant Physiol 136:257–263

Patnaik S, Saran BL, Patnaik SN (1984) Effect of Zarba (processed tobacco leaf) extract on the chromasomes of *Allium cepa* Linn. Cytologia 49:807–814

Paull RE, Johnson CM, Jones RL (1975) Studies on the secretion of maize root cap slime. 1. Some properties of the secreted polymer. Plant Physiol 56:300–306

Pauly G, Douce R, Carde JP (1984) Effect of β-pinen on spinach chloroplasts photosynthesis. Z Pflanzenphysiol 104:199–206

Peiser GD, Lizanda MAE, Shang Fa Yang (1982) Sulfite-induced lipid peroxidation in chloroplasts as determined by ethane production. Plant Physiol 70:994–998

Pell EJ, Puente M (1986) Emission of ethylene by oat plants treated with ozone and simulated acid rains. New Phytol 103:709–715

Perata P, Alpi A, Loschiavo F (1986) Influence of ethanol of plant cells and tissues. J Plant Physiol 126:181–188

Perez FJ, Ormeno-Nunez J (1991) Root exudates of wild oats: allelopathic effect on spring wheat. Phytochemistry 30:2199–2202

Perretto R, Perrotto S, Faccio A, Bonfante-Fasolo P (1990) Cell surface in *Calluna vulgaris* L hair roots. In situ localization of polysaccharidic components. Protoplasma 155:1–18

Perrin A, Karlin R (1966) Presence de vitamines hygrosolubles dans le liquide de guttation de jeunes plantules de mais. C R Acad Sci D263: 366–367

Perry SF (1967) Inhibition of respiration by juglone in *Phaseolus* and *Lycopersicon*. Bull Torrey Bot Club 94:26–30

Peterson RL, Vermeer J (1984) Histochemistry of trichomes. In: Rodriguez E, Healey PL, Mehta I (eds) Biology and chemistry of trichomes. Plenum Press, New York, pp 71–94

Petril E, Chrastil J (1955) Excretion of a flavonone by explants of pea nuts. Cesk Biol 4: 162–164

Petruzzelli L (1984) The effect of ethanol on germination of aleppo pine seeds (*Pinus halepensis* Mill). Biol Plant 26:235–238

Pham T, Homeyer BC, Roberts MF (1989) *Papaver somniferum* latex vesicle alkaloid sequestration mechanism. J Pharm Pharmacol 41:66

Philosoph-Hadas S, Anaroni N, Yang SF (1986) Carbon dioxide enhances the development of the ethylene-forming enzyme in tobacco leaf discs. Plant physiol 82:925–929

Philosoph-Hadas S, Meir S, Pesis E, Reuveni A, Aharoni N (1989) In: Hormone-enhanced ethylene production in leaves. Biochemistry and physiology aspects of ethylene production in lower and higher plants. Proc Conf Diepenbeek, 22–27 August 1988, Dordrecht, Kluwer Academic Press, pp 135–142

Piazza GG, Saggese EJ, Spletzer KM (1987a) Triterpene biosynthesis in the latex of *Euphorbia lathyris*. Effect of calmodulin antagonists and chlorinated phenoxy compounds. Plant Physiol 83:177–180

Piazza GG, Saggese EJ, Thompson MP (1987b) Regulation of terpenoid biosynthesis in tapped latex, In: Stumpf PK, Mudd JB, Nes WD (eds) The metabolism, structure and function of plant lipids. Plenum Press, New York, pp 111–113

Pickett JA, Griffith DC (1980) Composition of aphid alarm pheromones. J Chem Biol 6:349–360

Piumbe AM, Willmer CM (1985) Phytoalexins, water-stress and stomata. 1. Do phytoalexins accumulate in leaves under water-stress? New Phytol 101:269–274

Platt-Aloia KA, Bliss RD, Thomson WW (1983) Lipid-lipid interactions and membrane fusion in plant salt glands. In: Thomson WW, Mudd JB, Gibbs M (eds) Biosynthesis and function of plant lipids. Waverly Press, Baltimore (Md), pp 160–172

Plhak F (1972) Influence of volatile excretions by leaves of willow on the oxidative phosphorylation in rye seedlings. In: Grodzinsky AM (ed) physiologically biochemical base of plant interaction in phytocenosis. Naukova Dumka, Kiev, 3: 16–20

Pollak G, Waisel Y (1979) Ecophysiology of salt excretion in *Aeluropus litoralis* (Gramineae). Physiol Plant 47:177–184

Pollard AJ, Briggs D (1984) Genecological studies of *Urtica dioica* L. III. Stinging hairs and plant-herbivore interactions. New Phytol 97:507–522

Polova M, Vicherkova M (1986) Leaf diffusibility changes of bean and sunflower plants treated with essential oil vapours. Scr Fac Sci Nat UIEP Brun 16:119–127

Porutskii GV, Luchko AS, Matkovskii KI (1962) About the content of the ethylene hydrocarbons in volatile plant excretions. Fiziol Rast 9:482–485

Postek MT, Tucker SC (1983) Ontogeny and ultrastructure of secretory oil cells in *Magnolia grandiflora* L. Bot Gaz 144:501–512

Poulton JE (1983) Cyanogenic compounds in plants and their toxis effects. In: Keeler RF, Tu AT (eds) Handbook of natural toxins, vol 1. Dekker, New York, pp 117–157

Poulton JE (1990) Cyanogenesis in plants. Plant Physiol 94:401–405

Pribella A, Vasatko J (1967) Stanovenie prchvych aldehydov a ketonov v ovoci.. Biologia (Czech) 22:96–107

Pridgeon AM, Stern WL (1983) Ultrastructure of osmophores in Restrepi (Orchidaceae). Am J Bot 70:1233–1243

Pridgeon AM, Stern WL (1985) Osmophores of *Scaphosepalum* (Orchidaceae). Bot Gaz 146:115–123

Prokhorchik RA, Volinets AP (1973) Action of flavonoids on the chloroplasts activity of lupin and pea. Fiziol Biokhim Kul't Rast 5:623–628

Prokofyev AA (1948) Localization, formation and state of rubber in plants . AN SSSR, Moscow

Proksch P (1985) Vorkommen und biologische Bedeutung von Benzopyranen (Chrameonen) und Benzofuranen in den Araceae. Plant Syst Evol 150:89–100

Protacio CM, Dai Y-ren, Lewis EF, Flores HE (1992) Growth stimulation by catecholamines in plant tissue/organ cultures. Plant Physiol 98:89–96

Protopopov VV, Chernyaeva GN (1967) Quantity of volatile organic compounds in atmosphere of some forest types in Middle Siberia. In: Vasil'ev NG (ed) Results of study of the Far East forest. Vladivostok Publ, Vladivostok, pp 53–54

Prutenskaya NI, Einor LO, Grodzinskii AM (1967) About mechanism of action of volatile substances of rotted plant mass. Fiziol Rast 14:292–298

Punnet T (1965) Influence of growth conditions on the enhancement by carbon dioxide of photophosphorylation. Plant Physiol 40:1233–1284

Putnam AR (1983) Allelopathic chemicals: nature herbicides in action. Chem Eng News 61: 34–45

QiW, Fong C, Lamport DTA (1991) Gum arabic glycoprotein is a twisted hairy rope. Plant Physiol 96:848–855

Rabinowitch E (1951) Photosynthesis and related processes vol 2, part 1. Interscience, New York

Radin IW, Loomis RS (1969) Ethylene and carbon dioxide in the growth and development of cultured radish roots. Plant Physiol 44:1584–1589

Rakhteenko IN, Min'ko IF, Kaurov IA, Budkevich TA (1977) Amino acid composition of root excretions of some agricultural plants. In: Grodzinskii AM (ed) Physiologo-biochemical base in interactions between plants in phytocenosis. Naukova Dumka, Kiev, pp 97–103

Rakitin VYu (1967) Ethylene as high-effective defoliant. Fiziol Rast 14:936–951

Rakitin VYu, Rakitin LYu (1986) The determination of gas-exchange and content of ethylene, carbon dioxide and oxygen in plant tissues. Fiziol Rast 33:403–413

Rakitina TYa, Kefeli VI (1989) The ethylene release and pea growth in different light conditions. Fiziol Rast 36:789–793

Rakitina ZG (1970) Influence of ice crust on the gaseous composition of internal atmosphere of winter wheat. Fiziol Rast 17:907–912

Rao AS (1990) Root flavonoids. Bot Rev 56:1–90

Rao VSN, Menezes AMS, Viana GSB (1990) Effect of myrcene on nociception in mice. J Pharm Pharmacol 42:877–878

Raschke K (1979) Movements of stomata. In: Haupt W, Felnleib ME (eds) Physiology of movements. Encyclopedia of plant physiology. New Series, vol 7. Springer, Berlin Heidelberg New York, pp 383–441

Raskin I, Kende H (1984) Regulation of growth in stem sections of deepwater rice. Planta 160:66–72

Rasmussen RA (1970) Isoprene: indentified as a forest-type emission to the atmosphere. Environ Sci Technol 4:667–671

Rasmussen RA, Jones CA (1973) Emission of isoprene from leaf discs of *Hamamelis*. Phytochemistry 12:15–19

Rasmussen RA, Khalil MAK (1988) Isoprene over the Amazon basin. J Geophys Res 93:1417–1421

Razumovskii SD (1979) Elementary forms of oxygen and their features. Khimiya. Moscow

Reddy AR, Das VSR (1988) Enhanced rubber accumulation and rubber transferase activity in guayule under water stress. J Plant Physiol 133:152–155

Reddy AR, Suhasini M, Das VSR (1987) Impairment of photorespiratory carbon flow into rubber by the inhibition of the glycolate pathway in guayule (*Parthenium argentatum* Gray). Plant. Physiol 84:1447–1450

Reger BJ (1989) Stigma surface secretions of *Pennisetum americanum*. Am J Bot 76:1–5

Regula I, Kolevska-Pleticapic B, Krsnik-Rasol M (1989) Presence of serotonin in *Juglans ailanthifolia* var. *ailanthifolia* Carr and its physiological effects on plants. Acta Bot Croat 48:57–62

Reicher F, Leitner SCS, Sierakowski MR, Fontana JD, Correa JBC (1991) Seed gum of *Stryphnodendron barbatiman* (Barbatimao). Appl Biochem Biotechnol 28/29:353–351

Rembold H (1989) Kairomones – chemical signals related to plant resistance against insect attack. In: Wickens GE, Haq N, Day P (eds) New crops for food and industry, Int Symp Southampton 1986. Chapman and Hall, London, pp 352–364

Rennenberg H, Huber B, Schröder P, Stahl K, Haunold W, Georgii HW, Slovik S, Pflanz H (1990) Emission of volatile sulfur compounds from spruce trees. Plant Physiol 92:560–564

Rhodes M, Wooltorton LSC (1978) The biosynthesis of phenolic compounds in wounded plant storage tissues. In: Günder K (ed) Biochemistry of wounded plant tissues. De Gruyter, Berlin, pp 243–286

Ribeiro OA, Pais MS, Salome S (1988) Lipophilic secretion in the reticulum of *Ophrys lutea* (Orchidaceae): Ultrastructural aspects. Cienc Biol Mol Cell Biol 13:120

Rice El (1974) Allelopathy. Academic Press. New York, San Francisco, London

Rice El (1983) Pest control with nature's chemicals. Allelochemics and pheromones in gardening and agriculture. University of Oklahoma Press, Norman Oklahoma

Rice EL (1984) Allelopathy. Academic Press, Orlando

Richardson DP, Messer AC, Newton BA, Lindeman NI (1991) Identification and preparation of antiinsectan dienol from *Dipterocarpus kerrii* tree resins. J Chem Ecol 17:663–685

Riesenberg LH, Soltis DE, Arnold D (1987) Variation and localization of flavonoid aglycones in *Helianthus annuus* (Compositae). Am J Bot 74:224–233

Riffle MS, Waller GR, Murray DS, Sgaramello RP (1990) Devil's-claw (*Proboscidea louisianiaca*) essential oil and its components. Potential allelochemical agents on cotton and wheat. J Chem Ecol 16:1927–1934

Righetti B, Magnanini E, Infante R, Predieri S (1990) Ethylene, ethanol, acetaldehyde and carbon dioxide released by *Prunus avium* shoot cultures. Physiol Plant 78:507–510

Riov J, Jaffe MJ (1973) Cholinesterase from mung bean roots and its inhibition by plant growth retardants. Experientia 29:264–265

Robacker DC, Mecuse BJD, Erickson EH (1988) Floral aroma. Bioscience 36:390–398

Robards AW, Stark M (1988) Nectar secretion in abutilon: a new model. Protoplasma 142 : 79–91

Roberts JA, Osborne DJ (1981) Auxin and the control of ethylene production during the development and senescence of leaves and fruits. J Exp Bot 32:875–887

Roberts MF (1986) Papaver latex and alkaloid storage vacuoles. In: Martin B (ed) Plant vacuoles. Plenum Press, New York, pp 513–528

Roberts MF, Homeyer BC, Pham TDT (1991) Further studies of sesquestration of alkaloids in *Papaver Somniferum* L. latex vacuoles. Z. Naturforsch 46c: 377–388

Robertson PA, Macfarlane WV (1957) Pain-producing substances from the stinging bush *Laportea moroides*. Aust J Exp Biol 35:381–394

Robins RJ, Juniper BE (1980) The secretory cycle of *Dionaea muscipula* Ellis. III. The mechanism of release of digestive secretion. New Phytol 86:313–327

Rodecar KD, Tingey DT (1986) Ozone-induced ethylene release from leaf surfaces. Plant Sci 44:73–76

Rodriguez E (1983) Cytotoxic and insecticidal chemicals of desert plants. In: Hedin PA (ed) Plant resistance to insects. 183rd Meet Am Chem Soc, Las Vegas, Nevada, March 28-April 2, 1982 ACS Symp Ser vol 208. Am Chem Soc. Washington, pp 291–302

Rolfe BG, Sargent CL, Weinmann JJ, Djordievic MA, Mclver J, Redmond JW, Batley M, Yuan DC, Sutherland MW (1990) Signal exchange between *R. trifolii* and clovers. In: Lugtenberg BJJ (ed) Signal molecules in plants and plant microbe interactions. Springer, Berlin Heidelberg New York, pp 303–310

Roos W, Luckner M (1986) The spatial organization of secondary metabolism in microbial and plant cells. In: Subramanian TAV (ed) Cell metabolism, growth and environment, vol 1. CRC Press, Boca Raton, pp 45–73

Rorchert R (1984) Functional anatomy of the calcium-excreting system of *Gleditsia triacanthos* L. Bot Gaz 145:474–482

Rosenthal GA (1983) L-Canavanine and L-canaline: protective allelochemicals of certain leguminous plants. In: Hedin PA (ed) Plant resistance to insects. 183rd Meet Am Chem Soc, Las Vegas, Nevada, March 28- April 2, 1982, ACS Symp Ser. vol 208. Chem Soc, Washington, pp 279–290

Roshchina VD (1965) Phytoncides and chloroplast movement. Fiziol Rast 12:489–493

Roshchina VD (1971) On the composition of transpiration water in plants. Fiziol Rast 18:433–435

Roshchina VD (1972a) Use of gas chromatography in allelopathic study. Lesn N 3:29–31

Roshchina VD (1972b) The study of the intracellular water state under treatment by phenols. Sel'skokhoz Biol 7:554–558

Roshchina VD (1972c) Mechanism of the phytoncides action on the motility of cytoplasmic structures in cell. In: Aizenman BE (ed) Phytoncides. Naukova Dumka, Kiev, pp 59–61

Roshchina VD (1973a) Some components of the transpiration water and extracts from leaves of woody plants. Biol Nauki N 4:94–98

Roshchina VD (1973b) Seasonal dynamics of internal carbon dioxide in shoots of woody plants. Lesn N 4:24–31

Roshchina VD (1973c) Reducing power of woody plant leaf infusions and the inhibition of the Hill reaction with 2,6-dichlorophenol indophenol by them. In: Grodzinskii (ed) Physiologically biochemical base of interactions between plants in phytocenosis, vol 4. Naukova Dumka, Kiev, pp 10–15

Roshchina VD (1974a) Volatile and water-soluble metabolites of woody species leaves. In: Grodzinskii AM (ed) Physiologically biochemical base of interactions between plants in phytocenosis, vol 5. Naukova Dumka, Kiev, pp 36–40

Roshchina VD (1974b) The toxicity of gaseous substances released by woody species buds. In: Belov SV (ed) Forest geobotany and biology of woody species, vol 2. Priokskoe, Bryansk, pp 114–117

Roshchina VD (1974c) Exometabolites of woody plants and mechanisms of their action on plant cells. PhD Thesis, Institute of Plant Physiology of Ukrainien SSR, Kiev

Roshchina VD (1975) Certain physico-chemical and biological methods for study of plant excrets. In: Grodzinskii AM (ed) Physiologically biochemical base of interaction between plants in phytocenosis, vol.6 Naukova Dumka, Kiev, pp 109–114

Roshchina VD (1976) Physico-chemical mechanisms of phytoncides action. In: Grodzinskii AM (ed) Problems of allelopathy. Naukova Dumka, Kiev, pp 8–9

Roshchina VD (1981) Possible mechanisms of phytoncides action. In: Aizenman BE (ed) Phytoncides' role in biogeocenosis, significance for medicine. Naukova Dumka, Kiev, pp 80–82

Roshchina VD, Roshchina VV(1970) influence of water-soluble excreta from the woody species leaves on cytoplasm permeability for anthocyanin. In: Grodzinskii AM (ed) Physiologically biochemical base of interaction between plants in phytocenosis, vol 1. Naukova Dumka, Kiev, pp 257–262

Roshchina VD, Roshchina VV (1983) Aldehydes as components of plant excreta. In: Grodzinsky AM (ed) Role of plant and microbial toxins in allelopathy. Naukova Dumka, Kiev, pp 122–127

Roshchina VD, Roshchina VV (1989) The excretory function of higher plants. Nauka, Moscow.
Roshchina VD, Roshchina VV, Kotova IN (1979) The effect of extracts from *Cicuta virosa* on chloroplast movement and on some photosynthetic reactions. Fiziol Rast 26:147–152
Roshchina VV (1981) Phytoncides of *Cicuta* and *Malus* and photochemical reactions of chloroplasts. In: Aizenman BE (ed) Phytoncides. Role in biogeocenosis, significance for medicine. Naukova Dumka, Kiev, pp 177–179
Roshchina VV (1986) Cholinesterase from pea chloroplasts. Dokl AN SSSR 290:486–489
Roshchina VV (1987) Action of acetylcholine agonists and antagonists on reactions of photosynthetic membranes. Photosynthetica 21:296–300
Roshchina VV (1988a) Characterization of pea chloroplast cholinesterase: effect of inhibitors of animal enzymes. Photosynthetica 22:20–26
Roshchina VV (1988b) Cholinesterases from chloroplasts of higher plants. Fiziol Rast 35:899–906
Roshchina VV (1989a) Biomediators in chloroplasts of higher plants. I. The interaction with photosynthetic membranes. Photosynthetica 23:197–206
Roshchina VV (1989b) Cell membranes as a target of allelopathic and phytoncidic action. In: Grodzinsky AM (ed) Methodological problems of allelopathy. Naukova Dumka, Kiev, pp 122–127
Roshchina VV (1989c) Reactions of chloroplasts on allelopathically active alkaloids and polyacetylenes. In: Grodzinski Am (ed) Methodological problems of allelopathy. Naukova Dumka, Kiev, pp 128–131
Roshchina VV (1990a) Biomediators in chloroplasts of higher plants. 2. The acetylcholine-hydrolyzing proteins. Photosynthetica 24:110–116
Roshchina VV (1990b) Biomediators in chloroplasts of higher plants. 4. Reception by photosynthetic membranes. Photosynthetica 24:539–549
Roshchina VV (1990c) Regulation of chloroplast's reactions by secondary metabolites acetylcholine and biogenic amines. Acta Bot Croat 49:29–35
Roshchina VV (1991a) Biomediators in plants. Acetylcholine and biogenic amines. Department of Scientific and Technical Information, Pushchino (USSR)
Roshchina VV (1991b) Neurotransmitters catecholamines and serotonin in plants. Uspekhi Sovremennoi Biologii 111:631–645
Roshchina VV (1992) Action of neurotransmitters on the seed germination. Biologicheskie Nauki 9: 124–129
Roshchina VV Akulova EA (1978) Effect of phloridzin on photosynthetic electron transport. Biokhimiya 43:899–903
Roshchina VV, Mukhin EN (1984) The acetylcholinesterase activity in chloroplasts of higher plants. Dokl AN SSSR 278:754–757
Roshchina VV, Mukhin EN (1985a) Acetylcholinesterase activity in chloroplasts and acetylcholine effects on photochemical reactions. Photosynthetica 19:164–171
Roshchina VV, Mukhin EN (1985b) Acetylcholine action on the photochemical reactions in chloroplasts. Plant Sci 42:95–98
Roshchina VV, Mukhin EN (1986) Acetylcholine, its role in plant life. Usp Sovrem Biol (Trends Modern Biol – Rev) 101:265–274
Roshchina VV, Mukhin EN (1987a) Acetylcholine-acetylcholinesterase system in isolated pea chloroplasts and regulation of some photosynthetic reactions. Fiziol Rast 34:67–73
Roshchina VV, Mukhin EN (1987b) Acetylcholine changes photochemical reactions and stimulates Na^+ and K^+efflux from chloroplasts. Fiziol Rast 34:907–911
Roshchina VV, Mukhin EN (1989) Reactions of chloroplasts on allelopathically active alkaloids and polyacetylenes. In: Grodzinsky AM (ed) Methodological problems of allelopathy. Naukova Dumka, Kiev, pp 128–131
Roshchina VV, Roshchina VD (1977) Some aspects of mechanism involved in the movement of chloroplasts in the flow of cytoplasm. Fiziologia rastenii 24: 261–266
Roshchina VV, Roshchina VD, (1983) Role of allelopathically active substances in regulation of light stage of photosynthesis. In: Grodzinsky AM (ed) Role of plant and microbial toxins in allelopathy. Naukova Dumka, Kiev, pp 127–133

Roshchina VV, Semenova MN (1990) Plant cholinesterases: activity and substrate-inhibitory specificity. J Evol Biochem Physiol 26:644–651

Roshchina VV, Solomatkin VP, Roshchina VD (1980) Cicutotoxin as an inhibitor of electron transport in photosynthesis. Fiziol Rast 27:704–709

Roshchina VV, Ruzieva RKh, Mukhin EN (1986) Capsaicin from the fruits of red pepper *Capsicum annuum* L. as a regulator of the photosynthetic electron transport. Prikl Biokhim Mikrobiol 22:403–409

Rost TL (1969) Vascular pattern and hydathodes in leaves of *Crassula argentea* (Crassulaceae). Bot Gaz 130:267–270

Rost TL, Jones T, Robbins JA (1986) The role of ethylene in the control of cell division in cultured pea root tips: a mechanism to explain the excision effect. Protoplasma 130:68–72

Rougier M (1981) Secretory activity of root cap. In: Tanner W, Loewus FA (eds) Plant carbohydrates. II. Encyclopedia of plant physiology, New Ser. vol. 13B. Springer, Berlin Heidelberg New York, pp 542–574

Rougier M, Chaboud A (1985) Mucilages secreted by roots and their biological function. Isr J Bot 34:129–146

Rovira AD (1959) Root excretions in relation to the rhizosphere effect. 4. Influence of plant species, age of plant, light, temperature and calcium nutrition on exudation. Plant Soil 11:53–64

Rovira AD (1969a) Diffusion of carbon compounds away from wheat roots. Aust J Biol Sci 22:1287–1290

Rovira AD (1969b) Plant root exudates. Bot Rev 35:35–59

Rovira AD (1971) Plant root exudates. In: US Nat Comm for IBP (eds) Biochemical interactions among plants. Natt Acad Sci, Washington, pp 19–24

Rozema J, Gude H, Pollak G (1981) An ecophysiological study of the salt secretion of four halophytes. New Phytol 89:201–217

Rozen VB (1986) The hormones' receptors, their structure, characteristics and regularity of function in cells. In: Shalyapina VG (ed) Physiology of hormonal reception. Nauka, Leningrad, pp 5–33

Rubin BA, Panasenko NP (1956) To the question of daily dynamics of the content of carbon dioxide and oxygen in leaves of sugar beet. Izv AN SSSR Ser Biol N 1:55–62

Rüedi P, Wollenweber E, Marx D, Scheele C (1989) Kaurane-type diterpenes from fern frond exudates. J Biosci (Z Naturforsch) 44c:901–904

Russin WA, Uchytil F, Feistner G, Durbin RD (1988) Developmental changes in content of foliar secretory cavities of *Tagetes erecta* (Asteraceae). Am J Bot 75:1787–1793

Sabinin DA (1955) Physiological basis of plant nutrition. AN USSR Publ. Moscow

Sakai WS (1974) Scanning electron microscopy and energy dispersive X-ray analysis of chalk secreting leaf glands of *Plumbago capensis.* Am J Bot 61: 94–99

Salatino A, Monteiro WR, Bomtempi N Jr (1988) Histochemical localization of phenolic deposits in shoot apices of common species of Asteraceae. Ann Bot 61:557–559

Salyaev RK (1969) The absorption of substances by the plant cell. Nauka, Moscow

Salyaev RK (1985) Problems and perspectives of the study of plant biological membranes in plants. In: Salyaev RK, Voinikov VK (eds) Structure and function of biological membranes in plants. Nauka, Novosibirsk, pp 5–14

Samtsevich SA (1965) Active plant root excretions and their significance. Fiziol Rast 12:837–846

Sanadze GA (1961a) The excretion of volatile organic substances by plants. Issue of AN Gruzinskoi SSR, Tbilisi

Sanadze GA (1961b) The absorption of molecular hydrogen by green leaves on the light. Fiziol Rast 8:555–559

Sanadze GA (1964) About conditions of the diene C_5H_8 isoprene excretion from leaves. Fiziol Rast 11:49–52

Sanadze GA (1969) Light-dependent excretion of molecular isoprene. In: metzne (ed) Progress in photosynthetic research vol 2. pp 701–706 Tübingen

Sanadze GA (1990) The principle scheme of photosynthetic carbon conversion in cells of isoprene releasing plants. In: Baltschevffsky M (ed) Current research in photosynthesis, vol 4. Kluwer, Dordrecht pp 231–237

Sanadze GA, Dzhaiani GI, Tevzadze IM (1972) On the incorporation of ^{13}C-carbon from $^{13}CO_2$ assimilated during photosynthesis into the isoprene molecule. Fiziol Rast 19:24–27

Sanadze GA, Mgaloblishvili MP, Dalakishvili KG (1990a) Action of perulenin, organophosphorous compounds and synthetic analogs of carbohydrates on the CO_2-assimilation and isoprene effect of poplar isolated protoplasts. Fiziol Rast 37:6–11

Sanadze GA, Surmava MK, Dolidze ML, Aleksidze GYa (1990b) Biosynthesis of isoprene in vitro by chlorophyll-containing isolated culture of mesophyll leaf tissue of poplar, *Populus deltoides* Marsh. Soobshch An Gruz SSR 137:581–584

Sanchez-Alonso F, Lachica M (1988) Oxalate salts in the leaves of plum (*Prunus salicina* L.) and Cherry (*P. avium* L.). New Phytol 108:505–508

Sanders IO, Robertson DR, Smith AR, Hall MA (1987) Ethylene binding and metabolism in *Pisum sativum* L. cv. Alaska'. In: Klämbt D(ed) Plant Hormone Receptors. Springer, Berlin Heidelberg New York, pp 289–295

Sand-Jensen K, Prahl C (1982) Oxygen exchange with the lacunae and across leaves and roots of the submerged vascular macrophyte, *Lobelia dormanna.* New Phytol 91:103–120

Sandmann G, Boger P (1982) Volatile hydrocarbons from photosynthetic membranes containing different fatty acids. Lipids 17:35–41

Sarkar, Kumar A (1974) Influence of plant extracts on chromosome metabolism of mature nuclei. Bull Bot Soc Bengal 28:113–115

Sastry BVR, Sadavongvivad C (1979) Cholinergic systems in non-nervous tissues. Pharmacol Rev 30:65–132

Satcliffe JF (1962) Mineral salt absorption in plants. Pergamon Press, Oxford

Satler SO, Kende H (1985) Ethylene and the growth of rice seedlings. Plant Physiol 79:194–198

Sato H, Kobayashi Y, Fukui H, Tabata M (1990) Specific differences in tolerance to exogenous berberine among plant cell cultures. Plant Cell Rep 9:133–136

Sauth GR, Whittick A (1987) Introduction to phycology. Blackwell Oxford

Saxena PR, Pant MC, Kishor K, Bhargava KP (1965) Identification of pharmacologically active substances in the Indian stinging nettle *Urtica parviflora* Roxb. Can J Physiol Pharmacol 43: 869–876

Saxena PR, Tangri KK, Bhargava KP (1966) Identification of acetylcholine, histamine and 5-hydroxytrpytamine in *Girardinia heterophylla* (Decne). Can J Physiol Pharmacol 44: 621–627

Schauenstein E, Esterbauer H, Zollner H (1977) Aldehydes in the biological system: their natural occurrence and biological activities. Pion, London

Scheel D, Colling C, Keller H, Parker J, Schulte W, Hahlbrock K (1990) Studies on elicitor recognition and signal transduction in host and non-host plant/fungus pathogenic interactions. In: Lugtenberg BJJ (ed) Signal molecules in plants and plant microbe interactions. Springer, Berlin Heidelberg New York, pp 214–218

Scheffer F, Stricker G, Kickuth R (1965) Organische in der Guttationflüssingkeit einiger Wild und Kulturweizen. Z. Pflanzenernähr Bodenkd 109:240–248

Schildknecht Hvon Rauch G (1961) Die chemische Natur der Luftphytoncide von Blattpflanzen in besondere von *Robinia pseudoacacia.* 11. Mitteilung über Abwehrstoffe von Pflanzen, Z Naturforsch 16B:422–429

Schloß P, Walter C, Mäder M (1987) Basic peroxidases in isolated vacuoles in *Nicotiana tabacum* L. Planta 170:225–229

Schnabl H, Elbert C, Youngman RJ (1983) Release of ethane from immobilized plant cell protoplasts in response to chemical treatment. Physiol Plant 59:46–47

Schnabl H, Künne J, Florschütz K, Youngman RJ (1984) Differential effects of ecotoxicological substances on lipid peroxidation of plant cell protoplasts. Plant Cell Environ 7:499–505

Schnepf E (1964) Über Zellwandstrukturen bei der Köpfchen-drüsen der Schuppenblatter von Lathraea clandestina L. Planta 60:473–482

Schnepf E (1969) Sekretion und Exkretion bei Pflanzen. Ser Proto-plasmatologia, Bd 8. Springer, Berlin Heidelberg New York

Schnepf E (1974) Gland cells. In: Robards AW (ed) Dynamic aspects of plant ultrastructure. McGraw-Hill, London, pp 331–357

Schnepf E, Deichgraber G (1983a) Structure and formation of fibrillar mucilage in seeds epidermis cells. 1. *Collomia grandiflora* (polemoniaceae). Protoplasma 114:210–221

Schnepf E, Deichgraber G (1983b) Structure and formation of fibrillar mucilage in seed epidermis cells. 11. Ruellia (Acanthaceae). Protoplasma 114:222–234

Schonwitz R, Steinbrecher R, Ziegler H (1988) Emission of monoterpenes from needles of *Picea abies* (L.) Karst under natural conditions. Abstr 6th Congr FESPP, Split, 4–10 September 1988 p 1518

Schötz F, Diers L, Bathelt H (1970) Zur Feinstruktur der Raphidenzellen. 1. Die Entwicklung der Vakuolen und der Raphiden. Z Pflanzenphysiol 63:91–113

Seamon FC (1982) Sesquiterpene lactones as taxonic characters in the Asteraceae. Bot Rev 48:121–195

Seigler D, Price PW (1976) Secondary compounds in plants, primary functions. Am Nat 110:101–105

Selye H (1936) A syndrome produced by diverse nocuous agents. Nature 138:32

Selye H (1956) The stress in life. McGraw-Hill, New York

Senalik D, Simon PW (1986) Relationship between oil ducts and volatile terpenoid content in carrot roots. Am J Bot 73:60–63

Setter TL, Kupkanchanakul T, Kupkanchanakul K, Bhekasut P, Wiengweera A, Greenway H (1987) Concentrations of CO_2 and O_2 in floodwater and in internodal lacunae of floating rice growing at 1–2 metre water depth. Plant Cell Environ 10:767–776

Severson RF, Arrendale RF, Chortyk, OT, Green CR, Thome FA, Stewart JL, Johnson AW (1985) Isolation and characterization of the sucrose esters of the cuticular waxes of green tobacco leaf. J Agric Food Chem 35:870–875

Sfakiotakis EM, Dilley DR (1973) Inductionof autocatalytic ethylene production in apple fruits by propylene in relation to maturity and oxygen. J Am Soc Hortic Sci 98:504–508

Sguler SA, Taylor GE Jr, Selvidge WJ, Gunderson CA (1985) Effect of ethylene and related hydrocarbons on carbon assimilation and transpiration in herbaceous and woody species. Environ Sci Technol 19:432–437

Shaedle M (1975) Tree photosynthesis. Annu Rev Plant Physiol 26:101–115

Shah JJ, Setia RC (1976) Histological and histochemical changes during the development of gum canals in *Sterculia urens*. Phytomorphology 26:151–158

Shapovalov AA (1973) The release of organic substances of plant cells in connection with the functional state of plasmatic membranes. Usp Sovrem Biol 76:82–95

Sharma R, Singh G (1987) Effect of the phenolic compounds on Hill activity in rice (*Oryza sativa*). Ann Bot 60:189–190

Sharma R, Sharma B, Singh G (1984) Phenols as regulators of nitrate reductase activity in *Cicer arietinum* L. Phyton 44:185–188

Sharon N, Lis H (1989) Lectins. Chapman and Hall. London

Shavarda AL, Telepova MN, Budantsev AL (1990) Comparative study of the essential oils content and ultrastructure of glandular leaf trichomes in some species *Dracocephalus* L. Rastit Resur 26:252–362

Shchiparev SM, Chuprova GV, Polevoi VV (1976) Secretion of acids by isolated maize small shields. Vestn Leningr Univ Ser Biol 21:130–133

Shehab AS, Adam ZM (1983) Cytological effects of medicinal plants in Qatar. III. Mitotic effect of water extract of *Anastatica hierochuntica* L. on *Allium cepa*. Cytologia 48:343–348

Shimokawa K, Sakanoshita A, Horiba K (1978) Ethylene-induced changes of chloroplast structure in *Satsuma mandarin* (Citrus unshiu Marc). Plant Cell Physiol 19:229–236

Shimomura K, Sudo H, Saga H, Kamada H (1991) Shikonin production and secretion by hairy root cultures of *Lithspermum erythrorhizon*. Plant Cell Reports 10:282–285

Shipilova SV, Zaprometov MN (1980) Localization of phenylalanin-ammonium lyase in chloroplasts of maize (*Zea mays* L.). Fiziol Rast 27:67–73

Shpilevaya SP, Il'chenko KV (1984) Secretion of proteins during growth of pollen tubes of *Nicotiana alata* Link et Otto in vitro. Dokl AN Ukr SSR Ser Biol N 8:82–84

Shropshire W Jr (1979) Stimulus perception. In: Haupt W, Feinleib ME (eds) Physiology of movement. Encyclopedia of plant physiology. New Ser, vol 7. Springer, Berlin Heidelberg New York, pp 10–41

Shubnikova EA (1967) Cytology and cytophysiology of the secretory process. Moscow University Press. Moscow

Siegel SM, Siegel BZ (1987) Biogenesis of carbon monoxide: production by fungi and seed plants in the dark. Phytochemistry 26:3117–3119

Simpson FJ, Talbort G, Westlake DWS (1960) Production of carbon monooxide in the enzymatic degradation of rutin . Biochem Biophys Res Commun 2: 15–18

Sinyukhin AM (1973) Functional activity of action potential in ferns and mosses at fertilization. Biophyzika 18:477–482

Sinyukhin AM, Britikov EA (1967) Action potential in the reproductive system of plants. Nature 215:1278–1279

Siqueira JO, Safir GR, Nair MG (1991) Stimulation of vesicular-arbuscular mycorrhiza formation and growth of white clover by flavonoid compounds. New Phytol 118:87–93

Siriphanich J, Kader AA (1986) Changes in cytoplasmic and vacuolar pH in harvested lettuce tissue as influenced by CO_2. J Am Soc Hortic Sci 111:73–77

Sisler EC (1987) Purification of the ethylene-binding component from mung bean sprout and seeds. In: Klämbt D (ed) Plant hormone receptors. Springer, Berlin Heidelberg New York, pp 297–301

Sisler EC, Wood C (1988) Interaction of ethylene and CO_2. Physiol Plant 73:440–443

Sisler EC, Yang S Fa (1984) Ethylene – the gaseous plant hormone. BioScience 34:234–238

Sjöstrand T (1952) The formation of carbon monooxide by in vitro decomposition of haemoglobin in bile pigments. Acta Physiol Scand 26:328–333

Skrukrud CL, Taylor SE, Hawkins DR, Nemethy EK, Calvin M (1988) Subcellular fractionation of triterpenoid biosynthesis in *Euphorbia lathyris* latex. Physiol Plant 74:306–316

Skvortsov SS, Smirnova GN (1972) The season changes in biological activity and composition of volatile compounds of cowberry. In: Aizenmann BE (ed) Phytonicides: results, perspectives and problems of the study. Naukova Dumka, Kiev, pp 137–139

Slatyer B, Daday A, Smith GD (1983) The effects of acetaldehyde on nitrogenase, hydrogenase and photosynthesis in the cyanobacterium *Anaebaena cylindrica*. Biochem J 212:755–758

Smallman BN, Maneckjee A (1981) The synthesis of acetylcholine by plants. Biochem J 194:361–364

Smirnoff N, Stewart GR (1985) Stress metabolites and their role in plants. Vegetatio 62: 273–278

Smith DA (1982) Toxicity of phytoalexins. In: Bailey JA, Mansfield JW (eds) Phytoalexins. Blackie, Glasgow, pp 218–252

Smith BN, Meeuse BJD (1966) production of volatile amines and skatole at anthesis in some Arum lily species.Plant Physiol 41:343–347

Smith MM, Hartley RD (1983) Occurrence and nature of ferulic acid substitution of cell-wall polysaccharides in graminaceous plants. Carbohydr Res 118:65–80

Smith WH, Parker JG (1966) Prevention of ethylene injury to carnations by low concentrations of carbon dioxide. Nature 211:100–101

Smucker AJM, Erickson AE (1987) Anaerobic stimulation of root exudates and desease of peas. Plant Soil 99:423–433

Sokolov AA (1972) Chemical composition of the atmosphere precipitations dropped through cover of spruce and brich trees. Lesovedenie N 3:103–106

Soldatenkov SV (1971) Exchange of organic acids in plants. 30th Timiryazev Lecture. Nauka, Leningrad

Soldatenkov SV, Chirkova TV (1963) About role of leaves in respiration of roots with the oxygen deficit. Fiziol Rast 10:535–543

Solomonson LP (1981) Cyanide as a metabolic inhibitor. In: Vennesland B, Conn EE, Knowles CJ, Westley J, Wissing F (eds) Cyanide in biology. Academic Press, New york, pp 11–28
South GR, Whittick A (1987) Introduction to phycology. Blackwell, Oxford
Spilator SR, Mahlberg PG (1985) Composition and structure of nonutilizable laticifer starch grains of *Euphorbia pulcherrima* Wild. Bot Gaz 146:26–31
Spring O, Bienert U, Klemt V (1987) Sesquiterpene lactones in glandular trichomes of sunflower leaves. J Plant Physiol 130:433–439
Spring O, Benz T, Ilg M (1989) Sesquiterpene lactones of the capitate glandular trichomes of *Helianthus annuus*. Phytochemistry 28:745–749
Spring O, Vargas D, Fisher NH (1991) Sesquiterpene lactones and benzofuranes in glandular trichomes of three *Pappobolus* species. Phytochemistry 30:1861–1867
Sriram N, Ram MHY (1984) Sex-associated differences in peroxidases and ethylene production and their modification by ethephon treatment in the flowers of *Cannabis sativa* L. Curr Sci (India) 53:735–739
Stahl E (1952) Nachweis der Vorstufe des Azulens in den Drüsenhaaren der Schafgarbe. Naturwissenschaften 39:551–552
Stahl E (1957) Über Vorgänge in den Drüsenhaaren der Schafgarbe. Z. Bot 45:297–315
Stalfelt MG (1967) The components of CO_2-induced stomatal movements. Physiol Plant 20:634–642
Stanier RY, Adelberg EA, Ingraham JL (1976) The microbial world. Prentice-Hall, Engelwood Cliffs
Steelink C, Yeung M, Caldewell RL (1967) Phenolic constituents of healthy and wound tissues in the giant cactus (*Carnegiea gigantea*). Phytochemistry 6:1436–1440
Steer MW (1988) The role of calcium in exocytosis and endocytosis in plant cells. Physiol Plant 72:213–220
Stenlid G (1970) Flavonoids as inhibitors of the formation of adenosine triphosphate in plant mitochondria. Phytochemistry 9:2251–2256
Stepanov EV (1972) Volatile substances and phytoncideness of forest communities. PhD Thesis, Institute of Forest and Wood, Sverdlovsk, USSR
Stepanova AA (1983a) Ultrastructure of salt glands of *Plumbago capensis* (Piumbaginaceae) leaf. Bot Sh (USSR) 68:818–827
Stepanova AA (1983b) Change of ultrastructure of salt glands of *Limonium platyphylium* (Plumbaginaceae) leaf under induction of the sodium chloride secretion. Bot Zh (USSR) 68:1003–1011
Stepen' RA, Sukhinin AI (1985) Excretion of volatile compounds by conifer plantations and possibility of their inflamination. Izv Sib Dep AN SSSR Ser Biol Sci 13:47–52
Stephen AM (1980) Plant carbohydrates. In: Bell EA. Charlwood BV (eds) Secondary plant products. Encyclopedia of plant physiology, vol 8. Springer, Berlin Heidelberg New York, pp 555–584
Stern WL, Curry KJ, Pridgeon AM (1987) Osmophores of *Stanhopea* (Orchidaceae). Am J Bot 74:1323–1331
Stipanovic RD (1983) Function and Chemistry of plant trichomes and glands in insect resistance. Protective chemicals in plant epidermal glands and appendages. In: Hedin PA (ed) Plant resistance to insects 183rd Meet Am Chem Soc, Las Vegas Nev, March 28 -April 2 1982, ACS Symp Ser, vol 208 Am Chem Soc Washington, pp 69–100
Stobart AK, Stymne S, Hoglund S (1986) Safflower microsomes catalyse oil accumulation in vitro: a model system. Planta 169:33–37
Stoessl A 91982) Biosynthesis of phytoalexins. In: Bailey JA, Mansfiled Jw (eds) Phytoalexins. Blackie, Glasgow, pp 133–323
Stom DI (1979) The movement of phenols through tonoplast ans its physiological significance. Usp Sovrem Biol 87:78–92
Stuhlfauth T, Klug K, Fock HP (1987) The production of secondary metabolites by *Digitalis lanata* during CO_2 enrichment and water stress. Phytochemistry 26:2735–2739
Stünzi JT, Kende H (1989) Gas composition in the internal air spaces of deepwater rice in relation to growth induced by submergence. Plant Cell Physiol 30 :49–56

Suga T, Iwata N, Asakawa Y (1972) Chemical constituents of the male flower of *Alnus pendula* (Betulaceae). Bull Chem Soc Jpn 45:2058

Suslova TA, Sharygina IS (1968) To study of localization and the essential oils content in *Asarum europaeum* L. Rastite Rcsur 4:173–179

Suttle TC (1986) Cytokinin-induced ethylene-biosynthesis in nonsenescing cotton leaves. Plant Physiol 82:930–935

Suzuki T, Waller GR (1987) Allelopathy due to purine alkaloids in tea seeds during germination. Plant Soil 98:131–136

Sviridova IK (1960) Results of the study of washing out of nitrogen and the ashes elements by rain from tops of woody plants. Dokl AN SSSR 133:706–708

Tabata M, Nakagawa K, Yamamoto H, Kobayashi Y, Suzuki M, Fukui H (1987) Formation and secretion of alkaloids in *Thalictrum* cell cultures. Abstr 14th Int Bot Congr,Berlin-Dahlem Bot Garden West Berlin, p 54

Takahama U (1983) Redox reactions between kaempferol and illuminated chloroplasts. Plant Physiol 71:598–601

Takeoka Y, Kaufman PB, Matsumura O (1979) Comparative microscopy of idioblasts in lemma epidermis of some C_3 and C_4 grasses (Poaceae) using sump method. Phytomorphology 29:330–337

Tamarin AL (1986) Thin structure of surface of traping vesicles of *Utricularia vulgaris* L. (Lent). Bull Mosc Soc Nat Res Biol Dep 91:33–46

Tambiev AK (1984) The reaction capability of plant exometabolites. Moscow University Press, Moscow

Tan KS, Kubo I (1990a) Release of oxidases from the roots of plants. Experientia 46:478–481

Tan KS, Kubo I (1990b) Inhibitor of the oxidation of catechin released from the roots of corn seedlings. Experientia 46:971–972

Tanaka S, Yamaura T, Shigemoto R, Tabata M (1989) Phytochrome-mediated production of monoterpenes in thyme seedlings. Phytochemistry 28:2955–2957

Tanasienko FS (1985) Essential oils: content and composition in plants. Naukova Dumka, Kiev

Tang CS, Wat CK, Towers GHN (1987) Thiophenes and benzofurans in the undescribed rhizosphere of *Tagetes patula* L. Plant Soil 98:93–97

Tang ZC, Kozlowski TT (1984) Ethylene production and morphological adaptation of woody plants to flooding. Can J Bot 62:1659–1664

Tanowitz BD, Koehler DL (1986) Carbohydrate analysis of floral and extrafloral nectars in selected taxa of *Sansevieria* (Agavaceae). Ann Bot 58:541–545

Tata SJ, Beintema JJ, Balabaskaran S (1983) The lysozyme of *Hevea brasiliensis* latex: isolation, purification, enzyme kinetics and a partial amino-acid sequence. J Rubber Res Inst Malay 31:35–48

Taylor GE Jr, Gunderson CA (1986) The response of foliar gas exchange to exogenously applied ethylene. Plant Physiol 82:653–657

Taylorson RB (1979) Response of weeds to ethylene and related hydrocarbons. Weed Sci 27:7–10

Taylorson RB (1984) Prevention of action of far-red-absorbing phytochrome in *Rumex crispus* L. seeds by ethanol. Plant Physiol 74:223–236

Tester M (1990) Plant ion channels: whole cell and single-channel studies. New Phytol 114 :305–340

Thomas DS, Murashige T (1979) Volatile emission of plant tissue culture. I. Identification of the major components. In Vitro 15:654–658

Thomas V (1991) Structural, functional and phylogenetic aspects of the colleters. Ann Bot 68:287–305

Thomson CJ, Greenway H (1991) Metabolic evidence for stellar anoxia in maize roots exposed to low concentrations. Plant Physiol 96:1294–1301

Thomson WW (1975) The structure and function of salt glands. In: Poljakoff-mayber A, Gale J (eds) Plant and saline environments. Ecol Stud, vol 15. Springer, Berlin Heidelberg New York, pp 118–146

Thomson WW, Healey PL (1984) Cellular basis of trichome secretion. In: Rodriguez E, Healey PL, Mehta I (eds) Biology and chemistry of plant trichomes. Plenum Press, New York, pp 95–111

Thurnston EL (1974) Morphology, fine structure and ontogeny of the stinging emergence in *Urtica dioica*. Am J Bot 61 :809–817

Tian Lan-xin (Tian Lan-hsing), Lu Min, Hu Zheng-hai (1990) Studies on the initiation and development of gutta-containing cells of *Eucommia* ulmoides Oliv. Acta Bot Sin 32:1–6

Tietjen KG, Matern U (1984) Induction and suppression of phytoalexin. Biosynthesis in cultured cells of safflower *Carthamus tinctorius* L. by metabolites of *Alternaria carthami* Chewdhury. Arch Biochem Biophys 229:136–144

Tiezzi A, Cresti M, Ciampolini F (1983) The stigmatic exudate of *Citrus limon* (L.) BURM. in relation to some stigmatic ultrastructural features. In: Fertility and embryogenesis evoluted plants. Proc 7th Int Cytoembryol Symp High Tatra (Rackova delina), June 14–17 1982, Bratislava. pp 221–226

Ting JP, Loomis WE (1965) Further studies concerning stomatal diffusion. Plant Physiol 40:220–228

Tingey DT, Evans RC, Bates EH, Gumpertz ML (1987) Isoprene emissions and photosynthesis in three ferns. The influence of light and temperature. Physiol Plant 69:609–616

Tkachenko KG (1987) Essential oils from fruits of species of *Heracleum* L. grown in the Leningrad region. Rastit Resur N 3:429–436

Tokin BP (1957) About role of phytoncides in nature. In Tokin BP (ed) Phytoncides, their role in nature. Leningrad University Press, Leningrad, pp 5–21

Tokin BP (1980a) Medicinal toxins of plants. Leningrad University Press, Leningrad

Tokin BP (1980b) The phenomena of phytoncides – ecological and evolutional problem. Biol Nauki (USSR) N 5:5–17

Tomoda M, Gonda R, Shimizu N, Nakanishi S, Hikino H (1987) A mucilage from *Hibiscus moscheutos* leaves. Phytochemistry 26:2297–2300

Toong YC, Schooley DA, Baker FC (1988) Isolation of insect juvenile hormone III from a plant. Nature 333:170–171

Tooze SA, Weiss U, Huttner WB (1990) Requirement for GTP hydrolysis in the formation of secretory vesicles. Nature 347:207–208

Towers GHN (1980) Photosensitizers from plants and their photodynamic action. In: Reinhold L, Harborne JB, Swain T (eds) Progress in phytochemistry, vol 6. Pergamon Press, Oxford, pp 183–202

Toxobaeva GA, Ponomarenko SV, Rakhimbaev IR, Ivashchenko AT (1987) Modification of the ATPase activity of subchloroplasts particles by alcohols and ketones. Biol Nauki (USSR) N 5:24–27

Trainer M, Williams EG, Parrish DD, Buhr MP, Allwine EJ, Westberg HH, Fehsenfeld FC, Liu SC (1987) Models and observations of the impact of natural hydrocarbons on rural ozone. Nature 329:705–707

Tretyn A, Kendrick RE (1991) Acetylcholine in plants: presence, metabolism and mechanism of action. Bot Rev 57:33–73

Tretyn A, Kopcewicz J, Slezak E (1988) Interaction of light and the cholinergic system in the regulation of seed germination. Biol Plant 30:338–342

Troxler RF, Dokos JM (1973) Formationof carbon monoxide and bile pigment in red and blue-green algae. Plant Physiol 51:72–75

Tucker ML, Latis GG (1984) Comparative effects of ethylene and cyanide on respiration, polysome prevalence and gene expression in carrot roots. Plant Physiol 75:343–348

Tukey HB Jr (1966) Leaching of metabolites from above-ground plant parts and its implications. Bull Torrey Bot Club 93:385–401

Tukey HB Jr (1970) The leaching of substances from plants. Annu Rev Plant Physiol 21: 305–324

Tukey HB Jr, Morgan JW (1963) Injury to foliage and its effect upon the leaching of nutrients from above-ground plant parts. Physiol Plant 16:557–564

Tukey HB Jr, Morgan JW (1964) The occurrence of leaching from above-ground plant parts and the nature of the material leached. Proc XVIth Int Horticultural Congr 1962, vol 4. pp 153–160, Brussels

Tukey HB Jr, Tukey HB, Witwer SH (1958) Loss of nutrients by foliar leaching determined by radioisotopes. Proc Am Soc Hortic Sci 71:496–506

Tunmann O (1910) Untersuchungen über Sekretbehälter drüsen einiger Myrtaceen, speziell über ihren Entleerungsapparat. Arch Pharm 248:23–42

Turnel JC, Hemphill JK, Mahlberg PG (1978) Quantitative determination of cannabinoides in individual glandular trichomes of *Cannabis sativa* L, (Cannabiaceae). Am J Bot 65:1103–1106

Turner C (1983) *Cannabis*: the plant, its drugs and their effects. Aviat Space Environ Med 54:363–368

Uchiyama Y, Sugimura Y (1985) Salt-excreting function of vesiculated hairs of *Atriplex nummularia*. Jpn J Crop Sci 54:39–46

Uegaki R, Fujimori T, Kubo S, Kato K (1985) Stress compounds from *Nicotiana rustica* inoculated with TMV. Phytochemistry 24:2445–2447

Ueyama Y, Hashimoto S, Nii H, Furukawa K (1990a) The volatile constituents of the flower concrete of *Chimonanthus praecox* Link from China. Flav Fragr J 5:85–88

Ueyama Y, Hashimoto S, Nii, H, Furukawa K (1990b) The volatile constituents of shimei (*Rosa davurica* Pall) flower concrete from China. Flav Fragr J 5:115–120

Umarov MM (1986) Associative nitrogen-fixation. Moscow University Press, Moscow

Urbasch I (1984) Produktion pflanzlicher C6-Wundgase und ihre Wirkung auf einige phytopathogene Pilze. Z Naturforsch 39C:1003–1007

Uribe S, Pena A (1990) Toxicity of allelopathic monoterpene suspension on yeast. Dependence on droplet size. J. Chem Ecol 16:1399–1408

Valant-Vetschera KM, Wollenweber E (1989) Flavonoid aglycones inthe leaf exudate of the genus *Leucocyclus* (Compositae). Z. Naturforsch 44C:323–324

Vallance KB, Coult DA (1951) Observation on the gaseous exchanges which take place between *Menyanthes trifoliata* L. and its environment. The composition of the internal gas of the plant. J Exp Bot 2:212–222

Vancura V (1964) Root exudates of plants. 1. Analysis of root exudates of barley and wheat in their initial phases of growth. Plant Soil 21:231–248

Vancura V, Stotzky G (1976) Gaseous and volatile exudates from germinating seeds and seedlings. Can J Bot 54:518–532

Vartapetyan BB (1985) Anaerobios and structurally-functional reorganization of cell. In: Kursanov AL (ed) New directions in plant physiology. Nauka, Moscow, pp 175–199

Vartapetyan BB, Agapova LP, Aver'yanov AA, Veselovskii VA (1974) The study of the molecular oxygen transport from above-ground parts to plant roots using the chemoluminescence method. Fiziol 21:502–505

Vasilyev AE (1977) Functional morphology of plant secretory cells. Nauka, Leningrad

Vasilyev AE, Muravnik LE (1988) The ultrastructure of the digestive glands in *Pinguicula vulgaris* L. (Lentibulariaceae) relative to their function. 1. The changes during maturation. Ann Bot (USA) 62:329–341

Vasilyev AE. Stepanova AA (1990) The ultrastructure of ion-secreting and non-secreting salt glands of *Limonium platyphyllum*. J Exp Bot 41:41–46

Vaverka S (1986) Indukce a biosynteza fytoalexin v rostlinach soje Sb ref Ceckosl Konf Ochr Rostl. Brno, 2–5 zazi 1986. Praha, pp 79–80

Venis M (1985) Hormone binding sites in plants. Logman, New York

Venkataramaiah C, Reddy KH Rao KN (1984) Phenolic compounds in seed and leachates of red sanders (*Pterocarpus santalinus* L.) Indian J Forest 7:19–21

Vennesland B, Pistorius EK, Gewitz Hs (1981) HCN production by unicellular algae. In: Vennesland B, Conn EE, Knowles CJ, Westly J, Wissing F (eds) Cyanide in biology. Academic Press, London, pp 349–361

Vennesland B, Castric PA, Conn EE, Solomonson LP, Volini M. Westley J (1982) Cyanide metabolism. Fed Proc 41:2639–2648

Verderevski DD (1959) Plant immunity to parasitar diseases. Selkhozgiz. Moscow

Verzilin NN, Stepanova AM, Mashinskii AL, Panova SA, (1985) Influence of CO_2 level in root zone on radish and wheat plants in hydroponic culture. Vestn Leningr Univ N 3:75–83

Vichkanova SA (1981) Inhibitors of microorganisms among natural substances of plants origin. In: Aizenman BE (ed) Phytoncides: role in biogeocenosis, significance for medicine. Naukova Dumka, Kiev, pp 210–216

Villalobos J, Ramirez F, Moussatche H (1974) Some observation on the presence of acetylcholine and histamine in plants. Soc Bras Para Progr Cienc Cult 26:690–693

Vinkler C, Apelbaum A (1985) Ethylene formation in plant mitochondria. Dependence on the transport of 1-aminocyclopropane-1-carboxylic acid across the membrane. Physiol Plant 63:387–392

Vinogradova VM (1991) The secretory conceptacles on the flower petales being a new diagnostic feature in the family Umbelliferae. Bot Zn (USSR) 76:1069–1072

Viswanathan K, Lakshmanan KK (1984) Phytoallelopathic effect on in vitro pollinial germination of *Colothropis gigantea* R. Br. Indian J Exp Bot 22:544–547

Vlasyuk PA, Porutskii G (1964) About selected accumulation of metals and phytoncides in plants. In: Drobod'ko VG (ed) Phytoncides in national economy. Naukova Dumka, Kiev, pp 63–67

Vogelmann AF, Turner JC, Mahlberg PG (1987) Cannabinoid occurrence in seedlings of *Cannabis sativa* L. quantitation in seedlings of known age and primary leaf length. Bot Gaz 148:468–474

Vogelsang R, Barz W (1990) Elicitation of β-1, 3-glucanase in chitinase activities in cell suspension cultures of *Ascochyta rabiei* resistant and susceptible cultivars of chickpea (*Cicer arietinum*). Z. Naturforsch 45C:233–239

Vogt T, Proksch P, Gülz PG (1987) Epicuticular flavonoid aglycones in the genus *Cistus*, Cistaceae. J Plant Physiol 131:25–36

Vovides AP (1991) Cone idioblasts of eleven cycad genera: morphology, distribution and significance. Bot Gaz 152:91–99

Wagner GJ (1991) Secreting glandular trichomes; more than just hairs. Plant Physiol 96 :675–679

Waisel Y, Eshel A, Agami M (1986) Salt balance of leaves of the mangrove *Avicennia marina*. Physiol Plant 67:67–72

Wallbank BE, Wheatley GA (1976) Volatile constituents from cauliflower and other crucifers. Phytochemistry 15:763–766

Walther HF, Hoffman GM, Elstner EF (1981) Ethylene formation by germinating, *Drechslero gramineas*-infected barley (*Hordeum sativum*) grains: a simple test for fungicides. Planta 151:251–255

Walther K, Rimpler H, Leuckert C (1966) Vorkommen von Flavonoides in den Exkretäumen von Citrusfruchten. Planta Med 14:453–459

Wang CY (1987) Changes of polyam,ines and ethylene in cucumber seedling response to chilling stress. Physiol Plant 69:253–257

Wang Y, Toyota M, Krause F, Hamburger Mn, Hostettmann K (1990) Polyacetylenes from *Artemisia borealis* and their biological activities. Phytochemistry 29:3101–3105

Wardle K, Short KC (1982) Effect of isoprenoid alcohols and fatty acids on root elongation, germination and their association with stomatal activity. Biochem Physiol Pflanz 117:210–215

Wasicky R (1959) Herausschwemmen von Substanzen aus Blättern und seine eventuelle biologische Bedeutung. Naturwissenschaften 46:172–180

Wat CK, Prasad SK, Graham EA, Partington S, Arnason T, Towers GHN, Lam J (1981) Photosensitization of invertebrates by natural polyacetylenes. Biochem Syst Ecol 9:59–62

Watson DG, Rycroft DS, Freer IM, Brooks CIW (1985) Sesquiterpenoid phytoalexins from suspended callus cultures of *Nicotiana tabacum*. Phytochemistry 24:2195–2200

Webb JK, Banthorpe DW, Watson DG (1984) Monoterpene synthesis in shoots regenerated from callus cultures. Phytochemistry 23:903–904

Wei L, Kogan M, Fisher D (1987) Trapping and identification of nitrogen oxides from soybean leaves. Phytochemistry 26:2397–2398

Weiglin C, Winter E (1988) Studies on the ultrastructure and development of the glandular trichomes of *Cressa cretica* L. Flora 181:19–27

Weissenböck G, Hedrich R, Sachs G (1986) Secondary phenolic products in isolated quard cells, epidermal cell and mesophyll cell protoplasts from pea (*Pisum sativum* L.) leaves: distribution and determination. Protoplasma 134:141–148

Went FW (1960) Organic manner in the atmosphere and its possible relation to petroleum formation. Proc Natl Acad Sci USA 46: 212–221

Werker E, Ravid U, Putievsky E (1985) Glandular hairs and their secretions in the vegetative and reproductive organs of *Salvia sclarea* and *S. dominica*. Isr J Bot 34: 239–252

Werner C, Hohl Hr (1987) Effects of cations on stress metabolite accumulation and secretion by soybean roots. Abstr 14th Int Bot Congr Berlin-Dahlem Botanical Garden, West Berlin. p52

West WC (1969) Ontogeny of oil cells in the woody ranales. Bull Torrey Bot Club 96: 329–344

Whitefield FB, Shea SR, Gillen KJ, Shaw KJ (1981) Volatile components from the roots of *Acacia pulchella* R. Br. and their effect of *Phytophthora cinnamomi* Rand. Am J Bot 29: 195–208

Whitehead CS, Halevy AH, Reid MS (1984) Roles of ethylene and 1-aminocyclo-propane-1-carboxylic acid in pollination and wound-induced senescence of *Petunia hybrida* flowers. Physiol Plant 61: 643–648

Wieneke J, Sarwar G, Roeb M (1987) Existence of salt glands on leaves of Kallar grass (*Leptochloa fusca* L. Kunth). J Plant Nutr 10: 805–820

Wiermann R (1981) Secondary plant products and cell and tissue differentiation. In: Conn EE (ed) The biochemistry of plants: a comprehensive treatise, vol 7. Academic Press, New York, pp 85–116

Wildman HG, Parkinson D (1981) Seasonal changes in water-soluble carbohydrates of *Populus tremuloides* leaves. Can J Bot 59: 862–869

Wilks SS (1959) Carbon monooxide in green plants. Science 129: 964–966

Willaman II, Brown WR (1930) Carbon dioxide dissolved in plant sap and its effect on respiration measurements. Plant Physiol 5: 535–542

Williams NH, Whitten WM (1983) Orchid floral fragrances and male euglossine bees: methods and advances in the last sesquidecade. Biol Bull 164: 355–395

Williams SE, Spanswick RM (1976) Propagation of the neuroid action potential of the carnivorous plant *Drosera*. J Comp Physiol 108: 211–223

Williamson GB, Fischer NH, Richardson DR, De La Pena A (1989) Chemical inhibition of fire-prone grasses by fire-sensitive shrub, *Conradina canescenes*. J chem Ecol 15: 1567–1577

Winget GD, Izawa S, Good NE (1969) The inhibition of photophosphorylation by phlorizin and closely related compounds. Biochemistry 8: 2067–2074

Wink M (1985) Metabolism of quinolizidine alkaloids in plant and cell suspension cultures: induction and degradation. In: Neumann KH (ed) Primary and secondary metabolism in plant cell culture. Springer, Berlin Heidelberg New Yor, pp 107–116

Wittenberg JB (1960) The source of carbon monoxide inthe float of the Portuguese man-of-war, *Physalia physalis* L. J Exp Biol 37: 698–705

Wolfenden J, Robinson DC, Cape JN, Paterson S, Francis BJ, Mehlhorn H, Wellburn AR (1988) Use of carotenoid ratios, ethylene emissions and buffer capacities for the early diagnosis of forest decline. New Phytol 109: 85–95

Wollenweber E (1984) The systematic implication of flavonoids secreted by plants. In: Rodriquez E, Healey PL, Mehta I (eds) Biology and chemistry of plant trichomes. Plenum Press, New York, pp 53–69

Wollenweber E (1974) Methylierte Flavonol-Aglykabei Aesculus. Z Pflanzenphysiol. 73: N 3, pp 277–280

Wollenweber E (1975) Flavonoidmuster als systematisches Mer kmal in der Gattung *Populus*. Biochem Syst Ecol 3: N 1, pp 35–45

Wollenweber E (1985) Exkret-Flavonoide bei höheren Pflanzen arider Gebiete. Plant Syst Evol 150: 83–88

Wollenweber E, Dietz VH (1981) Occurrence and distribution of free flavonoid aglycones in plants. Phytochemistry 20: 869–932

Wollenweber E, Mann K (1988) 2′5′-Dihydroxylflavone and its 5′-acetate-novel compounds from the farinose exudate of *Primula*. Z. Naturforsch 43c: 305–307

Wollenweber E, Roitman JN (1991) New frond exudate flavonoids from Cheilanthoid ferns. Z Naturforsch 46c: 325–330

Wollenweber E, Lebreton P, Chadenson M (1972) Flavonoide im Knospen-Exkret von Prunus-und Rhamnus-Arten. Z. Naturforsch 27b: 567–570

Wollenweber E, Asakawa Y, Schilo D, Lehmann U, Weigel H (1987a) A novel caffeic acid derivative and other constituents of *Populus* bud excretion and propolis (bee-glue). Z Naturforsch 42c: 1030–10434

Wollenweber E, Hradetzky D, Mann K, Roitman JN, Yatskievych G, Proksch M, Proksch P (1987b) Exudate flavonoids from aerial parts of five *Ambrosia* species. J Plant Physiol 131: 37–43

Wollenweber E, Arriage-Giner FJ, Rumbero A, Greenaway W (1989a) New phenolics from *Baccaris* leaf exudate. Z. Naturforsch 44c: 727–730

Wollenweber E, Schober I, Schilling G, Arriaga-Giner FJ, Roitman JN (1989b) A geranyl α-pyrone from the leaf resin of *Diplacus aurantiacus*. Phytochemistry 28: 3493–3496

Wollenweber E, Stern S, Roitman JH, Yatskievych G (1989c) External leaf flavonoids of *Polanisia trachysperma*. Phytochemistry 28: 303–305

Wollenweber E, Mann K, Roitman JN (1991a) Flavonoid aglycones from the bud exudates of three Betulaceae. Z. Naturforsch 46c: 495–497

Wollenweber E, Mayer K, Roitman JN (1991b) Exudate flavonoids of *Inula viscosa*. Phytochemistry 30: 2445–2446

Woltering EJ, Somhorst D (1990) Regulation of anthocyanin synthesis in *Cymbidium* flowers: effects of emasculation and ethylene. J Plant Physiol 136: 295–299

Woods FW (1960) Biological antagonisms due to phytotoxic root exudates. Bot Rev 26: 546–569

Wurzer-Fassnacht U, Hoffman GM (1986) Äthylenbildung von Weizen-Keimlingen bei Saatgutbefall durch Fusarium-Arten und die Wirkung von Beizmitteln. Z Pflanzenkr Pflanzenchutz 93: 337–346

Yokovleva OV (1988) Slime cells of the leaf epidermis in the dycotyledonous plants (Electron Microscopy Data). Bot Zh (USSR) 73: 977–987

Yalpany N, Silverman P, Wilsonn TMA, Kleier DA, Raskin I (1991) Salicylic acid is a systemic signal and an inducer or pathogenesis-related proteins in virus-infected tobacco. Plant Cell 3: 809–818

Yamamoto F, Kozlowski TT (1987) Effect of flooding, tilting of stems and ethrel application of growth, stem anatomy and ethylene production of *Pinus densiflora* sedlings. J Exp Bot 38: 293–310

Yamamoto H, Suzuki M, Suga Y, Fukui H, Tabata M (1987) Participation of an active transport system in berberine-secreting cultured cells of *Thalictrum minus*. Plant Cell Rep 6: 356–359

Yamashita I, Naito S, Iino K, Yoshikawa S (1978) Formation of aldehydes and alcohols from α-keto acids in strawberries. Nippon Shokuhin Kogyo Gakkaishi 25: 378–382

Yamaura T, Tanaka S, Tabata M (1989) Light-dependent formation of glandular trichomes and monoterpenes in thyme seedlings. Phytochemistry 28: 741–744

Yang SFa (1985) Biosynthesis and action of ethylene. Hortic Sci 20: 41–45

Yang SFa, Hoffman NE (1984) Ethylene biosynthesis and its regulation in higher plants. Annu Rev Plant Physiol 35: 155–189

Yatsenko-Khmelevskii AA (1954) Base and methods of anatomic study of wood. Issue of USSR Academy of Sciences, Moscow

Yatsenko-Khmelevskii AA (1980) Family Cannabiaceae. In: Takhtadjyan AL (ed) Plant life vol 5. Pt 1. Prosveschenie. Moscow, pp 279–282

Yokouchi Y, Ambe Y (1984) Factors effecting the emission of monoterpenes from red pine (*Pinus densiflora*). Plant Physiol 75: 1009–1012

Yokouchi Y, Okaniwa M, Ambe Y, Fuwa K (1983) Seasonal variation of monoterpenes in the atmosphere of a pine forest. Atmos Environ 17: 743–750

Yung-jen-Chou, Zhenguo Li, Shyung Yang, Ziwen Yu, Shuwen Yu (1985) Evolution of propylene and propane from tobacco leaves exposed to H_2SO_3. Environ Pollut A 37: 117–129

Yurin VM (1990) Electrophysiological aspects of xenobiology of plant cell. 3. Non-state ionic flows. Vestn AN BSSR Ser Biol N 5: 7–10

Yurin VM, Ivanchenko VM, Galactionov SG (1979) Regulation of functions of plant cell membranes. Nauka i Technika, Minsk

Zaat SAJ, Spaink HP, Wijffelman CA, van Brussel AAN, Okker RJH, Lugtenberg BJJ (1988) Flavonoid compounds as molecular signals in Rhizobium-legume symbiosis. In: Scannerini S (ed) Cell to cell signals in plant, animal and microbial symbiosis. NATO ASI Ser, vol H17. Springer Berline Heidelberg New York, pp 191–205

Zador E, Jones D (1986) The biosynthesis of a noval nicotine alkaloid in the trichomes of *Nicotiana stocktonii*. Plant Physiol 82:479–484

Zafra-Polo MC, Blasquez MA, Villar A (1988a) Volatile constituents of *Thymus webbianus*. Plant Med Phytother 22:184–188

Zafra-Polo MC, Blazsquez MA, Villar A (1988b) Variation in the composition of the essential oils from *Thymus leptophyllus* Lange and *Thymus webbianus* Rouy. Plant Med Phytother 22 : 189–194

Zamski E, Shoham O, Palevitch D, Levy A (1987) Ultrastructure of capsaicinoid-secreting cells in pungent and nonpungent red pepper (*Capsicum annuum* L.) cultivars. Bot Gaz 148 : 1–6

Zaprometov MN (1985) Phenolic compounds of plants: biosynthesis, conversions and function. In:Kursanov AL (ed) New directions in plant physiology. Nauka, Moscow, pp 143–162

Zaprometov MM, Kolonkova SV (1967) About biosynthesis of phenolic substances in chloroplasts. Dokl AN SSSR 176:470–473

Zauralov OA (1985) Formation and excretion of nectar. Usp Sovrem Biol 99:303–312

Zauralov OA, Chernavina MV, Tsang GD (1986) Phosphorus compounds in nectaries of gourd in connection with the nectar excretion. Fiziol Rast 33:593–597

Zelawski W, Riech FP (1970) Assimilation and release of internal carbon dioxide by woody plant shoots. Can J Bot 48:1351–1354

Zeroni M, Jerie PH, Hall MA (1977) Studies on the movement and distribution of ethylene in *Vicia faba*. Planta 134:119–125

Zimmerman DC, Vick BA (1988) Lipid peroxidation in plants – products and physiological roles. In: Sevanian A (ed) Lipid peroxidation in biological systems. Am Oil Chem Soc, Champain, Illinois, pp 196–212

Zimmerman PR, Chatfield RB, Fishman J, Crutzen PJ, Hanst PL (1978) Estimates on the production of CO and H from the oxidation of hydrocarbon emission from vegetation. Geophys Res Lett 5:679–682

Zindler-Frank E, Wichmann E, Korneli M (1988) Cells with crystals of calcium oxalate inthe leaves of *Phaseolus vulgaris* – a comparison with those in *Canavalia ensiformis*. Bot Acta 101 : 246–253

Zlobina ES (1987) Field and laboratory observation of guttation and transpiration. In: Prokof'ev AA (ed) Problems of plant drought-resistance. Nauka, Moscow, pp 234–238

Zobel AM (1986) Localization of phenolic compounds in tannin-secreting cells from *Sambucus racemosa* L. Shoots. Ann Bot 57:801–810

Zobel AM (1989) Cytoplasmic and apoplastic location of phenolic compounds in the covering tissue of the *Brassica napus* radicle between embryogenesis and germination. Ann Bot 64: 149–157

Zolkevich VN, Sinitsyna ZA, Peisakhzon BI (1979) On the nature of root pressure. Fiziol Rast 26:978–993

Zweig G, Hitt IE, Cho DH (1969) Possible mechanisms of the mode of action of quinone herbicides. In: Metzner H (ed) Progress in photosynthesis research vol 3. Springer-Verlag, Tübingen, Vol 3 pp 1728–1736

Index of Latin Names

Plants and Microorganisms

Animals

Subject Index

Springer-Verlag and the Environment

MIX
Papier aus verantwortungsvollen Quellen
Paper from responsible sources
FSC® C105338
FSC
www.fsc.org

If you have any concerns about our products,
you can contact us on
ProductSafety@springernature.com

In case Publisher is established outside the EU,
the EU authorized representative is:
Springer Nature Customer Service Center GmbH
Europaplatz 3, 69115 Heidelberg, Germany

Printed by Libri Plureos GmbH
in Hamburg, Germany